Apollo 17

The NASA Mission Reports
Volume One

Compiled from the archives & edited
by Robert Godwin
with an Introduction by
Dr Harrison H. Schmitt

Special thanks to:
Gene Cernan
Dr. Harrison Schmitt
Jim Busby

All rights reserved under article two of the Berne Copyright Convention (1971).
We acknowledge the financial support of the Government of Canada through the
Book Publishing Industry Development Program for our publishing activities.
Published by Apogee Books an imprint of Collector's Guide Publishing Inc., Box 62034, Burlington, Ontario, Canada, L7R 4K2
Printed and bound in Canada
Apollo 17 - The NASA Mission Reports - Volume One
by Robert Godwin
ISBN 1-896522-59-9
©2002 Apogee Books
All photos courtesy of NASA

Apollo 17
The NASA Mission Reports
Volume One
(from the archives of the National Aeronautics and Space Administration)

INTRODUCTION BY DR HARRISON SCHMITT

PRESS KIT

PRE-MISSION OPERATION REPORT

POST MISSION OPERATION REPORT

TECHNICAL CREW DEBRIEFING

RETURN TO THE MOON: 30 YEARS AND COUNTING

"Oh, hey!…Wait a minute!" I exclaimed[1], hesitating on my way back to the Rover from the big, shattered boulder on the rim of Shorty Crater. My exclamation and obvious excitement had been brought on by noticing something different about the scuff-marks I had made as I went along the crater rim to take a first look at the boulder.

"Where are the reflections? I've been fooled once." Remembering that I earlier had almost sampled the sun's orange reflection off an orange thermal blanket at a previous exploration site, I didn't want to be fooled again.

"There is orange soil!" I yelled across 240,000 miles of space.

"Well, don't move it until I see it." My partner, Gene Cernan, said later that he thought I had been on the Moon too long[2].

"It's all over!!! Orange!…I stirred it up with my feet!" I insisted.

"Hey, it is!! I can see it from here!" Cernan was about 20 meters away at the Lunar Rover finishing up his checklist duties there. "Wait a minute, let me put my visor up. It's still orange!"

"Sure it is! Crazy!…I've got to dig a trench, Houston." was my call to Bob Parker, the CAPCOM for our work outside the Lunar Module.

"Copy that. I guess we'd better work fast," replied Parker.

Thus, we began possibly the most productive half-hour on the Moon since Neil Armstrong gathered samples at Tranquillity Base.

Waiting for Gene to finish routine duties at the Rover, I quietly took about a minute and a half for a black and white, 360° photographic panorama from the south rim of Shorty while I continued to plan how to approach this unique sampling opportunity.

"Hey, he's not going out of his wits. It really is [orange]." Gene had interrupted his work to look over again at what I had discovered.

"Is it the same color as cheese?" asked Parker, succumbing to his streak of facetiousness even in momentous times. On the other hand, Bob may have thought we were trying to pull his leg as we had been known to do[1] before.

"It's almost the same color as the LMP decal on my camera," I replied, refusing to rise to the bait.

"That is orange, Jack!" Gene could not resist looking again at what had been discovered and was not yet exposed except by scuff-marks.

"Fantastic, sports fans! It's trench time!" I continued. "You can see this in your color television, I'll bet you." And indeed they could.

Thirty years ago this coming December 13, while enjoying the excitement of geological discovery in the Valley of Taurus-Littrow, I was temporarily oblivious to the fact that Apollo 17 would end the human exploration of the Moon for the foreseeable future. That the United States abandoned the capability to explore deep space in 1972 may be difficult for future historians to fully comprehend. Explanations at the time are understandable however historically inadequate they ultimately may turn out to be. Certainly playing their part in this decision were a lack of media interest, a lack of political will, the continuing war in Vietnam, and a general lack of understanding of the historical importance of frontier exploration in the continuing advancement of American society. With the continued advancement of scientific knowledge about the Moon and with the hindsight of 30 years, on the other hand, we can now rationally consider what deep space has to offer in the future of our country and humankind.

Trips to the Moon figured prominently in the history of the world during the latter half of the 20th Century. A return to the Moon in the first decades of the 21st Century may be even more significant. My interest in this possibility stems from having participated in the exploration of the Moon's Taurus-Littrow Valley as the only scientist to go to the Moon, as the Lunar Module Pilot on the last of the Apollo Missions. This personal opportunity came as a result of President John F. Kennedy's 1963 challenge to Americans, "to go to the Moon and return safely to Earth." Kennedy's inspiration coincided with a remarkable superposition of four social phenomena – public concern about the future, a sufficient base of technology, a catalytic and focusing event, and a leader who recognized a unique opportunity. The coincidence of these phenomena in the America of the 1960s provided the foundation for the success of Apollo, just as it did earlier for Thomas Jefferson's Louis and Clark Expedition, Theodore Roosevelt's Panama Canal, and other critical endeavors in the history of the United States.

Not all, great undertakings are assured of success, however. Apollo 11 succeeded in landing on the Moon on July 20, 1969, because competitive bidding brought the best of industry to the job, conservative engineering established strong margins of performance and safety, and highly flexible but disciplined management kept the ultimate objectives in perspective. Most importantly, nearly 500,000 highly motivated men and women, mostly in their mid-twenties, believed that meeting President Kennedy's challenge was the most important historical contribution they could make with their lives. Ten years of 16-hour days, seven day weeks, and the inevitable wear and tear on families could not have been sustained without such a belief. No amount of money would have bought us the quality control, attention to detail, teamwork, and spur of the moment innovation that became the hallmarks of Apollo. And, most of this Apollo team were in their mid-twenties, just out of the great engineering schools of the country, with only the astronauts and senior managers significantly older.

As President Kennedy appears to have anticipated, Apollo's success contributed in profound ways to the successful conclusion of the Cold War. Émigré reports and post-Cold War examination of Soviet records indicate that Apollo created a belief in the minds of the leadership of the Soviet Union that President Ronald Reagan's 1983 Strategic Defense Initiative probably would be successful as well, ultimately leading to a break-up of Soviet communism. Although not discussed at the time of Kennedy's decision, Apollo also established for all human beings a new evolutionary status in the Solar System. As a consequence, young people alive today realistically can think about living in settlements on the Moon and Mars. They can anticipate helping their home planet survive itself as Americans helped former homelands in Europe and Asia defeat oppression in the 20th Century and terrorism in the 21st. All in all, we have had an unprecedented and continuing return on the 1960's investment in a Cold War "race to the Moon." Both Americans and Russians can be proud of the results of their competition in that race.

Assistance to the Earth from future settlers of the Moon will come as a direct result of the scientific discoveries of Apollo, echoing earlier events in the history of the United States. The explorations of President Thomas Jefferson's 1803 Louisiana Purchase by Meriwether Lewis and William Clark lay the foundations for the growth of the economy and power of the United States. Theodore Roosevelt's project to construct a canal and lock system across Panama made the United States a naval power on two oceans and produced an explosion in medical, construction, and electrical technology. Similarly, the exploration of the Moon by Apollo astronauts created the Earth's first pre-eminent spacefaring nation and stimulated rapid advances in most fields of engineering. Additionally, Apollo lay the foundations for future terrestrial energy alternatives to fossil fuels, the growth of a lunar economy, and the settlement of the solar system by humans.

Further, as a consequence of lunar exploration by twelve Americans, and the more recent robotic exploration and scientific analysis built on that foundation, we have detailed first and second order understandings of the nature and history of the Moon, the smallest of the terrestrial planets. By extrapolation, we gained vastly improved insights about the history of other terrestrial planets - Earth, Venus, Mars and Mercury. We have scientific visibility into the first one and a half billion years of the geologically clouded history of these planets, including aspects related to the origin of life, not accessible by any other means. For example, it is now clear that the origins of the Moon and our home planet were closely related. Not yet certain, however, is whether the Moon's presence around Earth resulted from a giant impact on the Earth or by the Earth's capture of a small planet co-orbiting the sun. After its accretion from the solar nebula about 4.57 billion years ago, an ocean of hot magma existed on the Moon for about 50 million years during which mineral crystallization and density separation in that ocean caused the chemical differentiation of a crust and a mantle. For the next 700 million years, intense bombardment by asteroids and/or comets cratered and pulverized the lunar surface and the surfaces of other terrestrial planets, ending about 3.8 billion years ago. Great surface eruptions of lava then dominated the next billion years, gradually dying out as the Moon cooled. Many other details of lunar history are known, but what may be most interesting to note is the match between the appearance of isotopic evidence of life on the water-rich Earth with the end of the great bombardment, both occurring 3.8 billion years ago. The end of the extraordinary impact violence at this point in solar system evolution may have finally permitted simple, replicating life to form at the surface of the Earth, arising from a clay soup rich in water, complex organic molecules, and other necessary components. The same process may have begun on Mars only to be arrested later by the loss of its oceans and atmosphere.

The tie between preparation and planning on the one hand and scientific debate and understanding on the other is illustrated by my discovery of the Apollo 17 deposit of orange pyroclastic volcanic glasses. Data from these glasses have contributed significantly to the continuing debate over the origin of the Moon. The primary importance of pyroclastic glasses relative to lunar origin lies in their deep origins and in the composition of the adsorbed volatiles on the surfaces of the small glass beads and devitrified glass beads. The volatiles suggest that the lower mantle (below 550km) is largely undifferentiated from its primordial chemical nature, complicating the case for the Moon being the result of a giant impact on the Earth.

Most importantly, for future inhabitants of Earth and space, 15 years after Apollo 11 several Wisconsin engineers realized that the Apollo lunar samples show us that fusion energy resources exist in the pulverized upper several meters of the Moon's surface. These potentially commercial energy resources (solar wind derived Helium-3, a light isotope of normal Helium) provide both a long-term alternative to our use of fossil fuels on Earth as well as the basis for future lunar and Martian settlement. Further, by-products of the extraction of Helium-3 from lunar materials can sustain the future travelers and settlers of deep space with water, oxygen, hydrogen fuel, and food.

I doubt, however, that the United States or any government will initiate or finance a return of humans to the Moon or a human expedition to Mars in the predictable future. Governments, particularly that of the United States, have a very pragmatic excuse for turning their financial backs on the multifaceted potential of a return to the Moon. I learned during a term as a U.S. Senator that it is very difficult, politically, for Congress to commit to the long-term allocation of the required taxpayer-provided funds to space exploration or to any other so-called "discretionary" activities. This would be true under any circumstances but is made impossible now by an extended and necessary war on terrorism. This new demand for resources and attention has compounded problems arising from the political inability of governments to fund retirement and health security for the elderly and the poor by means other than income transfer from one generation to the next. Income transfer will lead to higher and higher tax rates on the children of the World War II Baby Boom as their parents begin to retire early in the 21st Century. By training and inclination, most elected representatives work to treat the symptoms of social problems rather than to solve those problems. Few of us in the non-political world can afford this luxury.

The entrepreneurial private sector, on the other hand, may find a business rationale for a return to the Moon, based on the economic value in the extraction of lunar Helium-3 and its use as a fusion fuel on Earth as a future economic and environmental alternative to fossil fuels. This possibility has been studied extensively since 1985 by a team I am part of at the University of Wisconsin-Madison and continues to appear to be a feasible approach to a return to the Moon and to providing clean terrestrial energy in the future. In addition, by-products of Helium-3 extraction from the pulverized lunar surface soils will include hydrogen, oxygen, and water - valuable materials needed for consumption by humans elsewhere in space. Thus, those who next return to the Moon in this century probably will approach work on the lunar surface very pragmatically, with humans in the roles of exploration geologist, mining geologist/engineer, heavy equipment operator/engineer, heavy equipment/robotic maintenance engineer, mine manager, and the like. To be successful, of course, a lunar resource and terrestrial fusion power business must be based on competitive rates of return to investors, innovative management of financial and technical risk, and reasonable regulatory and treaty oversight by government.

The long-term business case for private sector involvement in a return to the Moon most directly relates to terrestrial needs for clean energy. The global demand for energy will likely increase by a factor of eight or more by 2050. This will be due to a combination of needs reflecting the doubling of world population, new energy intensive technologies, demands to avoid the adverse consequences of climate change, and aspirations for improved standards of living in the less-developed world. Lunar Helium-3, with a resource base in the titanium-rich basaltic soils of Mare Tranquillitatis of at least 10,000 tonnes, represents one of several potential energy sources to meet this rapidly escalating demand. The energy equivalent value of Helium-3 delivered to future fusion power plants on Earth would be about $3 billion per metric tonne relative to $21 per barrel crude oil. The domestic U.S. electrical power market is worth approximately $120 billion, annually, at these prices. 40 tonnes of Helium-3 contains enough energy to supply that market's needs for one year. These numbers illustrate the theoretical magnitude of the potential business opportunity in a return to the Moon. In addition, the technology and facilities required for success of a lunar commercial enterprise will make possible and reduce the cost of continued activities in space, including investigations on and from the Moon, space station re-supply, exploration and settlement of Mars, and asteroid interception and diversion to prevent catastrophic impacts on the Earth.

Mining, extraction, processing, and transportation of Helium-3 to Earth, and the use of by-products in space, will require new innovations in robotic and long-life engineering but no known new engineering concepts. A business enterprise based on lunar resources will be driven by cost considerations to minimize the number of humans required for the extraction of each unit of resource. Humans will be required on the Moon, on the other hand, to reduce the business risk of lunar operations; to prevent costly breakdowns of semi-robotic mining, processing, and delivery systems; to provide manual back-up to robotic or tele-robotic operation; and to support other lunar activities in general. The creation of capabilities to support mining operations also will provide the opportunity for lunar observatories, renewed scientific exploration, and potential tourism at much reduced expense with the cost of capital for launch and basic operations being carried by the business enterprise.

During the early years of operations the number of personnel at a lunar resource extraction settlement will be about six per mining/processing unit plus four support personnel per three mining/processing units. It is anticipated that one mining/processing unit would support the annual fuel needs of a 1000 megawatt fusion power plant on Earth. About 21 persons on the Moon, therefore, could support three such plants. Cost considerations also will drive business to encourage or require personnel to become settlers, provide them all medical care and recreation, and provide technical control for most or all operations on the Moon. It will always be important to remember Wallace Stiegner's reminder that "a place is not a place until people have lived and died there."

Investors and others probably will always ask: "Why humans in space or on the Moon? Wouldn't robots be cheaper, better and safer?" Setting aside the inherent desire for human beings to "be there" where ever "there" may be, I know from personal experience that on the Moon humans contribute to space operations in unique and valuable ways. As with the discovery of the orange soil, they will provide instantaneous observation, interpretation, and assimilation of the environment in which they work and a creative reaction to that environment. Human eyes, experience, judgement, ingenuity, and manipulative capabilities are unique in and of themselves and highly additive in synergistic and spontaneous interaction with instruments and robotic systems. Due to inherent communication delays and the cost of returning samples and providing mission support, the deeper into space human beings desire to go, say Mars for example, the more important will become these unique human attributes.

Near-term, the single most critical question is: "Can you cut the cost of access to deep space from the approximately $70,000 per kilogram to the Moon (including the additional cost of private capital) required by the Apollo Saturn V rocket?" Heavy lift launch constitutes the largest cost uncertainty facing initial business planning, however, many factors, particularly post-Apollo technology and long term production contracts, promise to lower these costs into the range of $1-2000 per kilogram. Also contributing to a reduction in launch costs will be nearly 40 years engineering experience with heavy lift rockets and a clear focus on a set of business and financial requirements. New technologies that have not been applied to Saturn V level and greater rocket capabilities (50-100 metric tonnes to the Moon) include light weight materials, micro-electronic and micro-mechanical devices, imbedded computer controls and diagnostics, new manufacturing approaches, and many others.

Another critical question relates to the technology base for the use of Helium-3 as a fusion fuel. Inertial electrostatic confinement (IEC) fusion technology appears to be the most attractive and least capital intensive approach to terrestrial fusion power plants. Although great amounts of public funds have been spent on fusion research over the last half-century, government funded research has almost exclusively focused on the technology of non-electrostatic confinement devices. These technologies suffer from numerous disadvantages in their possible application to commercial electrical power plants, including very high capital and operating costs, large minimum operating size, relatively low conversion efficiency through a heat cycle, need for radioactive fuel, and creation of radioactive waste products. In contrast, electrostatic technology inherently offers the potential for low capital costs, size flexibility, high electrical conversion efficiency through direct conversion of charged particles, non-radioactive fuel, and no radioactive waste. Over the last two decades, steady progress in the advancement of IEC fusion technology has been made by my colleagues at the University of Wisconsin-Madison's Fusion Technology Institute under the guidance of Professor G.L. Kulcinski.

A private enterprise approach to developing lunar Helium-3 and terrestrial IEC fusion power would be the most expeditious means of realizing this unique opportunity. In spite of the large, long-term potential return on investment, however, access to capital markets for a lunar Helium-3 and terrestrial fusion power business will require a near-term return on investment, based on early applications of IEC fusion technology. The most obvious such application will be in the low cost, point-of-use production of short half-life medical isotopes.

The international space treaty environment forms an important backdrop to a return to the Moon for its resources. The only space treaty related to the use of space resources to which the United States is a party, the 1967 "Treaty on Principles Governing the Activities of States in the Exploration and Use of Outer Space", or Outer Space Treaty. The Outer Space Treaty specifically provides a generally recognized legal framework for such use. The Treaty does not contain narrow rules relative to the extraction and use of lunar resources, however, its provisions imply certain appropriate guidelines for such activities. Compliance with these guidelines by a legal corporate entity under the laws of the United States would be straightforward and easily enforced.

Thus, a return to the Moon inherently has both great potential benefits as well as great challenges. The 20th Century, however, saw the beginning of the movement of the human species into space, best symbolized by the astronauts' photographs of the crescent Earth rising over the lunar horizon. A return to the Moon to stay early in the 21st Century will be the most logical next step in continuing the migration of our species out of Africa that began hundreds of thousands of years ago. This continuation of that migration, this time from the "cradle" of Earth, will provide for both the perpetuation of the human race and of human freedom.

Dr Harrison Schmitt
New Mexico October 2002

[1] Jones, E. "Apollo Lunar Surface Journal," Apollo 17, 145:26:32 Mission Elapsed Time, http://www.hq.nasa.gov/alsj/frame.html.

[2] Jones, E. "Apollo Lunar Surface Journal," Apollo 17, 145:27:07 Mission Elapsed Time, http://www.hq.nasa.gov/alsj/frame.html.

EDITOR'S INTRODUCTION

It hardly seems conceivable to me that, as I write this, nearly thirty years have passed since a human set foot on the Moon.

What happened? Where did the time go and why didn't we do more with that time?

The answers are not too hard to find. The world moved on; governments changed and so did priorities.

In 1972 a glitch in the pressurization of the S-IVB fuel tank delayed the launch of Apollo 17 just long enough that the local time at the Cape shifted to December 7th before the last manned Saturn V roared to life. It is a date which stands out in the minds of most Americans. It is that very same date in 1941 when an unprecedented aggressive attack took place on the US Naval Base at Pearl Harbour in Hawaii.

Shortly after that "day of infamy" the United States propelled itself into a World war which would terminate three and a half years later with the dawn of the nuclear age and the birth of a new super-power. The reason that I mention this is because when December 7th 2003 rolls around next year — as much time will separate us from Apollo 17 as separated Apollo 17 from Pearl Harbour. Perspective is critically important.

Early that morning in 1941 the leaders of the United States were completely blind-sided by the attacking forces of Imperial Japan. Every year since, on December 7th, the victims and heroes of that day are honored and remembered. Every year the black and white newsreels sputter to life and we relive those horrific moments as the USS Arizona belches forth smoke and fire. To be sure, it is a sobering memory which deserves every minute of annual attention it receives. In the last sixty years, the only unanswered question seems to have been instigated by those who would believe that President Roosevelt was forewarned about the attack and chose to ignore the warnings to the detriment of his fellow citizens in Honolulu.

None of the conspiracy-mongers have dared to suggest that Pearl Harbour was faked. They only suggest that it was not a surprise and was deliberately endured for political reasons. This harsh treatment of Roosevelt has been accepted by those who want to believe it, but surely everyone watches the newsreels and empathizes with those who suffered that day.

Thirty-one years later, on December 7th, the last great expedition was launched to the moon. Veteran astronaut Gene Cernan, accompanied by rookies Ron Evans and Harrison Schmitt, led the way to a perfect landing in the lunar valley of Taurus-Littrow.

Who could have believed in 1941 that only a few short decades later the people of the United States would be sending their ninth manned expedition to the moon? The world of the 1940's was so completely different as to be almost unrecognisable. The average person commuting around the world would be compelled to do so by sea. No one had flown a jet aircraft and the notion of breaking the sound barrier was considered science fiction. Over the next three decades the United States forged ahead to become the world's ascendant economic and technological power-house. The zenith of that success was surely the Apollo program.

We now find ourselves separated from Apollo by the same interval of time that divided Apollo from Pearl Harbour. What have we done with that time? Not much.

We live in a world where millions of people actually believe that America's greatest moment was a fraud. What can be said for a world that has let its education standards slip so badly that people who were alive at the time of Apollo 17 actually think that the moon landings were faked? Clearly the opportunists in Hollywood are making lots of money from their ridiculous documentaries, and they can hardly be expected to do otherwise. What is disconcerting is that so many people actually believe them.

It is not worth wasting paper addressing the so-called evidence that persuades these poor deluded souls.

However it is worth looking at the overall situation.

The Apollo program was the single greatest accomplishment in the history of mankind. It proved that we as a species were capable of securing our ultimate survival. Unlike every form of life before us, we are gifted enough to leave this speck of orbital flotsam and take our intelligence, our culture and our society to another world. That single accomplishment may be our only definitive claim to calling ourselves the paragon of terrestrial life. No other species in four billion years has gone so far — and yet some would have us believe that it was all a deception.

Thousands of people witnessed the triumphant night-launch of Apollo 17. Billions have since watched it on newsreels and television. Gene Cernan and Harrison Schmitt actually *resided* on another world for three complete days. For the first time ever, humankind sent a trained empirical scientist to study this new world and bring back his observations for our benefit. It was such an amazing accomplishment that not only were people totally astounded thirty years ago, but today people are so incredulous that they are beginning to give credence to the absurd accusations of trickery.

Some futurists have suggested that our era will ultimately be remembered for only one thing — Apollo. In a few thousand years no one will care about who fought who, over which tiny scrap of land. All that will be remembered is that in a short three year time-span, twelve men walked on another world for the first time. If the conspiracy theorists win the day it may be that our common heritage will be forgotten much sooner than we suspect. In another few decades every living person who witnessed or worked on Apollo will be dead. Without their eyewitness testimony what chance does the truth have? Who will remember humankind's greatest triumph?

Less than thirty years after America had its worst hour, it had its greatest triumph. Apollo elevated the United States to a prominence that few could have anticipated and none can measure. Surely it is now time to restore some vision and hope and to send a host of scientists in Harrison Schmitt's footsteps? To push the envelope and add to the treasure-trove of knowledge brought back by the crew of Apollo 17. At the very least we owe it to our progeny; who will inevitably face-off with some kind of earth-shattering asteroid that will make Pearl Harbour pale by comparison.

I am humbled and honored to have been alive at the time of the Apollo program. I am fortunate enough to have been given an opportunity to study in detail the amazing history of that time and to have met some of the heroes who contributed to its success. I was not born in time to have witnessed Pearl Harbour and yet I do believe it happened and I still honor those who were there that day. In the face of opportunistic commercialism we must not allow ourselves to become so cynical that we don't even believe common-sense. We must strive to keep these memories alive for future generations and revitalise that sense of wonder we all had as we watched Cernan and Schmitt searching for the origins of the universe.

And so, as we remember December 7th 1941 for its tragedy and its heroes we should not forget to remember December 7th 1972 as the culmination of America's triumph over the tyrannical forces which were arrayed against it and its allies. That terrible war brought with it the seeds of a new world here on Earth as well as the birth of a new era of exploration. These new possibilities were epitomised by Gene Cernan when he said, "As we leave the Moon at Taurus-Littrow, we leave as we came and, God willing, as we shall return, with peace and hope for all mankind."

The truth is out there. Let's go see it....again.

Robert Godwin
(Editor)

NASA NEWS
RELEASE NO: 72-220K

NATIONAL AERONAUTICS AND SPACE ADMINISTRATION

Washington, D. C. 20546
202-755-8370

P

R

E

S

S

FOR RELEASE: Sunday November 26, 1972
PROJECT: APOLLO 17 (To be launched no earlier than Dec. 6)
David Garrett (Phone: 202 755-3114)

GENERAL RELEASE

APOLLO 17 LAUNCH DECEMBER 6

The night launch of Apollo 17 on December 6 will be visible to people on a large portion of the eastern seaboard as the final United States manned lunar landing mission gets underway.

Apollo 17 will be just one and a third days short of the US spaceflight duration record of 14 days set in 1965 by Gemini VII, and will be the sixth and final Moon landing in the Apollo program. Two of the three-man Apollo 17 crew will set up the fifth in a network of automatic scientific stations during their three-day stay at the Taurus-Littrow landing site.

In addition to erecting the scientific data relay station, Apollo 17 has the objectives of exploring and sampling the materials and surface features at the combination highland and lowland landing site and to conduct several inflight experiments and photographic tasks.

The Taurus-Littrow landing site is named for the Taurus mountains and Littrow crater located in a mountainous region southeast of the Serenitatis basin. Dominant features of the landing site are three rounded hills, or "massifs" surrounding the relatively flat target point and a range of what lunar geologists describe as sculptured hills.

K

Apollo 17 will be manned by Eugene A. Cernan, commander, Ronald E. Evans, command module pilot, and Harrison H. Schmitt, lunar module pilot. Cernan previously flew in space aboard Gemini 9 and Apollo 10, while Apollo 17 will be the first flight into space for Evans and Schmitt. Civilian astronaut Schmitt is also a professional geologist. Cernan holds the rank of Captain and Evans is a Commander in the US Navy.

I

During their 75 hours on the lunar surface, Cernan and Schmitt will conduct three seven-hour periods of exploration, sample collecting and emplacing the Apollo Lunar Surface Experiment Package (ALSEP). Four of the five Apollo 17 ALSEP experiments have never been flown before.

T

ALSEP, powered by a nuclear generator, will be deployed and set into operation during the first extravehicular activity (EVA) period, while the second and third EVAs will be devoted mainly to geological exploration and sample collection.

The crew's mobility on the surface at Taurus-Littrow again will be enhanced by the electric-powered Lunar Roving Vehicle (LRV). Attached to a mount on the front of the LRV will be a color television camera which can be aimed and focussed remotely from the Mission Control Center.

Cameras operated by Cernan and Schmitt will further record the characteristics of the landing site to aid in postflight geological analysis.

Data on the composition, density and constituents of the lunar atmosphere, a temperature profile of the lunar surface along the command module ground track and a geologic crosssection to a depth of 1.3 kilometers (.8 miles) will be gathered by instruments in the service module Scientific Instrument Module (SIM). Evans will operate the SIM bay experiments and mapping cameras while Cernan and Schmitt are on the lunar surface. During transearth coast, he will leave the spacecraft to recover film cassettes from the mapping cameras and the lunar sounder.

Apollo 17 will be launched from Kennedy Space Center Launch Complex 39 at 9:53 pm EST December 6. Lunar surface touchdown by the lunar module will be at 2:55 pm EST December 11, with return to lunar orbit scheduled at 5:56 pm EST December 14. After jettisoning the lunar module ascent stage to impact on the Moon, the crew will use the service propulsion system engine to leave lunar orbit for the return to Earth. Transearth injection will be at 6:32 pm EST on December 16. Command module splashdown in the Pacific, southeast of Samoa, will be at 2:24 pm EST December 19. There the spacecraft and crew will be recovered by the USS Ticonderoga.

Communications call signs to be used during Apollo 17 are America for the command module and Challenger for the lunar module. During docked operations and after lunar module jettison, the call sign will be simply "Apollo 17."

Apollo 17 backup crewmen are US Navy Captain John W. Young, commander; USAF Lieutenant Colonel Stuart A. Roosa, command module pilot; and USAF Colonel Charles M. Duke. All three have prior spaceflight experience: Young on Gemini 3 and 10, and Apollo 10 and 16; Roosa on Apollo 14; and Duke on Apollo 16.

Summary timeline of major Apollo 17 events:

Event	December Date	EST
Launch	6	9:53 pm
Translunar Injection	7	1:12 am
TV-Docking & LM extraction	7	2:05 am
Lunar Orbit Insertion	10	2:48 pm
Descent Orbit Insertion #1	10	7:06 pm
Descent Orbit Insertion #2	11	1:53 pm
Lunar Landing	11	2:54 pm
Start EVA 1 (7 hours)	11	6:33 pm
TV Camera on	11	7:48 pm
Start EVA 2 (7 hours)	12	5:03 pm
TV Camera on	12	5:31 pm
Start EVA 3 (7 hours)	13	4:33 pm
TV Camera on	13	4:58 pm
Lunar Liftoff (TV on)	14	5:56 pm
TV-LM & CSM Rendezvous	14	7:31 pm
TV Docking	14	7:54 pm
Transearth Injection	16	6:31 pm
TV- View of Moon	16	6:46 pm
Transearth Coast EVA (TV-1 hr)	17	3:18 pm
TV - Press Conference	18	6:00 pm
Splashdown	19	2:24 pm

(End of general release; background information follows.)

A COMPARISON OF LUNAR SCIENCE BEFORE AND AFTER APOLLO

The astronomical observations of the Moon prior to Apollo give us a very detailed picture of the surface of this planet. However, even the most sensitive telescopes were unable to furnish the variety of scientific data that is necessary to the understanding of the history and evolution of the planet. In particular, it was necessary to know something about the chemistry and something about the internal state or condition of the planet before we could do much more than speculate about the origin and past history of the Moon. The most important scientific observations concerning the Moon that existed prior to the direct exploration of the Moon by either manned or unmanned spacecraft are as follows:

1) The mean density of the moon is 3.34 gm/cc. When this number is compared to the density of other planets (this comparison involves a substantial correction for the effects of pressure in planets as large as the Earth and Venus), we see that the density of the Moon is less than that of any of the other terrestrial planets. If we accept the hypothesis that stony meteorites are samples of the asteroids, we also observe that the Moon is lower in density than the parent bodies of many meteorites. This single fact has been an enigma to anyone attempting to infer a chemical composition for the Moon. One thing can be clearly concluded from this fact — that is, that the Moon has less metallic iron than the Earth. The difference between the lunar density and that of chondritic meteorites is particularly puzzling because these objects have compositions that are similar to those of the Sun once one removes those elements which form gaseous compounds at modest temperatures (hydrogen, helium, nitrogen, carbon, neon, and the other rare gases).

2) The second major characteristic of the Moon goes back to Galileo, who observed that the Earth-facing side of the Moon consisted of mountainous regions that he designated terra, and smoother, physiographically lower regions which he designated mare by analogy with the terrestrial oceans and continents. The albedo or reflectivity of these two regions is markedly different — the mare regions being very dark when compared to the terra regions. Astronomical studies added a great deal of detail to Galileo's discovery, including some rather fine features such as the rilles which were just barely resolved by good telescopes. However, the cause of this fundamental physiographic difference was not well understood before the era of Apollo. The explanation of the relatively smooth mare basins ranged from the conclusion that they were very extensive lava fields to the hypothesis that they were, in fact, dust bowls — that is, extensive dust deposits. There were even some scientists who seriously suggested that they were filled by a type of sedimentary rock that was deposited at a very early stage in lunar history when the Moon had an atmosphere.

3) The origin of the circular depressions or craters, which are the most common physiographic feature of the lunar surface, was the basis of continual scientific controversy. Two types of explanations were offered — first, that they were volcanic features similar to terrestrial calderas or volcanic collapse features; secondly, that they were produced by projectiles impacting on the lunar surface in the way that meteorites had occasionally been observed to fall on Earth. In fairness to the proponents of the various theories, it should be recognized that no one ever claimed that all craters were either meteoritic or volcanic. Those scientists tending to favor the volcanic origin emphasized that large numbers — including some craters much larger than any terrestrial caldera — were volcanic in origin. Others favoring the impact origin also admitted that a few atypical craters such as Davy Rille and the dark halo craters of Alphonsus may, indeed, be evidence of minor volcanism on the lunar surface.

4) In parallel with the role of volcanism on the lunar surface, there were two schools of thought on the thermal history of the Moon. The first of these held that the Moon was a relatively inactive body which may have undergone some chemical differentiation which, in any event, took place very early in lunar history. The second expected that the Moon was similar to the Earth with a long and continuous record of volcanism and chemical differentiation. Some adherents to this school fully expected that some volcanism may have persisted to the most recent geologic epochs; that is, as recently as 10 million years ago.

5) The chemistry of the lunar surface was a total unknown before Surveyor V. Nevertheless, there were a number of definite suggestions — for example, it was at one time suggested that carbonaceous chondrites were derived from the dark mare regions of the Moon. Others suggested that a type of meteorite known as eucrites was representative of the lunar surface. Still others suggested that a very silica-rich glass found in mysterious terrestrial objects called tektites must represent parts of the lunar surface. One could not even be sure that these hypotheses were all inconsistent with each other. At this point in time, we will

never know the extent to which the Surveyor analyses may have affected our understanding of the Moon. The data returned from these analyses were of surprisingly high quality. They were, however, so quickly superseded by the analyses of the returned samples that there was never sufficient time for them to be completely integrated into scientific thinking on the Moon.

6) Several other results obtained by unmanned spacecraft helped set the stage for Apollo. They are the discovery of the mascons, which require a remarkably rigid or strong lunar shallow interior — the determination (by Explorer 35) that the Moon had a very weak, perhaps nonexistent, magnetic field; and finally, the observation (by both Russian and American spacecraft) that the lunar backside was very different from the frontside in that dark mare regions were essentially absent from the backside of the Moon.

As we anticipate the sixth manned landing on the lunar surface, we are infinitely richer in facts concerning the Moon. Many of the facts and observations have already been tentatively assembled into theories and models which are leading us to a genuine understanding of the Moon's history. In other cases, it is proving extremely difficult to come up with an explanation that accounts for all of these facts in a self-consistent way. The major areas of understanding which have come out of the unmanned exploration and five manned landings are briefly outlined here:

1) We now have a rather definite and reliable time scale for the sequence of events in lunar history. In particular, it has been established with some confidence that the filling of the mare basins largely took place between 3.1 and 3.8 billion years ago. Since these surfaces represent the major physiographic features on the lunar surface, we can immediately infer that the bulk of lunar history recorded on the surface of the Moon (that is, the time of formation of more than 90 percent of the craters) took place before 4 billion years ago. This is quite different from the terrestrial situation where most of the Earth's ocean basins are younger than 300 million years, and rocks older than 3 billion years make up an almost insignificant proportion of the surface of the Earth. One of the major objectives frequently stated by groups of scientists involved in planning the Apollo science activities was to find rocks that might date back to the formation of the Moon, itself. Underlying this objective was the hope that one might find a primitive or predifferentiation sample of the planet. Up until now, this objective has eluded us in the sense that none of the samples returned to date is unmodified by younger events in lunar history. There is strong circumstantial evidence (for example, the apparent age of some soils) that rocks dating back to 4.5 or 4.6 billion years must exist on the lunar surface. However, it now appears that the intense bombardment of the lunar surface by projectiles that range in size up to tens of kilometers in diameter was rather effective in resetting most of the clocks used to determine the absolute ages of rocks. The widespread occurence of highland material with an apparent age of 3.8 - 4.1 billion years is today associated by some scientists with the formation of the Imbrium basin, which is thought to be produced by the collision of a 50 kilometer projectile with the lunar surface.

We should recognize, however, that more detailed studies of some of the returned samples may change the interpretation of these ages. In other words, we cannot be nearly as sure that we know the age of formation of the Imbrium basin as we are of the time of crystallization of the mare volcanic rocks.

2) The relative importance of volcanic and impact-produced features on the lunar surface is today rather well established. There is almost unanimous agreement that the dark mare regions are, indeed, underlain by extensive lava flows. This is shown both by the rocks returned from the Apollo 11, 12, and 15 sites and by the high resolution photographs which give us very convincing pictures of features comparable to terrestrial lava flows. On the other hand, almost all craters appear to be caused by impacting projectiles. The occurrence of volcanic rocks in the terra regions is an open question. Preliminary interpretations of the Apollo 16 samples suggest that volcanic activity in the highland region may be highly restricted or virtually nonexistent.

3) A major objective underlying many of the Apollo experiments was the investigation of the lunar interior. Most of our information concerning the interior of the Earth derives from a knowledge of the way in which the velocity of acoustic waves varies with depth. The study of terrestrial earthquakes has provided a detailed picture of these variations within the Earth. The Apollo 11 seismograph indicated that when

compared to the Earth the Moon is seismically very quiet. This result: is, of course, consistent with the conclusion that volcanism and other types of tectonic activity have been rare or absent from the lunar scene for the last 2-3 billion years. It was a disappointment in that it indicated that information regarding the interior would be sparse. However, the use of SIVB impacts and the very fortunate impact of a large meteorite on May 13, 1972, have today shown a remarkable structure for the upper 150 kilometers of the Moon. In particular, we have learned from lunar seismology that the Moon has a crust more than 60 kilometers thick. More precisely, one should say that there is a seismic discontinuity where the velocity of sound increases suddenly from 7 kilometers per second to 8 kilometers per second at about 65 kilometers depth. The precise origin of this discontinuity is still a subject for debate. The most commonly held explanation suggests that it is due to the chemical differentiation of the upper part of the Moon in particular, that an extensive, partial melting of this region produced low-density liquids which arose to cover the lunar surface, leaving a high-density residue that accounts for the high velocity material below 65 kilometers. A minority opinion holds that the velocity contrast at 65 kilometers is due to a pressure-induced phase change.

4) We now have a much more detailed understanding of the Moon's present magnetic field. It is clearly not negligible as was thought prior to the Apollo missions. The magnetometers emplaced on the lunar surface reveal a surprisingly strong, but variable, field. Both the direction and the intensity of the magnetic field vary. This heterogeneity in the local field is, of course, smoothed out when one moves away from the lunar surface. So the field seen by an orbiting spacecraft is much lower than that recorded on the surface. We have also determined that the mare lava flows crystallized in a magnetic field which was much stronger than that of the present Moon. This raises the very interesting possibility that during its early history, the Moon was either embedded in a relatively strong interplanetary magnetic field or had a magnetic field of its own which has since disappeared. Either possibility presents very serious problems in the sense that we are forced to make assumptions which are not entirely consistent with what some scientists hypothesize we know about the Sun or the early history of the Moon.

5) The fluctuation in the magnetic fields measured at the lunar surface is a function of the flux of incoming charged particles (solar wind) and the internal electrical conductivity of the Moon. Careful study of these fluctuations shows that the Moon has a relatively low conductivity. To a first order, the conductivity of most silicates is a function of temperature and chemical composition. It is particularly sensitive to the abundance of ferrous and ferric iron. At present, one cannot completely sort out these two parameters. If the interior of the Moon has a "normal" iron concentration, the conductivities appear to place upper limits of 1200-1500°C on the temperature of the deep interior.

6) The heat escaping by conduction from the interior of a planet depends on the amount of heat produced by the decay of radioactive elements, the thermal conductivity of the deep interior, and the initial temperature of the deep interior. Using rather imprecise models and making reasonable assumptions concerning the abundance of the radioactive elements potassium, uranium, and thorium, it was expected that the energy flux from the interior of the Moon would be substantially lower than that for the Earth simply because the Moon is much smaller.

The first measurement of this quantity at the Apollo 15 site indicates that this is not the case. If this measurement is characteristic of the whole Moon, the only plausible explanation that has been put forth to date requires: First, that the Moon is richer in the radioactive elements uranium and thorium than the Earth; and secondly, that these elements are strongly concentrated into the upper parts of the Moon. When combined with the observations on the volcanic history of the Moon and the present-day internal temperatures, the energy flux leads to two current pictures of lunar evolution. The first assumes that the variation in radioactivity with depth is a primary characteristic of the planet; that is, the planet was chemically layered during its formation. In this case, the initial temperature of the lunar interior below 500 kilometers was relatively low, and the deep interior of the Moon gradually became hotter, perhaps reaching the melting point during the last billion years. Volcanism can be entirely accounted for by early melting in the outer 400 kilometers of the Moon which were formed at a higher temperature than the central core of the Moon. The alternative model of thermal evolution assumes that the Moon was chemically homogeneous when it formed and underwent extensive chemical differentiation to bring radioactivity to the surface very shortly after its formation - in other words, we begin with a molten Moon. Each of these models has some problems. The objectives are, in a sense, esthetic. Some scientists object to the hypothesis that the Moon was initially heterogeneous on the grounds that such structure requires special assumptions regarding the processes that formed planets. Others object to the idea of a molten Moon that conveniently differentiates to bring the radioactivity to the surface as an equally arbitrary idea.

7) The most extensive and diverse data obtained on the lunar surface are those concerned with the chemistry and mineralogy of the surface materials. The study of samples from the six Apollo sites and two Luna sites reveals a number of chemical characteristics that are apparently moonwide. There is, nevertheless, some hesitancy to generalize from these relatively minute samples to the whole lunar surface. Fortunately, two experiments carried out in lunar orbit provided excellent data regarding the regional distribution of various rock types.

The x-ray fluorescence experiment very convincingly defined the prime difference between the chemistry of the mare and highland regions. It showed that the highland regions are unusually rich in aluminum — much richer, in fact, than most terrestrial continents. This observation, along with the ubiquitous occurence of fragments which show an apparent enrichment in the mineral plagioclase, leads to the strong hypothesis that the regolith and soil of highland regions is underlain by a "crust" similar to the terrestrial rock designated anorthosite.

The x-ray fluorescence results show that mare regions have aluminum concentrations 2-3 times lower than those of the terra or highland regions, along with magnesium concentrations that are 1½ - 2 times greater than those of the terra regions. These differences are totally consistent with the chemistry of the returned samples. When combined with data from the returned samples, these observations provide an excellent explanation of the morphological and albedo differences. We have, for example, determined that all mare basalts are unusually rich in iron and sometimes rich in titanium. The high iron concentration of the mare vis-a-vis the low concentration of the highlands is the basic explanation of the albedo differences since both glass and mineral substances rich in iron and titanium are usually very dark. A second experiment carried from lunar orbit shows that the region north and south of the crater Copernicus is remarkably rich in radioactive elements. A band going north from the Fra Mauro site to a region west of the Apollo 15 site contains soil that must have 20 times more uranium and thorium than most of the mare or terra in other parts of the Moon. The existence of a rock rich in these elements was also inferred from samples from the Apollo 12, 14 and 15 sites. The uneven distribution of this rock — commonly designated KREEP basalt — is a major enigma in the early evolution of the Moon. The time of formation of rocks with this chemistry is not well determined. There is, however, strong circumstantial evidence that some of the uranium-rich KREEP basalts were originally formed between 4.3 and 4.4 billion years ago. Both the samples and orbital geochemical experiments indicate that the three most common rocks in the lunar surface are plagioclase - or aluminum-rich anorthosites; uranium, thorium-rich "KREEP" basaltic rocks; and iron-rich mare basalts. With the exception of the mare basalts, we do not have well documented, unambiguous theories or models that explain the chemical or mineralogical characteristics of these rocks. Nevertheless, the differences between the lunar rocks and terrestrial rocks are so marked that we can conclude that the Moon must be chemically different from the Earth. The Moon appears to be much richer in elements that form refractory compounds at temperatures of 1600-1800°K. Thermodynamic considerations show that calcium,aluminum, and titanium silicates are the most refractory compounds that exist in a solar dust cloud. Many scientists are now coming to the conclusion that the chemistry of the lunar surface is telling us that some separation of solid material and gas in this dust cloud took place at temperatures in excess of 1600°K. The Moon is also strongly depleted in elements that are volatile at high temperatures. This is, of course, consistent with the enrichment in refractory elements.

None of the three theories regarding the origin of the moon — that is, separation from the Earth, capture from a circumsolar orbit, or formation from a dust cloud surrounding the Earth — can be absolutely ruled out from the present data. The chemical difference between the Earth and the Moon must, however, be explained if the Moon was torn out of the Earth. The depletion in volatiles and enrichment in refractories place a constraint on this theory that will be very difficult to account for.

APOLLO 17 MISSION OBJECTIVES

The final mission in the Apollo lunar exploration program will gather information on yet another type of geological formation and add to the network of automatic scientific stations. The Taurus-Littrow landing site offers a combination of mountainous highlands and valley lowlands from which to sample surface materials. The Apollo 17 Lunar Surface Experiment Package (ALSEP) has four experiments never before flown, and will become the fifth in the lunar surface scientific station network. Data continues to be relayed to Earth from ALSEPs at the Apollo 12, 14, 15 and 16 landing sites.

The three basic objectives of Apollo 17 are to explore and sample the materials and surface features at Taurus-Littrow, to set up and activate experiments on the lunar surface for long-term relay of data, and to conduct inflight experiments and photographic tasks.

The scientific instrument module (SIM) bay in the service module is the heart of the inflight experiments effort on Apollo 17. The SIM Bay contains three experiments never flown before in addition to high-resolution and mapping cameras for photographing and measuring properties of the lunar surface and the environment around the Moon.

While in lunar orbit, command module pilot Evans will have the responsibility for operating the inflight experiments during the time his crewmates are on the lunar surface. During the homeward coast after transearth injection, Evans will perform an in-flight EVA hand-over-hand back to the SIM Bay to retrieve film cassettes from the SIM Bay and pass them back into the cabin for return to Earth.

The range of exploration and geological investigations made by Cernan and Schmitt at Taurus-Littrow again will be extended by the electric-powered lunar roving vehicle. Cernan and Schmitt will conduct three seven-hour EVAs.

Apollo 17 will spend an additional two days in lunar orbit after the landing crew has returned from the surface. The period will be spent in conducting orbital science experiments and expanding the fund of high-resolution photography of the Moon's surface.

SITE SCIENCE RATIONALE

	APOLLO 11	APOLLO 12	APOLLO 14	APOLLO 15	APOLLO 16	APOLLO 17
TYPE	MARE	MARE	HILLY UPLAND	MOUNTAIN FRONT/ RILLE/MARE	HIGHLAND HILLS AND PLAINS	HIGHLAND MASSIFS AND DARK MANTLE
PROCESS	BASIN FILLING	BASIN FILLING	EJECTA BLANKET FORMATION	• MOUNTAIN SCARP • BASIN FILLING • RILLE FORMATION	• VOLCANIC CONSTRUCTION • HIGHLAND BASIN FILLING	• MASSIF UPLIFT • LOW LAND FILLING • VOLCANIC MANTLE
MATERIAL	BASALTIC LAVA	BASALTIC LAVA	DEEP-SEATED CRUSTAL MATERIAL	• DEEPER-SEATED CRUSTAL MATERIAL • BASALTIC LAVA	VOLCANIC HIGHLAND MATERIALS	• CRUSTAL MATERIAL • VOLCANIC DEPOSITS
AGE	OLDER MARE FILLING	YOUNGER MARE FILLING	• EARLY HISTORY OF MOON • PRE-MARE MATERIAL • IMBRIUM BASIN FORMATION	• COMPOSITION AND AGE OF APENNINE FRONT MATERIAL • RILLE ORIGIN AND AGE • AGE OF IMBRIUM MARE FILL	• COMPOSITION AND AGE OF HIGHLAND CONSTRUCTION AND MODIFICATION • COMPOSITION AND AGE OF CAYLEY FORMATION	• COMPOSITION AND AGE OF HIGHLAND MASSIFS AND POSSIBLY OF LOW-LAND FILLING • COMPOSITION AND AGE OF DARK MANTLE • NATURE OF A ROCK LANDSLIDE

LAUNCH OPERATIONS

Prelaunch Preparations

NASA's John F. Kennedy Space Craft Center performs preflight checkout, test and launch of the Apollo 17 space vehicle. A government-industry team of about 600 will conduct the final countdown, 500 of them in Firing Room 1 in the Launch Control Center and 100 in the spacecraft control rooms in the Manned Spacecraft Operations Building (MSOB).

The firing room team is backed up by more than 5,000 persons who are directly involved in launch operations at KSC from the time the vehicle and spacecraft stages arrive at the Center until the launch is completed.

Initial checkout of the Apollo spacecraft is conducted in work stands and in the altitude chambers in the Manned Spacecraft Operations Building at Kennedy Space Center. After completion of checkout there, the assembled spacecraft is taken to the Vehicle Assembly Building (VAB) and mated with the launch vehicle. There the first integrated spacecraft and launch vehicle tests are conducted. The assembled space vehicle is then rolled out to the launch pad for final preparations and countdown to launch.

Flight hardware for Apollo 17 began arriving at KSC in October, 1970, while Apollo 14 was undergoing checkout in the VAB.

The command/service module arrived at KSC in late March, 1972, and was placed in an altitude chamber in the MSOB for systems tests and unmanned and manned chamber runs. During these runs, the chamber air was pumped out to simulate the vacuum of space at altitudes in excess of 200,000 feet. It is during these runs that spacecraft systems and astronauts' life support systems are tested.

The lunar module at KSC in June and its two stages were moved into an altitude chamber in the MSOB after an initial receiving inspection. It, too, was given a series of systems tests and unmanned and manned chamber runs. The prime and backup crews participated in the chamber runs on both the LM and the CSM.

In July, the LM and CSM were removed from the chambers. After installing the landing gear on the LM and the SPS nozzle on the CSM, the LM was encapsulated in the spacecraft LM adapter (SLA) and the CSM was mated to the SLA. On August 24, the assembled spacecraft was moved to the VAB where it was mated to the launch vehicle.

The Lunar Roving Vehicle (LRV), which the Apollo 17 crew will use in their exploratory traverses of the lunar surface, arrived at KSC on June 2. Following a series of tests, which included a mission simulation on August 9 and a deployment demonstration on August 10, the LRV was flight installed in the Lunar Module's descent stage on August 13.

Erection of the Saturn V launch vehicle's three stages and instrument unit on Mobile Launcher 3 in the VAB's High Bay 3 began on May 15 and was completed on June 27. Tests were conducted on individual systems on each of the stages and on the overall launch vehicle before the spacecraft was erected atop the vehicle on August 24.

Rollout of the space vehicle from the VAB to Pad A at KSC's Launch Complex 39 was accomplished on August 28.

Processing and erection of the Skylab 1 and Skylab 2 launch vehicles was underway in the VAB while preparations were made for moving Apollo 17 to the pad, giving KSC three Saturn space vehicles "in flow" for the first time since the peak of Apollo activity in 1969.

After the move to the pad, the spacecraft and launch vehicle were electrically mated and the first overall test (plugs-in) was conducted on October 11.

The plugs-in test verified the compatibility of the space vehicle systems, ground support equipment, and off-site support facilities by demonstrating the ability of the systems to proceed through a simulated countdown, launch and flight. During the simulated flight portion of the test, the systems were required to respond to both normal and emergency flight conditions.

The space vehicle Flight Readiness Test was conducted October 18-20. Both the prime and backup crews participate in portions of the FRT, which is a final overall test of the space vehicle systems and ground support equipment when all systems are as near as possible to a launch configuration.

After hypergolic fuels were loaded aboard the space vehicle and the launch vehicle first stage fuel (RP-1) was brought aboard, the final major test of the space vehicle began. This was the Countdown Demonstration Test (CDDT), a dress rehearsal for the final countdown to launch.

The CDDT for Apollo 17 was divided into a "wet" and a "dry" portion. During the first or "wet" portion, the entire countdown, including propellant loading was carried out down to 8.9 seconds, the time for ignition sequence start. The astronaut crew did not participate in the wet CDDT.

At the completion of the wet CDDT, the cryogenic propellants (liquid oxygen and liquid hydrogen) were off-loaded and the final portion of the countdown was re-run, this time simulating the fueling and with the prime astronaut crew participating as they will on launch day.

Apollo 17 will mark the 12th Saturn V launch from KSC and the eleventh from Complex 39's Pad A. Only Apollo 10 was launched from Pad B.

Because of the complexity involved in the checkout of the 110.6 meter (363-foot) tall Apollo/Saturn V configuration, the launch teams make use of extensive automation in their checkout. Automation is one of the major differences in checkout used in Apollo compared to the procedures used in earlier Mercury and Gemini programs.

Computers, data display equipment and digital data techniques are used throughout the automatic checkout from the time the launch vehicle is erected in the VAB through liftoff. A similar but separate computer operation called ACE (Acceptance Checkout Equipment) is used to verify the flight readiness of the spacecraft. Spacecraft checkout is controlled from separate rooms in the MSOB.

COUNTDOWN

The Apollo 17 precount activities will start at T-6 days. The early tasks include electrical connections and pyrotechnic installation in the space vehicle. Mechanical buildup of the spacecraft is completed, followed by servicing of the various gases and cyrogenics to the CSM and LM. Once this is accomplished, the fuel cells are activated.

The final countdown begins at T-28 hours when the flight batteries are installed in the three stages and instrument unit of the launch vehicle. At the T-9 hour mark, a built-in hold of nine hours and 53 minutes is planned to meet contingencies and provide a rest period for the launch crew. A one hour built-in hold is scheduled at T-3 hours 30 minutes.

Following are some of the highlights of the latter part of the count:

T-10 hours, 15 minutes	Start mobile service structure move to park site.
T-9 hours	Built-in hold for nine hours and 53 minutes. At end of hold, pad is cleared for LV propellant loading.
T-8 hours, 05 minutes	Launch vehicle propellant loading - Three stages (LOX in first stage, LOX and LH2 in second and third stages). Continues thru T-3 hours 38 minutes.

T-4 hours, 00 minutes	Crew medical examination.
T-3 hours, 30 minutes	Crew supper.
T-3 hours, 30 minutes	One-hour built-in hold.
T-3 hours, 06 minutes	Crew departs Manned Spacecraft Operations Building for LC-39 via transfer van.
T-2 hours, 48 minutes	Crew arrival at LC-39
T-2 hours, 40 minutes	Start flight crew ingress
T-1 hour, 51 minutes	Space Vehicle Emergency Detection System test (Young participates along with launch team).
T-43 minutes	Retract Apollo access arm to standby position (12 degrees).
T-42 minutes	Arm launch escape system. Launch vehicle power transfer test, LM switch to internal power.
T-37 minutes	Final launch vehicle range safety checks (to 35 minutes)
T-30 minutes	Launch vehicle power transfer test, LM switch over to internal power.
T-20 minutes to T-10 minutes	Shutdown LM operational instrumentation.
T-15 minutes	Spacecraft to full internal power.
T-6 minutes	Space vehicle final status checks.
T-5 minutes, 30 seconds	Arm destruct system.
T-5 minutes	Apollo access arm fully retracted.
T-3 minutes, 6 seconds	Firing command (automatic sequence).
T-50 seconds	Launch vehicle transfer to internal power.
T-8.9 seconds	Ignition start.
T-2 seconds	All engines running.
T-0	Liftoff.

NOTE: Some changes in the countdown are possible as a result of experience gained in the countdown demonstration test which occurs about two weeks before launch.

Launch Windows

The mission planning considerations for the launch phase of a lunar mission are, to a major extent, related to launch windows. Launch windows are defined for two different time periods: a "daily window" has a duration of a few hours during a given 24-hour period: a "monthly window" consists of a day or days which meet the mission operational constraints during a given month or lunar cycle.

Launch windows are based on flight azimuth limits of 72° to 100° (Earth-fixed heading east of north of the launch vehicle at the end of the roll program), on booster and spacecraft performance, on insertion tracking, and on Sun elevation angle at the lunar landing site. All times are EST.

Launch Windows

LAUNCH DATE	OPEN	CLOSE	SUN ELEVATION ANGLE
December 6, 1972	9:53 pm EST	1:31 am	13°
December 7, 1972	9:53 pm	1:31 am	16.9-19.1°
January 4, 1973*	9:51 pm EST	11:52 pm	6.8°
January 5, 1973	8:21 pm	11:51 pm	10.2-11.1°
January 6, 1973	8:28 pm	11:56 pm	20.3-22.4°
February 3, 1973	6:47 pm EST	10:13 pm	13.3-15.5°
February 4, 1973	6:58 pm	10:20 pm	13.5-15.5°

Launch azimuth limits for January are 84° to 100°.

Ground Elapsed Time Update

It is planned to update, if necessary, the actual ground elapsed time (GET) during the mission to allow the GET clock to coincide with the preplanned major flight event times should the event times be changed because of late liftoff or trajectory dispersions.

For example, if the flight plan calls for descent orbit insertion (DOI) to occur at GET 88 hours, 55 minutes and the flight time to the Moon is two minutes longer than planned due to trajectory dispersions at translunar injection, the GET clock will be turned back two minutes during the translunar coast period so that DOI occurs at the pre-planned time rather than at 88 hours, 57 minutes. It follows that the other major mission events would then also be accomplished at the pre-planned GET times.

Updating the GET clock will accomplish in one adjustment what would otherwise require separate time adjustments for each event. By updating the GET clock, the astronauts and ground flight control personnel will be relieved of the burden of changing their checklists, flight plans, etc.

The planned times in the mission for updating GET will be kept to a minimum and will, generally, be limited to three updates. If required, they will occur at about 63, 96 and 210 hours into the mission. Both the actual GET and the update GET will be maintained in the MCC throughout the mission.

Synchronization of Ground Elapsed Time (GET)

The realtime GET is synchronized with the Flight Plan GET. In TLC, the GET is synchronized at 63:30 if the time propagated ahead to start of Rev 2 is more than ±1 minute from the flight plan GET. In lunar orbit the GET is synchronized at 95:50 and 209:50 if the time propagated ahead to start of Rev 26 and Rev 64 respectively is more than ±2 minutes from the flight plan GET. The synchronization is performed by a V70 uplink from the ground followed by the crew synchronizing the mission timer to the CMC clock.

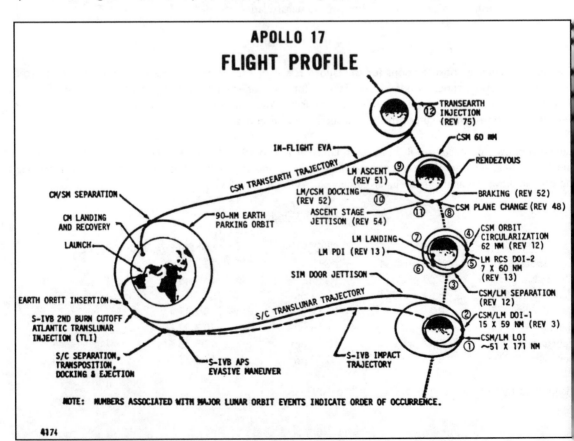

APOLLO 17 FLIGHT PROFILE

NOTE: NUMBERS ASSOCIATED WITH MAJOR LUNAR ORBIT EVENTS INDICATE ORDER OF OCCURRENCE.

4174

COMPARISON OF APOLLO MISSIONS

	PAYLOAD DELIVERED TO LUNAR SURFACE KG (LBS)		EVA DURATION (HR:MIN)	SURFACE DISTANCE TRAVERSED (KM)	SAMPLES RETURNED KG (LBS)	
APOLLO 11	104	(225)	2:24	.25	20.7	(46)
APOLLO 12	166	(365)	7:29	2.0	34.1	(75)
APOLLO 14	209	(460)	9:23	3.3	42.8	(94)
APOLLO 15	550	(1210)	18:33	27.9	76.6	(169)
APOLLO 16	558	(1228)	20:14	26.7	95.4	(210)
APOLLO 17 (PLANNED)	558	(1228)	21:00	32.9	95.4	(210)

LAUNCH AND MISSION PROFILE

he Saturn V launch vehicle (SA-512) will boost the Apollo 17 spacecraft from Launch Complex 39A at the ennedy Space Center at 9:53 p.m. EST December 6, 1972, on an azimuth of 72 degrees.

he first stage (S-IC) will lift the vehicle 66 kilometers (33 nautical miles) above the Earth. After separation, ne booster stage will fall into the Atlantic Ocean about 662 km (301 nm) downrange from Cape Kennedy pproximately nine minutes, 12 seconds after liftoff.

he second stage (S-II) will push the vehicle to an altitude of about 173 km (87 nm). After separation, the S- stage will follow a ballistic trajectory which will plunge it into the Atlantic about 4,185 km (2,093 nm) ownrange about 19 minutes, 51 seconds into the mission.

he single engine of the third stage (S-IVB) will insert the spacecraft into a 173—kilometer (93 nm) circular arth parking orbit before it is cut off for a coast period. When reignited, the engine will inject the Apollo pacecraft into a trans-lunar trajectory.

LAUNCH EVENTS

Time Hrs Min Sec Event	Vehicle Wt Kilograms (Pounds)*	Altitude meters (Feet)*	Velocity Mtrs/Sec (Ft/Sec)*	Range Kilometers (Naut Mi)*
00 00 00 First Motion	2,923,461 (6,445,127)	60 (198)	0 (0)	0 (0)
00 01 23 Maximum Dynamic Pressure	1,823,015 (4,019,061)	13,329 (43,731)	505 (1,658)	6 (3)
00 02 19 S-IC Center Engine Cutoff	1,080,495 (2,382,085)	46,703 (153,224)	1,717 (5,634)	52 (28)

Time Hrs Min Sec	Event	Vehicle Wt Kilograms (Pounds)*	Altitude meters (Feet)*	Velocity Mtrs/Sec (Ft/Sec)*	Range Kilometer (Naut Mi)*
00 02 41	S-IC Outboard Engines Cutoff	842,718 (1,857,873)	66,498 (218,169)	2,365 (7,760)	92 (50)
00 02 43	S-IC/S-II Separation	675,991 (1,490,306)	68,197 (223,744)	2,371 (7,779)	96 (52)
00 02 45	S-II Ignition	675,991 (1,490,306)	69,727 (228,763)	2,365 (7,758)	99 (54)
00 03 13	S-II Aft Interstage Jettison	643,675 (1,419,060)	93,755 (307,597)	2,470 (8,103)	162 (88)
00 03 19	Launch Escape Tower Jettison	633,126 (1,395,804)	98,095 (321,835)	2,499 (8,198)	176 (95)
00 07 41	S-II Center Engine Cutoff	303,587 (669,295)	173,099 (567,910)	5,179 (16,992)	1,094 (591)
00 09 20	S-II Outboard Engines Cutoff	216,768 (477,893)	173,593 (569,532)	6,540 (21,458)	1,654 (893)
00 09 21	S-II/S-IVB Separation	171,158 (377,337)	173,634 (569,664)	6,534 (21,466)	1,661 (897)
00 09 24	S-IVB First Ignition	171,117 (377,337)	173,731 (569,984)	6,534 (21,467)	1,680 (907)
00 11 50	S-IVB First Cutoff	139,217 (306,921)	172,882 (567,199)	7,400 (24,278)	2,668 (1,440)
00 12 00	Parking orbit Insertion	139,158 (306,791)	172,887 (567,215)	7,402 (24,284)	2,740 (1,479)
03 21 19	S-IVB Second Ignition	138,000 (304,237)	176,884 (580,326)	7,404 (24,291)	4,331 (2,339)
03 27 04	S-IVB Second Cutoff	65,156 (143,645)	298,545 (979,478)	10,440 (34,250)	7,228 (3,903)
03 27 14	Trans-Lunar Injection	65,092 (143,503)	312,029 (1,023,716)	10,432 (34,225)	7,326 (3,956)

*English measurements given in parentheses.

APOLLO 17 DESCENT ORBIT INSERTION MANEUVERS

DOI-1

PERILUNE ALTITUDE RAISED TO ~ 86, 000 FT VS ~ 54, 000 FT ON APOLLO 16
PERILUNE LOCATION SHIFTED TO 10° W. OF LANDING SITE VS 16° E ON APOLLO 16
(1) PROVIDES SUFFICIENT TIME FOR FLIGHT CONTROLLERS TO DETERMINE BURN CHARACTERISTICS
(2) REDUCES PROBABILITY OF NECESSITY FOR DOI-1 BAILOUT MANEUVER
(3) LANDMARK TRACKING ENHANCED BY HIGHER ALTITUDE SHOULD PRECLUDE EARLY CREW WAKEUP FOR A DOI TRIM MANEUVER

DOI-2

LOWERS PERILUNE FROM ~ 80,000 FT TO ~ 43,000 FT.
40 LBS OF LM RCS USED
NET GAIN IN HOVER TIME OF ~ 3 SEC.
SPS RESERVES INCREASED BY ~25 FPS

APOLLO 17
LUNAR ORBIT INSERTION

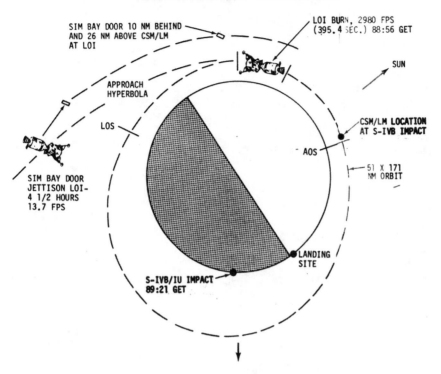

APOLLO 17
CSM/LM LANDING EVENTS

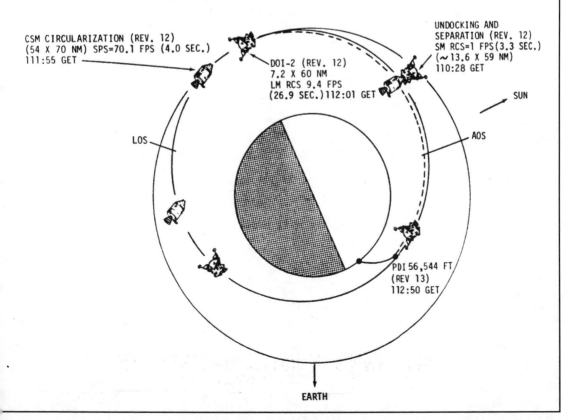

MISSION EVENTS

Events	GET hrs:min	Date CST	Velocity change m/sec (ft/sec)	Purpose and resultant orbit
Translunar injection (S-IVB engine start)	3:27	7/12:20 am	3,048(10,001)	Injection into translunar trajectory with 94km (51 nm) pericynthion
CSM separation, docking	4:02	7/12:55 am	———	Mating of CSM and LM
Ejection from SLA	4:47	7/1:40 am	.3 (1)	Separates CSM-LM from S-IVB/SLA
S-IVB evasive maneuver	5:10	7/2:03 am	3 (9.8)	Provides separation prior to S-IVB propellant and thruster maneuver to cause lunar impact
(S-IVB prop dump, APS burns from MSFC Launch Vehicle Trajectory documents)				
Midcourse correction 1	TLI+9 hr	7/9:20 am	0*	*These midcourse corrections have a nominal velocity change of 0 m/sec, but will be calculated in real time to correct TLI dispersions; trajectory remains within capability of a docked-DPS TEI burn should SPS fail to ignite.
Midcourse correction 2	TLI+32 hrs	8/8:20 am	0*	
Midcourse correction 3	LOI-22 hrs	9/3:48 pm	0*	
Midcourse correction 4	LOI-5 hrs	10/8:48 am	0*	
SIM door jettison	LOI-4.5 hrs	10/9:18 am	4.2 (13.7)	
Lunar orbit insertion	88:55	10/1:48 pm	-908.3 (-2980)	Inserts Apollo 17 into 94x316 km (51x170 nm) elliptical lunar orbit
S-IVB impacts lunar surface	89:21	10/2:14 pm		Seismic event for Apollo 12, 14, 15 and 16 passive seismometers. Target: 7 degrees south latitude by 8 degrees west longitude.
Descent orbit insertion No. 1	93:13	10/6:06 pm	-60.5 (-198.7)	SPS burn places CSM/LM into 25x109 km (15x59 nm) lunar orbit.
CSM/LM undocking	110:28	11/11:20 am	-	
CSM circularization burn	111:55	11/12:48 pm	21.4(70.1)	Inserts CSM into 99.9x129.6 km (54x70 nm) orbit (SPS burn)
Descent orbit insertion No. 2	112:00	11/12:53 pm	2.8 (9.4)	Lowers LM pericynthion to 12.9 km (7 nm)
LM powered descent	112:49	11/1:42 pm	-1,850 (6,701)	Three-phase DPS burn to brake LM out of transfer orbit, vertical descent & lunar touchdown
LM lunar surface contact	113:01	11/1:54 pm	———	Lunar exploration, deploy ALSEP, collect geological samples, photography.
EVA-1 begins	116:40	11/5:33 pm	———	See separate EVA timelines
EVA-2 begins	139:10	12/4:03 pm	———	See separate EVA timelines
EVA-3 begins	162:40	13/3:33 pm	———	See separate EVA timelines
CSM plane change	182:36	14/11:29 am	102.6 (336.7)	Changes CSM orbital plane by 3.6 degrees to coincide with LM orbital plane at time of LM ascent.
LM ascent	188:03	14/4:56 pm	1,847.7 (6,062)	Boosts ascent stage into 16.6 x 88.5 km (9x47.8 nm) lunar orbit for rendezvous with CSM
Lunar orbit insertion	188:10	14/5:03 pm		
Terminal phase initiate (TPI) LM APS	188:57	14/5:50 pm	16.7 (54.8)	Boosts ascent stage into 88x118.5 km (47x64 nm) catch-up orbit; LM trails CSM by 59.2 km (32 nm) and 27.7 km (15 nm) below at TPI burn time.
Braking: 4 LM RCS burns	189:39	14/6:32 pm	9.5 (31.2)	Line-of-sight terminal phase braking to place LM in 114.8x114.8 km (62x62 nm) orbit for final approach, docking.
Docking	190:00	14/6:53 pm	———	CDR and LMP transfer back to CSM
LM jettison, separation	194:09	14/11:02 pm	———	Prevents recontact of CSM with LM ascent stage for remainder of mission.
LM ascent stage deorbit	145:41	15/12:34 am	108.8 (-357)	ALSEP seismometers at Apollo landing sites record impact.
LM impact	195:58	15/12:51 am	———	Impact at about 1,643.9 m/sec (5,394 ft/sec) at -6.1 degree angle
Transearth injection (TEI)	236:39	16/5:32 pm	928.3 (3,045.7)	Injects CSM into transearth trajectory
Midcourse correction 5	253:40	17/10:35 am	0	Transearth midcourse corrections will be computed in real time for entry corridor control and recovery area weather avoidance.
Transearth EVA	257:25	17/2:18 PM	———	Retrieve SM SIM bay film canisters.
Midcourse correction 6	EI-22 hrs	18/3:11 pm	0	
Midcourse correction 7	EI-3 hrs	19/10:11 am	0	
CM/SM separation (EI-15 min)	304:03	19/12:56 pm	———	Command module oriented for Earth atmosphere entry
Entry interface	304:18	19/1:11 pm		Command module enters Earth atmosphere at 11,000 m/sec (36,090 fps)
Splashdown	304:31	19/1:24 pm		Landing 2,111 km (1,140 nm) downrange from entry; splash at 17.9 degrees south latitude by 166 degrees west longitude.

POWERED DESCENT VEHICLE POSITIONS

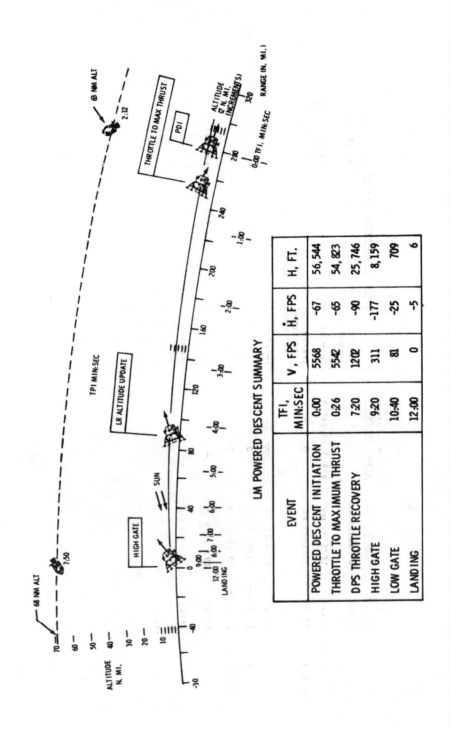

LM POWERED DESCENT SUMMARY

EVENT	TFI, MIN:SEC	V, FPS	Ḣ, FPS	H, FT.
POWERED DESCENT INITIATION	0:00	5568	-67	56,544
THROTTLE TO MAXIMUM THRUST	0:26	5542	-65	54,823
DPS THROTTLE RECOVERY	7:20	1202	-90	25,746
HIGH GATE	9:20	311	-177	8,159
LOW GATE	10:40	81	-25	709
LANDING	12:00	0	-5	6

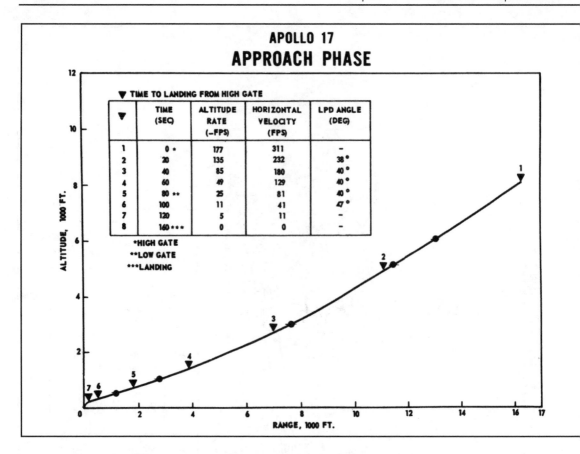

APOLLO 17
APPROACH PHASE

▼ TIME TO LANDING FROM HIGH GATE

▼	TIME (SEC)	ALTITUDE RATE (−FPS)	HORIZONTAL VELOCITY (FPS)	LPD ANGLE (DEG)
1	0 *	177	311	–
2	20	135	232	38°
3	40	85	180	40°
4	60	49	129	40°
5	80 **	25	81	40°
6	100	11	41	47°
7	120	5	11	–
8	160 ***	0	0	–

*HIGH GATE
**LOW GATE
***LANDING

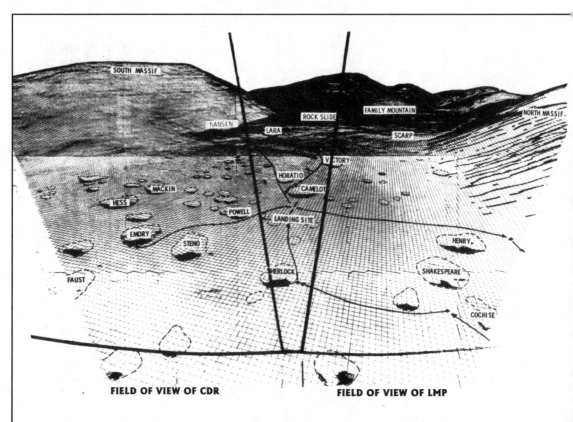

FIELD OF VIEW OF CDR   FIELD OF VIEW OF LMP

FILM RETRIEVAL FROM THE SIM BAY

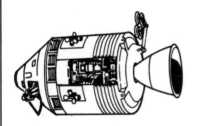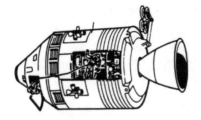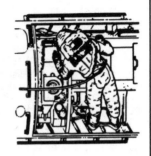

EVA TIMELINE

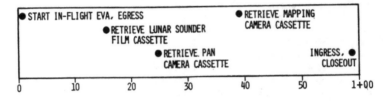

- START IN-FLIGHT EVA, EGRESS
- RETRIEVE LUNAR SOUNDER FILM CASSETTE
- RETRIEVE PAN CAMERA CASSETTE
- RETRIEVE MAPPING CAMERA CASSETTE
- INGRESS, CLOSEOUT

```
0        10        20        30        40        50      1+00
```

APOLLO 17
CREW POST LANDING ACTIVITIES

DAYS FROM RECOVERY	DATE		ACTIVITY
SPLASHDOWN	DEC	19	
R + 1	DEC	20	DEPART SHIP, ARRIVE HAWAII
R + 2	DEC	21	DEPART HAWAII, ARRIVE HOUSTON
R + 3/4	DEC	22,23	CREW TECHNICAL DEBRIEFING PERIOD
	DEC	24 THRU	NO DEBRIEFINGS SCHEDULED
	JAN	2	
	JAN	3	PICKUP CREW DEBRIEFINGS

EVA MISSION EVENTS

Events	hrs:min	Date CST GET	
Depressurize LM for EVA 1	116:40	11/5:33	pm
CDR steps onto surface	116:55	11/5:48	pm
LMP steps onto surface	116:58	11/5:51	pm
Crew offloads LRV	117:01	11/5:54	pm
CDR test drives LRV	117:20	11/6:13	pm
LRV parked near MESA	117:25	11/6:18	pm
LMP mounts geology pallet on LRV	117:29	11/6:22	pm
CDR mounts LCRU, TV on LRV	117:31	11/6:24	pm
LMP deploys United States flag	117:53	11/6:46	pm

		GET		
Events		hrs:min	Date CST	
CDR readies LRV for traverse		118:02	11/6:55	pm
LMP offloads ALSEP		118:07	11/8:00	pm
LMP carries ALSEP "barbell" to deployment site		118:20	11/7:13	pm
Crew begins ALSEP deploy		118:27	11/7:20	pm
ALSEP deploy complete		120:19	11/9:12	pm
Crew drive to Surface Electrical Properties (SEP) experiment site		120:38	11/9:31	pm
Crew arrives at SEP site and drops off transmitter		120:41	11/9:34	pm
Crew drives to station 1 Enroute, crew emplaces Lunar		120:46	11/9:39	pm
Seismic Profiling Experiment (LSPE) explosive package		120:55	11/9:48	pm
Crew arrives station 1 for documented/rake/soil samples, crater sampling, trench etc.		121:07	11/10:00	pm
Crew emplaces LSPE explosive package No. 5 at station 1		122:05	11/10:58	pm
Crew returns to SEP site		122:13	11/11:06	pm
LSPE explosive package No. 7 deployed enroute		122:27	11/11:20	pm
Crew arrives at SEP site		122:35	11/11:28	pm
SEP transmitter deploy completed		122:58	11/11:51	pm
Crew arrives at LM for EVA 1 closeout, sample packaging, load transfer bag		123:00	11/11:53	pm
LMP ingresses LM		123:24	12/12:17	am
CDR ingresses LM		123:36	12/12:29	am
Repressurize LM, end EVA 1		123:40	12/12:33	am
Depressurize LM for EVA 2		139:10	12/4:03	pm
CDR steps onto surface		139:26	12/4:19	pm
LMP steps onto surface		139:29	12/4:22	pm
Crew completes loading LRV for geology traverse		139:45	12/4:38	pm
Crewmen load geological gear on each other's PLSS		139:46	12/4:39	pm
LMP walks to SEP site, turns SEP on		139:50	12/4:43	pm
Crew drives toward station 2, deploys explosive package No. 4 enroute		140:02	12/4:55	pm
Crew arrives at station 2 for rake, core, documented samples and polarimetry at base of South Massif		141:08	12/5:01	pm
Crew drives toward station 3 collecting samples enroute with LRV sampling device		141:59	12/6:52	pm
Crew arrives at station 3 for rake, trench, and documented sampling of scarp and light mantle		142:28	12/7:21	pm
Crew drives toward station 4		143:13	12/8:06	pm
Crew arrives at station 4 for observations, rake & documented samples & a double core around dark halo crater	143:32	12:8:25	pm
Crew drives toward station 5, deploys explosive package No. 1 enroute		144:13	12/9:06	pm
Crew arrives at station 5 for observations, double core, rake & documented samples around 700m mantled crater	144:46	12/9:39	pm
Crew drives back to LM, deploys explosive package No. 8 enroute		145:16	12/10:09	pm
Crew arrives at LM for EVA closeout, packages samples, film mags		145:26	12/10:19	pm
LMP ingresses LM		145:57	12/10:50	pm
CDR ingresses LM		146:06	12/10:59	pm
Repressurize LM, end EVA 2		146:10	12/11:03	pm
Depressurize LM for EVA 3		162:40	13/3:33	pm
CDR steps onto surface		162:55	13/3:48	pm
LMP steps onto surface		162:59	13/3:52	pm
Crew completes loading LRV for geology traverse		163:06	13/3:59	pm
Crewmen load geological gear on each other's PLSS		163:10	13/4:03	pm
LMP walks to SEP site, turns SEP on		163:15	13/4:08	pm
CDR drives to SEP site, picks up LMP, depart for station 6		163:25	13/4:18	pm
Crew arrives at station 6 for rake and documented samples and polarimetry near base of North Massif		163:52	13/4:45	pm
Crew drives to station 7		164:39	13/5:32	pm
Crew arrives at station 7 for rake and documented sampling at base of North Massif		164:50	13/5:43	pm
Crew drives to station 8		165:37	13/6:30	pm
Crew arrives at station 8 for rake and documented samples at base of sculptured hills		165:50	13/6:43	pm
Crew drives to station 9		166:37	13/7:30	pm
Crew arrives at station 9 for radial, rake and documented samples at fresh 80-meter crater on dark mantle		166:53	13/7:46	pm
Crew drives to station 10		167:23	13/8:16	pm
Crew arrives at station 10, for documented samples, double core, on rim of blocky-rimmed crater		167:47	13/8:40	pm
Crew drives back to LM, deploys explosive package No. 2 enroute		168:23	13/9:16	pm
Crew arrives at LM and crew begins EVA 3 close-out, sample stowage etc.		168:42	13/9:35	pm
LMP hikes to ALSEP site to fetch neutron probe		168:59	13/9:52	pm
CDR deploys explosive package No. 3 near LRV final parking location		169:21	13/10:14	pm
LMP ingresses LM		169:32	13/10:25	pm
CDR ingresses LM		169:38	13/10:31	pm
Repressurize LM, end EVA 3		169:40	13/10:33	pm

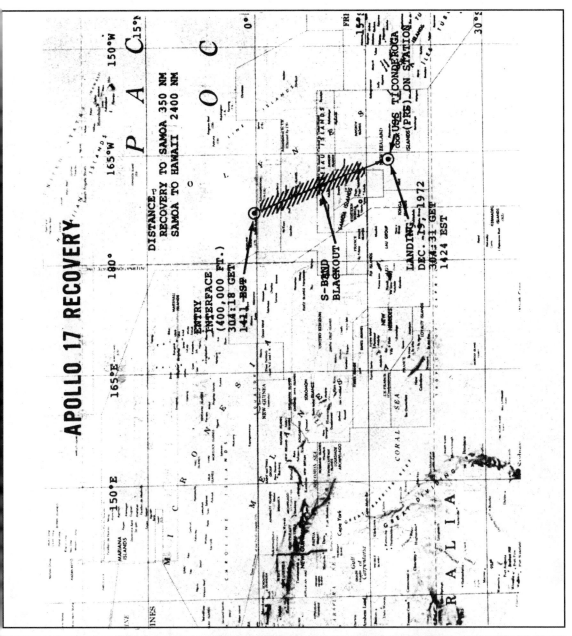

APOLLO 17 VS APOLLO 16 OPERATIONAL DIFFERENCES

ITEM	APOLLO 17	APOLLO 16
LAUNCH TIME	NIGHT	DAY
TRANSLUNAR INJECTION	ATLANTIC (3RD REV)	PACIFIC (2ND REV)
DESCENT ORBIT INSERTION	DOI-1 & 2 MANEUVERS	ONE DOI MANEUVER
PERILUNE LOCATION	10° W OF LANDING SITE	16° E OF LANDING SITE
LUNAR SURFACE STAY	75 HOURS	73 HOURS (PLANNED 71 HOURS (ACTUAL)
TRAVERSE DISTANCE	32.9 KM	25.2 KM (PLANNED) 26.7 KM (ACTUAL)
LUNAR ORBIT PLANE CHANGES	1	2 (PLANNED)
EARTH RETURN INCLINATION	66.5° DESCENDING	62° ASCENDING
TOTAL MISSION TIME	304:31 (PLANNED)	290:36 (PLANNED) 265:51 (ACTUAL)

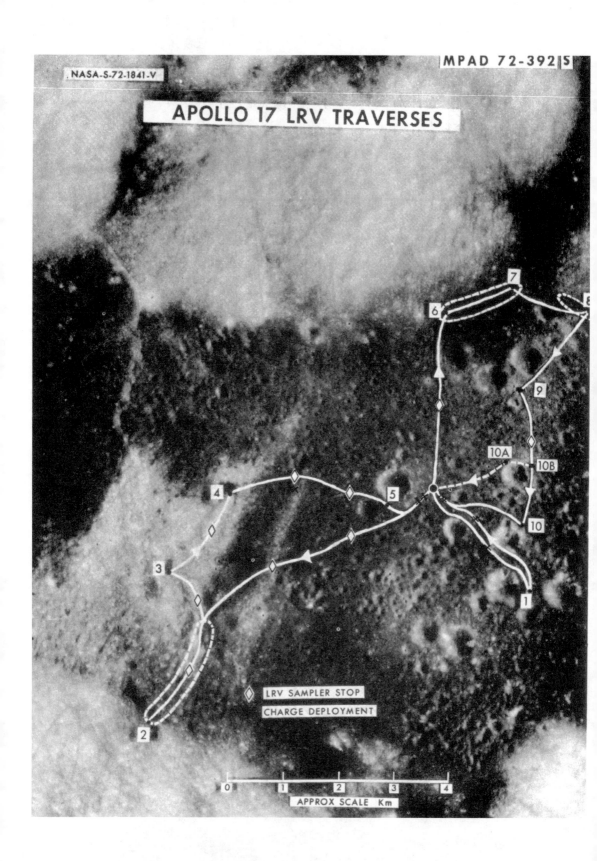

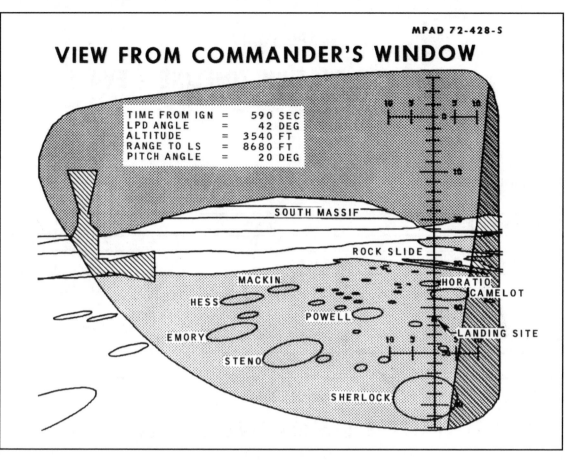

MPAD 72-428-S

VIEW FROM COMMANDER'S WINDOW

TIME FROM IGN	=	590 SEC
LPD ANGLE	=	42 DEG
ALTITUDE	=	3540 FT
RANGE TO LS	=	8680 FT
PITCH ANGLE	=	20 DEG

SOUTH MASSIF

ROCK SLIDE

MACKIN

HESS

POWELL

HORATIO
CAMELOT

EMORY

LANDING SITE

STENO

SHERLOCK

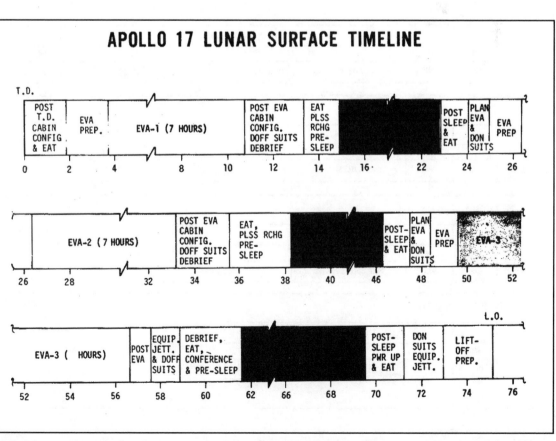

APOLLO 17 LUNAR SURFACE TIMELINE

APOLLO 17
TRAVERSE STATION TIMELINE - EVA 1

STATION 1: EMORY (1:06)

CDR	OVERHEAD	DESCRIPTION	SAMPLING		O/H
	:05	:05	:52		:04
LMP	O/H	PAN	DESCRIPTION	RAKE/SOIL SAMPLE AND SAMPLING	O/H

NOTES:
O/H = OVERHEAD

APOLLO 17
TRAVERSE STATION TIMELINE - EVA 2

STATION 2: NANSEN (:51)

CDR	O/H		DESCRIPTION AND SAMPLING		O/H
	:05	:05	:21	:16	:04
LMP	O/H	PAN	RAKE/SOIL SAMPLE/POLARIMETRY	RAKE SAMPLE, SAMPLING AND SINGLE CORE	O/H

STATION 3: LARA (:45)

CDR	O/H	DESCRIPTION AND SAMPLING	SAMPLING	O/H
	:05	:05	:13	:18	:04
LMP	O/H	PAN	RAKE/SOIL SAMPLING	SAMPLING AND EXPLORATORY TRENCH	O/H

STATION 4: SHORTY (:41)

CDR	O/H	DESCRIPTION AND SAMPLING	O/H
	:05	:05	0:11	0:16	:04
LMP	O/H	PAN	RAKE/SOIL SAMPLING	DOUBLE CORE	O/H

STATION 5: CAMELOT (:30)

CDR	O/H	DESCRIPTION AND SAMPLING	O/H
	:05	:05	:05	:11	:04
LMP	O/H	PAN	RAKE/SOIL SAMPLING	DOUBLE CORE	O/H

APOLLO 17

TRAVERSE STATION TIMELINE - EVA 3

STATION 6 & 7: N. MASSIF AREA (1:28)

CDR	O/H	DESCRIPTION	SAMPLING	PAN AND O/H
	:10	:10	:20 :40	:08
LMP	O/H	PAN	DESCRIPTION	RAKE/SOIL SAMPLING POLARIMETRY	SAMPLING	PAN AND O/H

STATION 8: SCULPTURED HILLS AREA (:44)

CDR	O/H	DESCRIPTION	SAMPLING	O/H
	:05	:05	:20 :10	:04
LMP	O/H	PAN	DESCRIPTION	RAKE/SOIL SAMPLE	SAMPLING	O/H

STATION 9: VAN SERG (:29)

CDR	O/H	DESCRIPTION	SAMPLING	O/H
	:05	:05	:15	:04
LMP	O/H	PAN	DESCRIPTION	RAKE/SOIL SAMPLE	O/H

STATION 10: SHERLOCK (:36)

CDR	O/H	DESCRIPTION	SAMPLING	O/H
	:05	:05	:12 :10	:04
LMP	O/H	PAN	DESCRIPTION	SAMPLING	DOUBLE CORE	O/H

APOLLO 17
EVA 1 TIMELINE

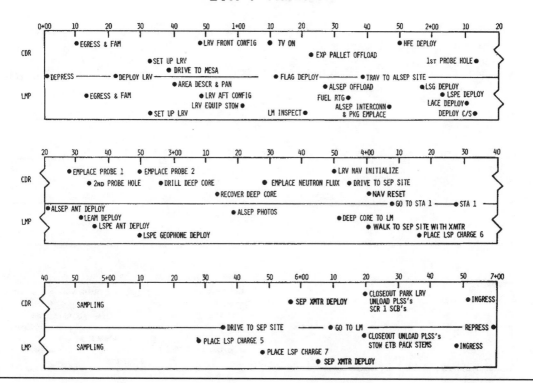

APOLLO 17
EVA 2 TIMELINE

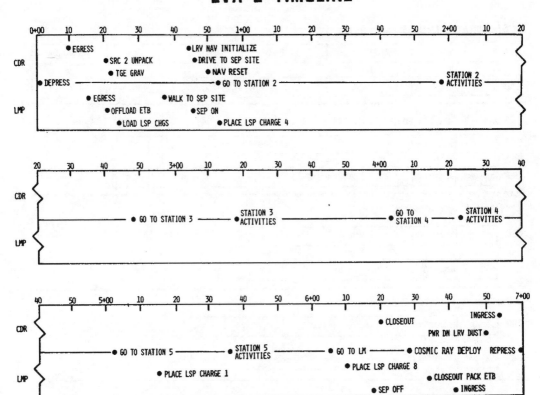

EVA 3 TIMELINE

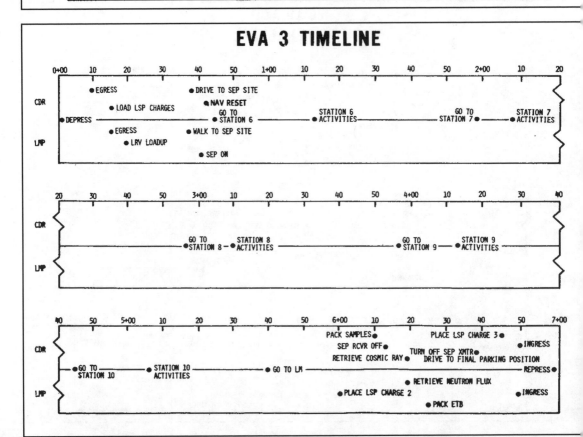

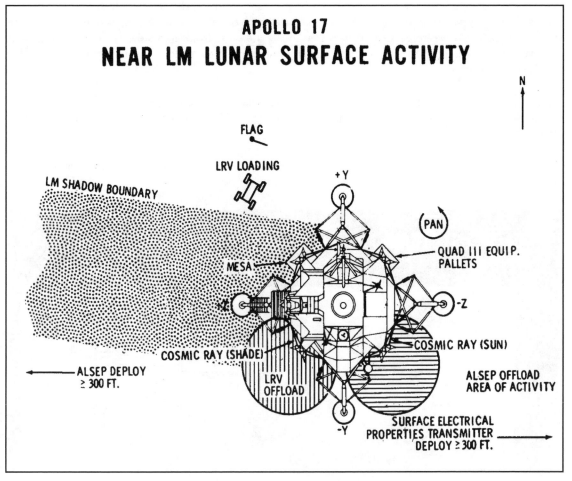

APOLLO 17
NEAR LM LUNAR SURFACE ACTIVITY

TAURUS-LITTROW — APOLLO 17 LANDING SITE

Landing site for the final Apollo lunar landing mission, Taurus-Littrow, takes its name from the Taurus mountains and Littrow crater which are located in a mountainous region on the southeastern rim of the Serenitatis basin.

The actual target landing site is at 30° 44'58.3" east longitude by 20°09'50.5" north latitude —- about 750 km east of the Apollo 15 landing site at Hadley Rille.

Geologists speculate that most of the landing site region probably consists of highland material which was uplifted to its present height at the time the Serenitatis basin was formed. The valley in which the landing site is located is covered by a fine-grained dark mantle that may consist of volcanic fragments. The site is surrounded by three high, steep massifs which likely are composed of breccia formed by the impacts that created some of the major mare basins —- probably pre-Imbrian in age.

A range of "sculptured hills" to the northeast of the landing site is believed to be of the same origin as the massifs, but probably having a different history of erosion and deformation. In gross morphology, the sculptured hills possess some of the characteristics of volcanic structures.

Most of the plain between the massifs is covered by a dark mantle which apparently has no large blocks or boulders, and which has been interpreted to be a pyroclastic deposit. The dark mantle is pocked by several small, dark halo craters that could be volcanic vents all near the landing site. The dark mantle material is thought to be younger in age than all of the large craters on the plain —- probably Eratusthenian Copernican age.

Extending northward from the south massif is a bright mantle with ray-like fingers which overlies the dark mantle. Geologists believe the light mantle is from an avalanche of debris down the slopes of the south massif, and that, it is of Copernican age. Craters near the landing site range from large, steepside craters-one-half to one kilometer in diameter that are grouped near the landing point to scattered clusters of craters less than a half-kilometer across.

Another prominent landing site feature is an 80-meter high scarp trending roughly north-south near the West side of the valley into the north massif. The scarp is thought to be a surface expression of a fault running through the general region.

The Apollo 17 lunar module will approach from the east over the 750 meter hills, clearing them by about 3000 meters.

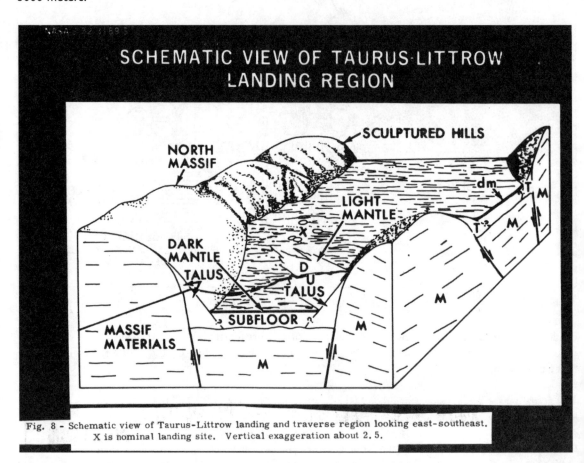

Fig. 8 - Schematic view of Taurus-Littrow landing and traverse region looking east-southeast. X is nominal landing site. Vertical exaggeration about 2.5.

LUNAR SURFACE SCIENCE

S-IVB Lunar Impact - The Saturn V's third stage, after it has completed its job of placing the Apollo 17 spacecraft on a lunar trajectory will be aimed to impact on the Moon. As in several previous missions, this will stimulate the passive seismometers left on the lunar surface in earlier Apollo flights.

The S-IVB, with instrument unit attached, will be commanded to hit the Moon 305 kilometers (155 nautical miles) southeast of the Apollo 14 ALSEP site, at a target point 7 degrees south latitude by 8 degrees west longitude, near Ptolemaeus Crater.

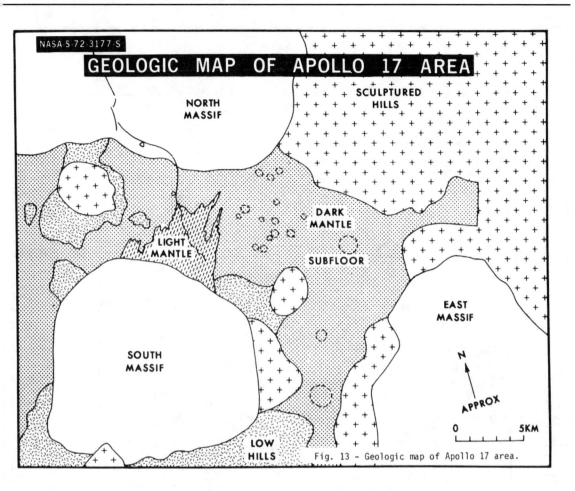

NASA-S-72-3177-S

GEOLOGIC MAP OF APOLLO 17 AREA

Fig. 13 - Geologic map of Apollo 17 area.

After the spacecraft is ejected from the launch vehicle, a launch vehicle auxiliary propulsion system (APS) ullage motor will be fired to separate the vehicle to a safe distance from the spacecraft. Residual liquid oxygen in the almost spent S-IVB/IU will then be dumped through the engine with the vehicle positioned so the dump will slow it into an impact trajectory. Mid-course corrections will be made with the stage's APS ullage motors if necessary.

The S-IVB/IU will weigh 13,931 kilograms (30,712 pounds) and will be traveling 9,147 kilometers an hour (4,939 nautical mph) at lunar impact. It will provide an energy source at impact equivalent to about 11 tons of TNT.

ALSEP Package

The Apollo Lunar Surface Experiments Package (ALSEP) array carried on Apollo 17 has five experiments: heat flow, lunar ejecta and meteorites, lunar seismic profiling, lunar atmospheric composition and lunar surface gravimeter.

Additional experiments and investigations to be conducted at the Taurus-Littrow landing site will include traverse gravimeter, surface electrical properties, lunar neutron probe, cosmic ray director, soil mechanics and lunar geology investigation.

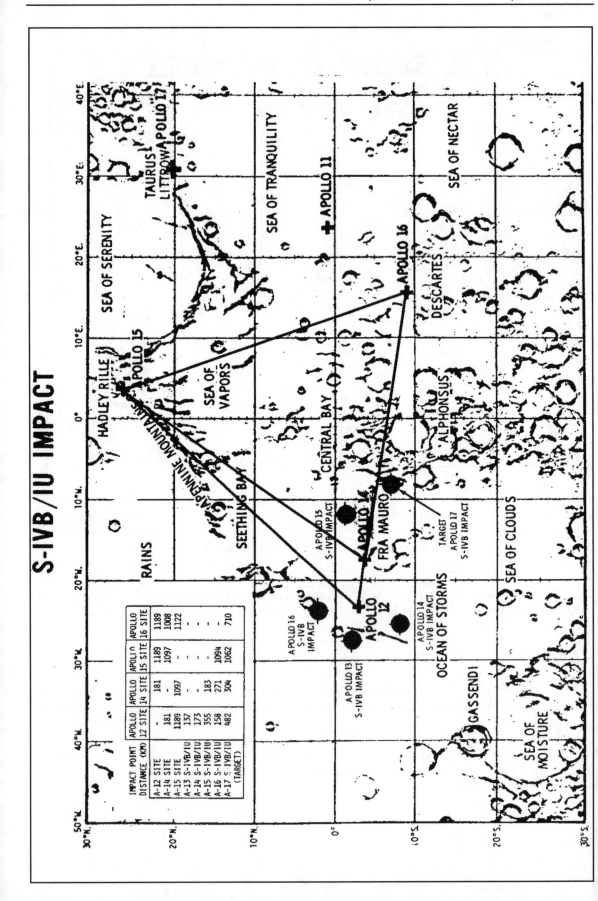

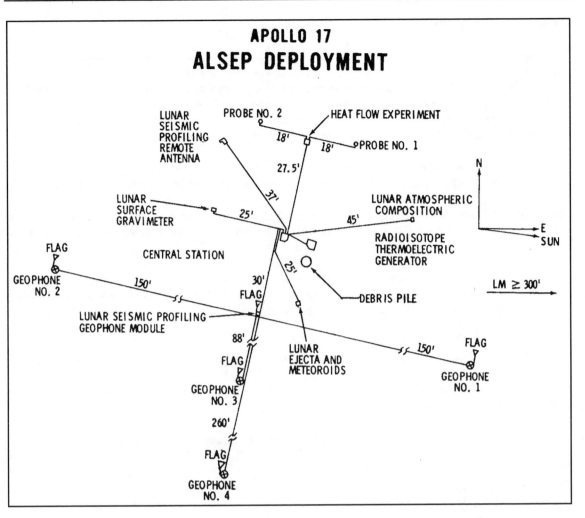

APOLLO 17
ALSEP DEPLOYMENT

SNAP-27— Power Source for ALSEP

A SNAP-27 nuclear generator, similar to four others deployed on the Moon, will provide power for the Apollo 17 ALSEP package. The continuous operation of the four nuclear generators makes possible uninterrupted scientific surveillance of the lunar surface by the instrument packages.

SNAP-27 is one of a series of radioisotope thermoelectric generators (atomic batteries) developed by the Atomic Enerqy Commission under its space SNAP program. The SNAP (Systems for Nuclear Auxiliary Power) program is directed at development of generators and reactors for use in space, on land and in the sea.

SNAP-27 on Apollo 12 marked the first use of a nuclear power system on the Moon. The nuclear generators are required to provide power for at least one year. Thus far, the SNAP-27 unit on Apollo 12 has operated over three years, the unit on Apollo 14 has operated almost two years, and the unit on Apollo 15 has operated for over a year. The fourth nuclear generator was deployed by the Apollo 16 astronauts in April 1972 and is functioning normally.

The basic SNAP-27 unit is designed to produce at least 63.5 watts of electrical power. The SNAP-27 unit is a cylindrical generator, fueled with the radioisotope plutonium-238. It is about 46 cm (18 inches) high and 41 cm (16 inches) in diameter, including the heat radiating fins. The generator, making maximum use of the lightweight material beryllium, weighs about 12.7 kilograms (28 pounds) without fuel. The fuel capsule, made of a superalloy material, is 42 cm (16.5 inches) long and 6.4 cm (2.5 inches) in diameter. It weighs about 7 km (15.5 pounds), of which 3.8 km (8.36 pounds) represent fuel. The plutonium-238 fuel is fully oxidized and is chemically and biologically inert.

The rugged fuel capsule is stowed within a graphite fuel cask from launch through lunar landing. The cask is designed to provide reentry heating protection and containment for the fuel capsule in the event of an aborted mission. The cylindrical cask with hemispherical ends includes a primary graphite heat shield, a secondary beryllium thermal shield, and a fuel capsule support structure. The cast is 58.4 cm (23 inches) long and 20 cm (eight inches) in diameter and weighs about 11 kg (24.5 pounds). With the fuel capsule installed, it weighs about 18 kg (40 pounds). It is mounted on the lunar module descent stage.

Once the lunar module is on the Moon, an Apollo astronaut will remove the fuel capsule from the cask and insert it into the SNAP-27 generator which will have been placed on the lunar surface near the module.

The spontaneous radioactive decay of the plutonium-238 within the fuel capsule generates heat which is converted directly into electrical energy — at least 63.5 watts. The units now on the lunar surface are producing 70 to 74 watts. There are no moving parts.

The unique properties of plutonium-238, make it an excellent isotope for use in space nuclear generators. At the end of almost 90 years, plutonium-238 is still supplying half of its original heat. In the decay process, plutonium-238 emits mainly the nuclei of helium (alpha radiation), a very mild type of radiation with a short emission range.

Before the use of the SNAP-27 system was authorized for the Apollo program a thorough review was conducted to assure the health and safety of personnel involved in the launch and of the general public. Extensive safety analyses and tests were conducted which demonstrated that the fuel would be safely contained under almost all credible accident conditions.

APOLLO LUNAR SURFACE SCIENCE MISSION ASSIGNMENTS

EXPERIMENT	11	12	14	15	16	17
S-031 PASSIVE SEISMIC	X	X	X	X	X	
S-033 ACTIVE SEISMIC			X		X	
S-034 LUNAR SURFACE MAGNETOMETER		X		X	X	
S-035 SOLAR WIND SPECTROMETER		X		X		
S-036 SUPRATHERMAL ION DETECTOR		X	X	X		
S-037 HEAT FLOW				X	X	X
S-038 CHARGED PARTICLE LUNAR ENVIRONMENT			X			
S-058 COLD CATHODE IONIZATION		X	X	X		
M-515 LUNAR DUST DETECTOR		X	X	X		
S-207 LUNAR SURFACE GRAVIMETER						X
S-202 LUNAR EJECTA AND METEORITES						X
S-203 LUNAR SEISMIC PROFILING						x
S-205 LUNAR ATMOSPHERIC COMPOSITION						X
S-201 FAR UV CAMERA/SPECTROSCOPE					X	
S-059 LUNAR GEOLOGY INVESTIGATION	X	X	X	X	X	X
S-078 LASER RANGING RETRO-REFLECTOR	X		X	X		
S-080 SOLAR WIND COMPOSITION	X	X	X	X	X	
S-184 LUNAR SURFACE CLOSE-UP CAMERA	X	X	X			
S-152 COSMIC RAY DETECTOR					X	X
S-198 LUNAR PORTABLE MAGNETOMETER			X		X	
S-199 LUNAR TRAVERSE GRAVIMETER						X
S-200 SOIL MECHANICS			X	X	X	X
S-204 SURFACE ELECTRICAL PROPERTIES						X
S-229 LUNAR NEUTRON PROBE						X

Heat Flow Experiment (S-037):

The thermal conductivity and temperature gradient of the upper 2.44 meters of the lunar surface will be measured by the HFE for gathering data on the Moon's internal heating process and for a basis for comparing the radioactive content of the Moon's interior with the Earth's mantle. Results from the HFE deployed on Apollo 15 show that the Moon's outward heat flow from the interior at the Hadley-Apennine site is about half that of the Earth. The HFE principle investigator believes that the relationship between Moon and Earth heat flow is an indication that the amounts of heat-producing elements (uranium, thorium and potassium) present in lunar soil are about the same as on Earth. Temperature variations over a lunar month range at the surface from -185°C to +86°C— a spread of 271°C— but at a depth of one meter, the variations are only a few thousandths of a degree.

The Apollo Lunar Surface Drill (ALSD) is used to drill two holes about 2.6 meters deep and 10 meters apart into which the two heat flow probes are lowered. The bore stems are left in the hole and serve as casings to prevent collapse of the walls when the probes are lowered. An emplacement tool aids the crewman in ramming the probes to the maximum depth. Flat cables connect the probes to the HFE electronics package and thence to the ALSEP central station. Radiation shields are placed over each bore hole after the data cables are connected.

During deployment of the Apollo 16 HFE, the cable connecting the experiment probe to the central station was broken when an astronaut caught his foot in the cable. A fix, consisting of a strain release device, has been installed on all of the cables to preclude a reoccurrence of such an accident. It was successfully deployed on Apollo 15, however.

Principal Investigator is Dr. Marcus E. Langseth of the Lamont-Doherty Geological Observatory, Columbia University.

Lunar Ejecta and Meteorites (S-202):

The purpose of the Lunar Ejecta and Meteorites Experiment is to measure the physical parameters of primary and secondary particles impacting the lunar surface.

The detailed objectives of the experiment are: determine the background and long-term variations in cosmic dust influx rates in cislunar space; determine the extent and nature of lunar ejecta produced by meteorite impacts on the lunar surface; determine the relative contributions of comets and asteroids to the Earth's meteorite flux; study possible correlations between the associated ejecta events and times of Earth's crossing of comet orbital planes and meteor streams; determine the extent of contribution of interstellar particles toward the maintenance of the zodiacal cloud as the solar system passes through galactic space; and investigate the existence of an effect called "Earth focusing of dust particles".

The experiment package is aligned and leveled by the crew about eight meters south of the ALSEP central station using a built-in bubble level and Sun-shadow gnomon. Detector plates on the surfaces of the experiment housing are protected from lunar module ascent dust particles by a cover device which is later jettisoned by ground command.

Using a sophisticated array of detectors, the experiment measures and telemeters information such as particle velocities from 1 to 75 km/sec, energy ranges from 1 to 1000 ergs and particle impact frequencies up to 100,000 impacts per square meter per second.

Dr. Otto E. Berg, NASA Goddard Space Flight Center, is the principal investigator.

Lunar Seismic Profiling (S-203):

A major scientific tool in the exploration of the Moon has been seismology. Through this discipline, much has been learned about the structure of the lunar interior — particularly through the passive seismic experiments carried as a part of each ALSEP array from Apollo 11's initial landing through Apollo 16. The passive seismic experiments provided data on the lunar interior. All but the Apollo 11 seismometer are still operating effectively.

On Apollo 17, the experiment's data-gathering network consists of four geophones placed in the center and at each corner of a 90-meter equilateral triangle. Explosive charges placed on the surface will generate seismic waves of varying strengths to provide data on the structural profile of the landing site. The triangular arrangement of the geophones allows measurement of the azimuths and velocities of seismic waves more accurately than was possible with the Active Seismic Experiments and their linear array of geophones using mortar-fired grenades emplaced on Apollos 14 and 16.

After the charges have been fired by ground command, the experiment will settle down into a passive listening mode, detecting Moonquakes, meteorite impacts and the thump caused by the lunar module ascent stage impact. Knowledge on the surface and subsurface geologic characteristics to depths of three kilometers could be gained by the experiment. The experiment will be deactivated after the LM impact.

Components of the lunar seismic profiling experiment are four geophones similar to those used in the earlier active seismic experiment, an electronics package in the ALSEP central station, and eight explosive packages which will be deployed during the geology traverse — the lightest charge no closer than 150 meters from the geophone triangle, and the heaviest charge no further than 2.5 kilometers. Each charge has two delay timers which start after a crewman pulls three arming pins. (Timer delays vary from 90-93 hours). A coded-pulse ground command relayed through the central station will detonate each charge. Two charges weigh 57g (1/8 lb), two weigh 113g (1/4 lb), and the remaining four charges weigh 227g (1/2 lb), 454g (1 lb), 1361g (3 lb) and 2722g (6 lb) respectively. Television observations of the LSPE charge detonations as well as the LM impact are planned during the post lift-off period.

Dr. Robert L. Kovach of the Stanford University Department of Geophysics is the experiment principal investigator.

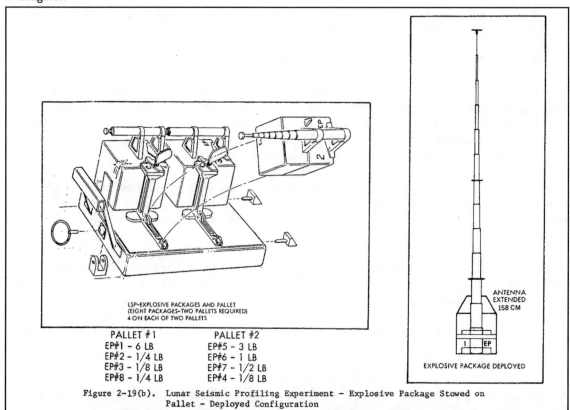

LSP-EXPLOSIVE PACKAGES AND PALLET
(EIGHT PACKAGES-TWO PALLETS REQUIRED)
4 ON EACH OF TWO PALLETS

PALLET #1	PALLET #2
EP#1 - 6 LB	EP#5 - 3 LB
EP#2 - 1/4 LB	EP#6 - 1 LB
EP#3 - 1/8 LB	EP#7 - 1/2 LB
EP#8 - 1/4 LB	EP#4 - 1/8 LB

ANTENNA
EXTENDED
158 CM

EXPLOSIVE PACKAGE DEPLOYED

Figure 2-19(b). Lunar Seismic Profiling Experiment - Explosive Package Stowed on Pallet - Deployed Configuration

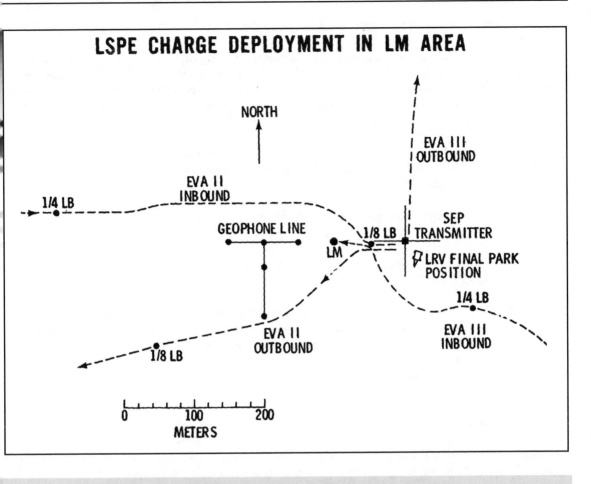

LSPE CHARGE DEPLOYMENT IN LM AREA

LSPE CHARGE DEPLOYMENT PLAN

CHARGE SIZE LB	CHARGE NO.	PALLET NO.	DEPLOYMENT LOCATION, RADIUS FROM ALSEP (KM)	DEPLOYMENT TIME EVA	HR:MIN FROM EVA START	TIME AFTER DEPLOYMENT OF DETONATION (HOURS)	TIME AFTER LIFTOFF FOR DETONATION* (HR:MIN)	TIME* GET (HR:MIN)	DATE/EST
1	6	2	1.3	1 (OUTBOUND)	4:40	91	24:17	212:20	12/15-1813
3	5	2	2.3	1 (STATION 1)	5:53	92	26:30	214:33	12/15-2026
1/2	7	2	.8	1 (INBOUND)	6:12	93	27:49	215:52	12/15-2145
1/8	4	2	.16	2 (OUTBOUND)	:55	91	43:02	231:05	12/16-1258
6	1	1	2.4	2 (INBOUND)	5:17	92	48:24	236:27	12/16-1820
1/4	8	1	.25	2 (INBOUND)	6:11	94	51:18	239:21	12/16-2114
1/4	2	1	.25	3 (INBOUND)	5:59	93	73:36	261:39	12/17-1932
1/8	3	1	.16	3 (LRV FINAL PARKING)	6:04	94	75:07	263:10	12/17-2113

*BASED ON THE FOLLOWING MISSION TIMES:

LANDING	113:02 GET	START EVA 3	162:40 GET
START EVA 1	116:40 GET	LIFTOFF	188:03 GET
START EVA 2	139:10 GET		

OTHER TIMES OF INTEREST:

LM IMPACT: 7:56 AFTER LIFT-OFF 12/15-0152 EST

TEI: 48:37 AFTER LIFT-OFF 12/16-1833 EST

Lunar Atmospheric Composition Experiment (LACE) (S-205):

This experiment will measure components in the ambient lunar atmosphere in the range of one to 110 atomic mass units (AMU). It can measure gases as thin as one billion billionth of the Earth's atmosphere. The instrument is capable of detecting changes in the atmosphere near the surface originating from the lunar module, identifying native gases and their relative mass concentrations, measuring changes in concentrations from one lunation to the next, and measuring short-term atmospheric changes.

It has been suggested that lunar volcanism may release carbon monoxide, hydrogen sulfide, ammonia, sulfur dioxide and water vapor, and that solar wind bombardment may generate an atmosphere of the noble gases - helium, neon, argon, and krypton.

The instrument can measure gases ranging from hydrogen and helium at the low end of the atomic mass scale to krypton at the high end.

The instrument is a Neir-type magnetic sector field mass spectrometer having three analyzers whose mass ranges are 1-4 AMU, 12-48 AMU and 40-110 AMU.

The experiment will be set up 15 meters northwest of ALSEP central station. After placement, it will be heated to drive off contaminating gases. The instrument will be turned on by ground command after lunar module ascent during the first lunar night and will be active thereafter.

Dr. John H. Hoffman of the Department of Atmospheric and Space Sciences of the University of Texas at Dallas is principal investigator.

Lunar Surface Gravimeter (S-207):

The major goal of the LSG is to confirm the existence of gravity waves as predicted by Einstein's general theory of relavity. Additional insights into the Moon's internal structure are expected to come from the measurement of the tidal deformation in the lunar material caused by the Earth and the Sun — much as the Earth's ocean tides are caused by the changing Earth-Moon-Sun alignment.

Additionally the experiment is expected to detect free Moon oscillations in periods of 15 minutes upward which may be caused by gravitational radiation from cosmic sources.

The device will also measure vertical components of natural lunar seismic events in frequencies up to 16 cycles per second. It will thus supplement the passive seismic network emplaced by Apollos 12, 14, 15 and 16.

The surface gravimeter uses a sensor based on the LaEvete-Romberg gravimeter widely used on Earth, modified to be remotely controlled and read. The package includes a sunshield, electronics and ribbon cable connecting it to the ALSEP central station. The crew will erect the experiment eight meters west of the central station.

Dr. Joseph Weber of the University of Maryland Department of Physics and Astronomy is the principal investigator.

Traverse Gravimeter (S-199):

Gravimetry has proven itself to be a valuable tool for geophysical measurements of the Earth, and the aim of the traverse gravimeter experiment is to determine whether the same techniques can also be used in making similar measurements of the Moon to help determine its internal structures. Many of the major findings in geophysics, such as variations in the lateral density in the Earth's crust and mantle, tectogenes, batholiths and isostasy, are a result of gravimetric investigations.

The Apollo 17 landing site gravitational properties will be measured in the immediate touchdown area as well as at remote locations along the geology traverse routes. The measurements of the Taurus-Littrow site will be related to geologically similar areas on Earth in an effort to draw parallels between gravimetry investigations on the two bodies.

In an extension of Earth gravimetry techniques to the lunar surface, it is felt that small-scale lunar features such as mare ridges, craters, rilles, scarps and thickness variations in the regolith can better be understood.

The traverse gravimeter instrument will be mounted on the LRV. Measurements can be made when the vehicle is not in motion or when the instrument is placed on the lunar surface. The crew, after starting the instrument's measurement sequence, will read off the numbers from the digital display on the air/ground circuit to Mission Control.

Dr. Manik Talwani of the Columbia University Lamont-Doherty Geological Observatory is the traverse gravimeter principal investigator.

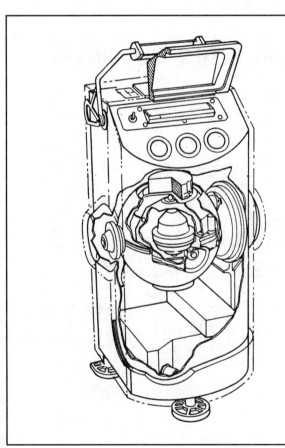

APOLLO 17

TRAVERSE GRAVIMETER

PURPOSE:

TO MAKE A RELATIVE SURVEY
OF THE LUNAR GRAVITATIONAL FIELD
IN THE LANDING AREA AND TO MAKE
AN EARTH-MOON GRAVITY TIE

Surface Electrical Properties (SEP) (S-204):

This experiment measures electromagnetic energy transmission, absorption and reflective characteristics of the lunar surface and subsurface. The instrument can measure the electrical properties at varying depths and the data gathered, when compared to the traverse gravimeter and seismic profiling data, will serve as a basis for a geological model of the upper layers of the Moon.

Frequencies transmitted downward into the lunar crust have been selected to allow determination of layering and scattering over a range of depths from a few meters to a few kilometers. Moreover, the experiment is expected to yield knowledge of the thickness of the regolith at the Taurus-Littrow site, and may provide an insight into the overall geological history of the outer few kilometers of the lunar crust.

Continuous successive waves at frequencies of 1, 2.1, 4, 8.1, 16 and 32.1 megahertz (MHz) broadcast downward into the Moon allow measurement of the size and number of scattered bodies in the subsurface. Also, any moisture present in the subsurface can easily be detected since small amounts of water in the rocks or subsoil would greatly change electrical conductivity.

The instrument consists of a deployable self-contained transmitter, having a multi-frequency antenna and a portable receiver with a wide-band three-axis antenna. The receiver contains a retrievable data recorder.

The transmitter with its antenna is deployed by the crew about 100 meters east of the lunar module, while the receiver-recorder is mounted on the LRV. The exact location of each reading is recorded on the recorder using information from the LRV's navigation system. After the final readings are made near the end of the third EVA, the recorder will be removed for return to Earth.

Dr. M. Gene Simmons of the Massachusetts Institute of Technology, Cambridge, Mass., is the principal investigator.

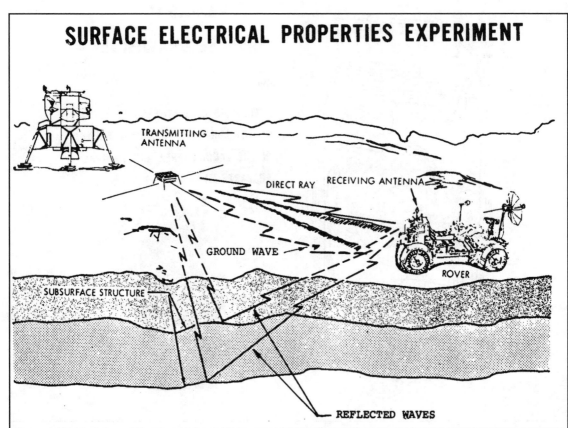

Lunar Neutron Probe (S-229):

Neutron capture rates of the lunar regolith and average mixing depths of lunar surface material are among the problems which it is hoped this experiment will help solve. The 2.4 meter long probe will be inserted into the hole left by the drill core sample gathered during the lunar geology experiment. It will measure the degree of present neutron flux in the top two meters of the regolith. Data from the instrument also will help determine the average irradiation depth for lunar rocks and yield information on the lunar neutron energy spectrum. The data from the probe will be compared with the distribution of gadolinium isotopes in the lunar drill core sample returned by Apollo 17 as an aid in interpreting neutron dosages on samples from previous missions.

The two-section cylindrical probe 2.4 meters long by 2 cm in diameter, is activated by the crew at the emplacement site. The crew inserts the probe in the drill core sample hole during the first EVA, and retrieves it at the end of the third EVA for deactivation and stowage for the trip home.

Principal investigator is Dr. Don S. Burnett of the California Institute of Technology, Division of Geology and Planetary Sciences, Pasadena, California.

Soil Mechanics (S-200):

While this experiment is officially listed as "passive," since no specific crew actions or hardware are involved, the wide range of knowledge gained of the physical characteristics and mechanical properties of the surface material in the landing site will make the experiment active from an information standpoint.

Crew observations and photography during the three EVAs will aid in the interpretation of lunar history and landing site geological processes, such as the form and compaction of surface layers, characteristics of rays, mares, slopes and other surface units, and deposits of different chemical and mineralogical compounds.

The experiment is further expected to contribute to knowledge on slope stability, causes of downslope movement and the natural angle of repose for different types of lunar soils.

The experiment data will also aid in predicting seismic velocities in different types of material for interpreting seismic studies.

Material density, specific heat and thermal conductivity data for heat-flow investigations are also expected to be supplemented by crew observations.

Among the specific items the crew will observe are the LM footpad impressions, any soil deposits on LM vertical surfaces and tracks made by the LRV during the geology traverses. These observations will aid in determining the in-place strength and compressibility of the surface material. Moreover, the observations will aid in defining conditions for simulations studies back on Earth of the returned sample density, porosity and confining pressures.

Dr. James K. Mitchell of the University of California at Berkeley is principal investigator.

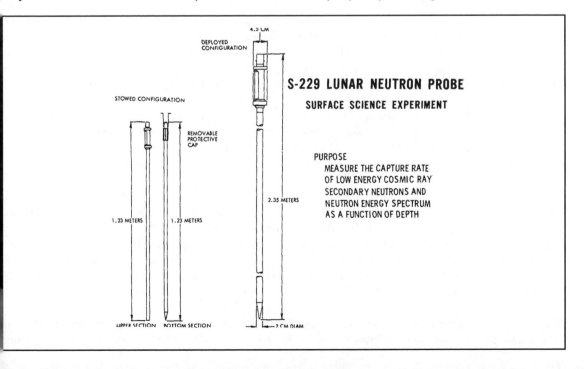

APOLLO 17
LRV TRAVERSES
(PRELIMINARY)

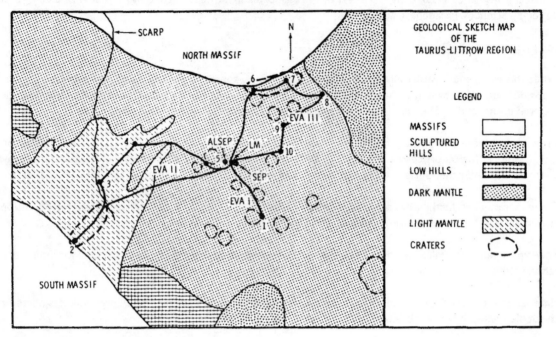

Lunar Geology Investigation (S-059):

The fundamental objective of the lunar geology investigation experiment is to provide data in the vicinity of the landing site for use in the interpretation of the geologic history of the Moon. Apollo lunar landing missions offer the opportunity to correlate carefully collected samples with a variety of observational data on at least the upper portions of the mare basin filling and the lunar highlands, the two major geologic subdivisions of the Moon. The nature and origin of the maria and highlands will bear directly on the history of lunar differentiation and differentiation processes. From the lunar bedrock, structure, land forms and special materials, information will be gained about the internal processes of the Moon. The nature and origin of the debris layer (regolith) and the land forms superimposed on the maria and highland regions are a record of lunar history subsequent to their formation. This later history predominately reflects the history of the extralunar environment. Within and on the regolith, there will also be materials that will aid in the understanding of geologic units elsewhere on the Moon and the broader aspects of lunar history.

The primary data for the lunar geology investigation experiment come from photographs, verbal data, and returned lunar samples. Photographs taken according to specific procedures will supplement and illustrate crew comments, record details not discussed by the crew, provide a framework for debriefing, and record a wealth of lunar surface information that cannot be returned or adequately described by any other means.

In any Hasselblad picture taken from the lunar surface, as much as 90 percent of the total image information may be less than 100 feet from the camera, depending on topography and how far the camera is depressed below horizontal. Images of distant surface detail are so foreshortened that they are difficult to interpret. Therefore, it is important that panoramas be taken at intervals during the traverse and at the farthest excursion of the traverse. This procedure will extend the high resolution photographic coverage to the areas examined and discussed by the astronaut, and will show the regional context of areas of specific interest that have been discussed and photographed in detail.

The polarizing filters will permit the measurement of the degree of polarization and orientation of the plane of polarization contained in light reflected from the lunar surface. Different lunar materials (i.e., fine-grained glass and/or fragments, strongly shocked rocks, slightly shocked rocks and shock-lithified fragmental material.) have different polarimetric functions, in other words, different polarimetric "signatures."

Comparison of the polarimetric function of known material, such as returned samples and close-up lunar surface measurements, with materials photographed beyond the traverse of the astronaut will allow the classification and correlation of these materials even though their textures are not resolvable. The polarimetric properties of lunar materials and rock types are a useful tool for correlation and geologic mapping of each landing site, and for extrapolation of geologic data from site to site across the lunar surface.

The "in situ" photometric properties of both fine-grained materials and coarse rock fragments will serve as a basis for delineating, recognizing, describing, and classifying lunar materials. The gnomon, with photometric chart attached, will be photographed beside a representative rock and, if practical, beside any rock or fine-grained material with unusual features.

The long focal length (500 mm) lens with the HEDC will be used to provide high resolution data. A 5 to 10 centimeter resolution is anticipated at a distance of 1 to 2 km (0.6 to 1.2 mi.).

Small exploratory trenches, several centimeters deep, are to be dug to determine the character of the regolith down to these depths. The trenches should be dug in the various types of terrain and in areas where the surface characteristics of the regolith are of significant interest as determined by the astronaut crew. The main purpose of the trenches will be to determine the small scale stratigraphy (or lack of) in the upper few inches of the regolith in terms of petrological characteristics and particle size.

An organic control sample, carried in each sample return case (SRC), will be analyzed after the mission to determine the level of contamination in each SRC.

In order to more fully sample the major geological features of the Apollo 17 landing site, various groupings of sampling tasks are combined and will be accomplished in concentrated areas. This will aid in obtaining vertical as well as lateral data to be obtained in the principal geological settings. Thus, some trench samples, core tube samples and lunar environmental soil samples will be collected in association with comprehensive samples.

In addition, sampling of crater rims of widely differing sizes in a concentrated area will give a sampling of the deeper stratigraphic divisions at that site. Repeating this sampling technique at successive traverse stations will show the continuity of the main units within the area.

Sampling and photographic techniques used to gather data in the landing site include:

Documented samples of lunar surface material which, prior to gathering, are photographed in color and stereo — using the gnomon and photometric chart for comparison of position and color properties — to show the sample's relation to other surface features.

Rock, boulder and soil samples of rocks from deep layers, and soil samples from the regolith in the immediate area where the rocks are gathered.

Radial sampling of material on the rim of a fresh crater — material that should be from the deepest strata.

Photopanoramas for building mosaics which will allow accurate control for landing site map correlation.

Polarimetric photography for comparison with known materials.

Double drive tube samples to depths of 60 cm (23.6 in.) for determining the stratigraphy in multi-layer areas.

Single drive tube samples to depths of 38 cm (15 in.) in the comprehensive sample area and in such target of opportunity areas as mounds and fillets.

Drill core sample of the regolith which will further spell out the stratigraphy of the area sampled.

Small exploratory trenches, ranging from 8 to 20 cm (2.4 to 9.7 in.) in depth, to determine regolith particle size and small-scale stratigraphy

Large equidimensional rocks ranging from 15 to 24 cm (6 to 9.4 in.) in diameter for data on the history of solar radiation. Similar sampling of rocks from 6 to 15 cm (2.4 to 6 in.) in diameter will also be made.

Vacuum-packed lunar environment soil and rock samples kept-biologically pure for postflight gas, chemical and microphysical analysis.

Dr. William R. Muehlberger of the US Geological Survey Center of Astrogeology, Flagstaff Ariz. is the lunar geology principal investigator.

<u>Lunar Geology Hand Tools:</u>

<u>Sample scale</u> - The scale is used to weigh the loaded sample return containers, sample bags, and other containers to maintain the weight budget for return to Earth. The scale has graduated markings in increments of 5 pounds to a maximum capacity of 80 pounds. The scale is stowed and used in the lunar module ascent stage.

<u>Tongs</u> - The tongs are used by the astronaut while in a standing position to pick up lunar samples from pebble size to fist size. The tines of the tongs are made of stainless steel and the handle of aluminum. The tongs are operated by squeezing the T-bar grips at the top of the handle to open the tines. In addition to picking up samples, the tongs are used to retrieve equipment the astronaut may inadvertantly drop. This tool is 81 cm (32 inches) long overall.

<u>Lunar rake</u> - The rake is used to collect discrete samples of rocks and rock chips ranging from 1.3 cm (one-half inch) to 2.5 cm (one inch) in size. The rake is adjustable for ease of sample collection and stowage. The tines, formed in the shape of a scoop, are stainless steel. A handle, approximately 25 cm (10 inches) long, attaches to the extension handle for sample collection tasks.

<u>Adjustable scoop</u> - The sampling scoop is used to collect soil material or other lunar samples too small for the rake or tongs to pick up. The stainless steel pan of the scoop, which is 5 cm (2 inches) by 11 cm (41/2 inches) by 15 cm (6 inches) has a flat bottom flanged on both sides and a partial cover on the top to prevent loss of contents. The pan is adjustable from horizontal to 55 degrees and 90 degrees from the horizontal for use in scooping and trenching. The scoop handle is compatible with the extension handle.

<u>Hammer</u> - This tool serves three functions; as a sampling hammer to chip or break large rocks, as a pick, and as a hammer to drive the drive tubes or other pieces of lunar equipment. The head is made of impact resistant tool steel, has a small hammer face on one end, a broad flat blade on the other, and large hammering flats on the sides. The handle, made of aluminum and partly coated with silicone rubber, is 36 cm (14 inches) long; its lower end fits the extension handle when the tool is used as a hoe.

<u>Extension handle</u> — The extension handle extends the astronaut's reach to permit working access to the lunar surface by adding 76 cm (30 inches) of length to the handles of the scoop, rake, hammer, drive tubes and other pieces of lunar equipment. This tool is made of aluminum alloy tubing with a malleable stainless steel cap designed to be used as an anvil surface. The lower end has a quick-disconnect mount and lock designed to resist compression, tension, torsion, or a combination of these loads. The upper end is fitted with a sliding "T" handle to facilitate any torqueing operation.

<u>Drive Tubes</u> - These nine tubes are designed to be driven or augured into soil, loose gravel, or soft rock such as pumice. Each is a hollow thin-walled aluminum tube 41 cm (16 inches) long and 4 cm (1.75 inch) diameter with an integral coring bit. Each tube can be attached to the extension handle to facilitate sampling. A deeper core sample can be obtained by joining tubes in series of two or three. When filled with sample, a Teflon cap is used to seal the open end of the tube, and a keeper device within the drive tube is positioned against the top of the core sample to preserve the stratigraphic integrity of the core. Three Teflon caps are packed in a cap dispenser that is approximately a 5.7 cm (2.25 inch) cube.

<u>Gnomon and Color Patch</u> - The gnomon is used as a photographic reference to establish local vertical Sun angle, scale, and lunar color. This tool consists of a weighted staff mounted on a tripod. It is constructed in such a way that the staff will right itself in a vertical position when the legs of the tripod are on the lunar surface. The part of the staff that extends above the tripod gimbal is painted with a gray scale from 5 to 35 percent reflectivity and a color scale of blue, orange, and green. The color patch, similarly painted in gray scale and color scale, mounted on one of the tripod legs provides a larger target for accurately determining colors in color photography.

<u>LRV Soil Sampler</u> - A scoop device attached to the end of the Universal Hand Tool for gathering surface soil samples and small rock fragments without dismounting from the LRV. The device has a ring 7.5 cm in diameter and a five-wire stiffening cage on the end that holds 12 telescoped plastic cap-shaped bags which are removed and sealed as each sample is taken. The sampler is 25 cm long and 7.5 cm wide.

<u>Sample Bags</u> - Several different types of bags are furnished for collecting lunar surface samples. The Teflon documented sample bag (DSB), 19 by 20 cm (7-1/2 by 8 inches) in size, is prenumbered and packed in a 20-bag dispenser that can be mounted on a bracket on the Hasselblad camera. Documented sample bags (120) will be available during the lunar surface EVAs. The sample collection bag (SCB), also of Teflon, has interior pockets along one side for holding drive tubes and exterior pockets for the special environmental sample container and for a drive tube cap dispenser. This bag is 17 by 23 by 41 cm (6-3/4 by 9 by 16 inches) in size (exclusive of the exterior pockets) and fits inside the sample return containers. During the lunar surface EVAs this bag is hung on the hand or on the portable life support system tool carrier. Four SCBs will be carried on Apollo 17. The extra sample collection bag (ESCB) is identical to the SCE except that the interior and exterior pockets are omitted. During EVAs it is handled in the same way as an SCB. Four ESCB bags will be carried on the mission. A sample return bag, 13 by 33 by 57 cm (5 by 13 by 22.5 inches) in size, replaces the third sample return container and is used for the samples collected on the third EVA. It hangs on the LRV pallet during this EVA.

ong Term Surface Exposure Experiment:

tudy of the abundance and composition of high energetic particles of solar and galactic origin as well as articulate interplanetary matter will yield important clues concerning the chemical composition of the Sun, he formation of the solar system, nuclear reactions in the Sun and galaxy as well as small scale geologic urface processes on planetary bodies.

variety of aspects concerned with the above particles are the subject of detailed analysis of returned lunar aterials, as well as specific instruments emplaced on the lunar surface, e.g., Solar Wind Composition xperiment and Cosmic Ray Detector and in particular some pieces of Surveyor III hardware returned uring the Apollo 12 mission. The lunar sample analysis and the above instruments have yielded significant esults. However, lunar samples are exposed, on the average too long to the space environment (in the illions of years) which make the interpretation of the experimental results very complex. The emplaced struments suffer from the handicap that they are exposed for too short a period of time (a few hours to few years) thus leaving some questions about the representative character of the results. It therefore is esirable to expose certain materials for an extended and known time period.

his is the objective of the so called "Long Term Surface Exposure Experiment", where existing flight ardware will either be used as is or deployed in a specific fashion by the Apollo 17 crew in the hope that these materials will be retrieved at some undetermined future time, i.e., in a few decades. To investigat anticipated long-term exposure effects on the retrieved materials, precise documentation of the fligh hardware has been obtained. This documentation includes detailed photography. Certain surfaces wer photographed with a resolution of 80 microns. The surfaces include the LCRU-mirrors, TV-camera mirror and lenses, Hasselblad camera lens, Heat Flow Electronics Box, LEAM Lunar Shield, and the ALSEP Heli Antenna Gimbal Housing. In addition, the chemical composition of various materials have been obtaine Most important, however, representative pieces of the above surfaces have been incorporated in th Curatorial Facilities at NASA-MSC, where they will be stored for the next few decades. These samples wi serve as reference materials to those which are emplaced by the Apollo 17 crew and which hopefully will b returned.

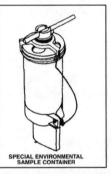

SPECIAL ENVIRONMENTAL SAMPLE CONTAINER

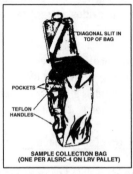

SAMPLE COLLECTION BAG (ONE PER ALSRC-4 ON LRV PALLET)

EXTRA SAMPLE COLLECTION BAG

SAMPLE RETURN BAG (BSLSS SAMPLE BAG)

SAMPLE RETURN CONTAINER

2-BAG DOCUMENTED SAMPLE BAG DISPENSER

DOCUMENTED SAMPLE BAG

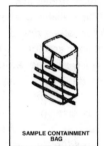

SAMPLE CONTAINMENT BAG

Lunar Geology Sample Containers

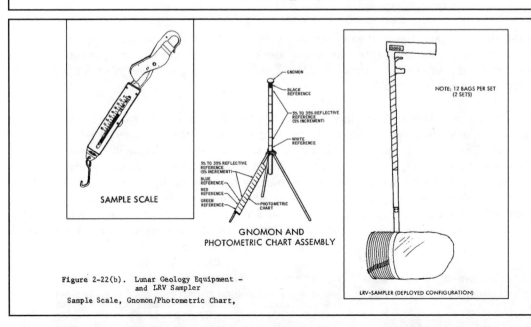

SAMPLE SCALE

GNOMON AND PHOTOMETRIC CHART ASSEMBLY

LRV-SAMPLER (DEPLOYED CONFIGURATION)

Figure 2-22(b). Lunar Geology Equipment - and LRV Sampler

Sample Scale, Gnomon/Photometric Chart,

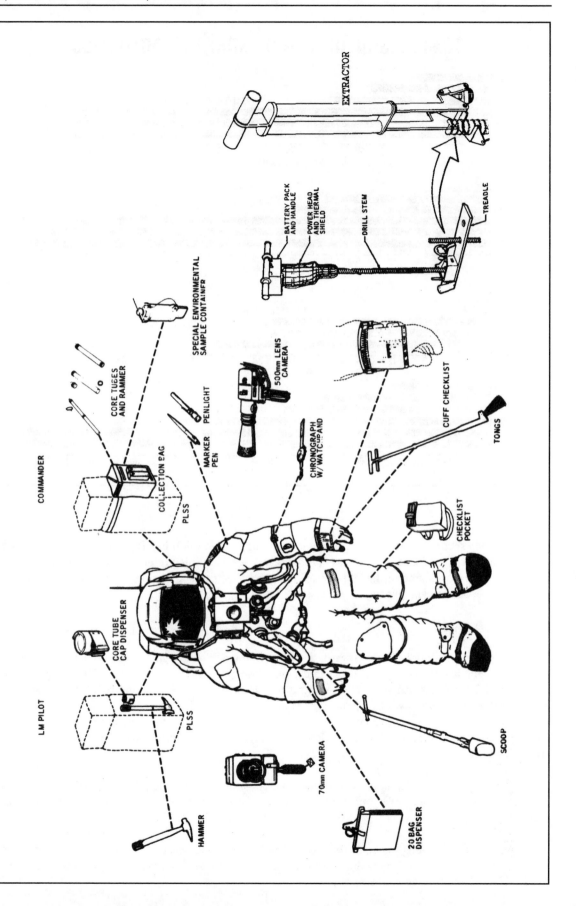

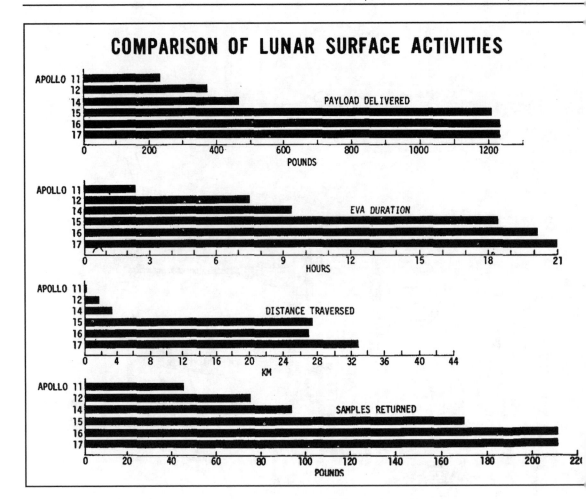

LUNAR ORBITAL SCIENCE

Service module sector 1 houses the scientific instrument module (SIM) bay. Three experiments are carrie in the SIM Bay: S-209 lunar sounder, S-171 infrared scanning spectrometer and S-169 far-ultraviole spectrometer. Also mounted in the SIM bay are the panoramic camera, mapping camera and laser altimete used in the service module photographic tasks.

Lunar Sounder (S-209):

Electromagnetic impulses beamed toward the lunar surface in the high frequency (HF) and very hig frequency (VHF) bands will provide recorded data for developing a geological model of the lunar interior t a depth of 1.3 km (4,280 ft.). In addition to stratigraphic, structural, tectonic and topographic data on th region of the Moon overflown by Apollo 17, the lunar sounder will measure the ambient electromagneti noise levels in the lunar environment at 5, 15 and 150 mHz and the occultation by the Moon c electromagnetic waves generated at the lunar surface by the surface electrical properties experimer transmitter.

Data from the lunar sounder coupled with information gathered from the SIM bay cameras and the lase altimeter, and from surface gravity measurements will allow experimenters to build an absolute topographi profile.

The experience in operating the lunar sounder and analyzing its data will contribute to the designs of futur instruments for detection of surface or near-surface water on mars, mapping of major geological units o Mars and Venus and topside sounding of Jupiter.

he lunar sounder has three major components: the coherent synthetic aperture radar (CSAR), the optical ecorder, and antennas — an HF retractable dipole and a VHF yagi. The radar and optical recorder are located n the lower portions of the SIM bay. The HF dipole antenna, which has a scan of 24.4m (80 feet), deploys om the base of the SIM. The VHF yagi antenna is automatically deployed after the spacecraft/LM adapter SLA) panels are jettisoned.

APOLLO ORBITAL SCIENCE MISSION ASSIGNMENTS

EXPERIMENT		11	12	14	15	16	17
ERVICE MODULE:							
-160	GAMMA-RAY SPECTROMETER				X	X	
-161	X-RAY SPECTROMETER				X	X	
-162	ALPHA-PARTICLE SPECTROMETER				X	X	
-164	S-BAND TRANSPONDER (CSM/LM)			X	X	X	X
-165	MASS SPECTROMETER				X	X	
-169	FAR UV SPECTROMETER						X
-170	BISTATIC RADAR			X	X		
-171	IR SCAN RADIOMETER						X
UBSATELLITE:							
-173	PARTICLE MEASUREMENT				X	X	
-174	MAGNETOMETER				X	X	
-164	S-BAND TRANSPONDER				X	X	
-209	LUNAR SOUNDER						X
M PHOTOGRAPHIC TASKS:							
4" PANORAMIC CAMERA					X	X	X
" MAPPING CAMERA					X	X	X
ASER ALTIMETER					X	X	X
COMMAND MODULE:							
CM PHOTOGRAPHIC TASKS:		X	X	X	X	X	X
-158	MULTI SPECTRAL PHOTOGRAPHY		X				
-176	CM WINDOW METEOROID			X	X	X	X
-177	UV PHOTO EARTH & MOON				X	X	
-178	GEGENSCHEIN			X	X		
1-211	BIOSTACK					X	X
1-212	BIOCORE						X

APOLLO 17
SIM BAY

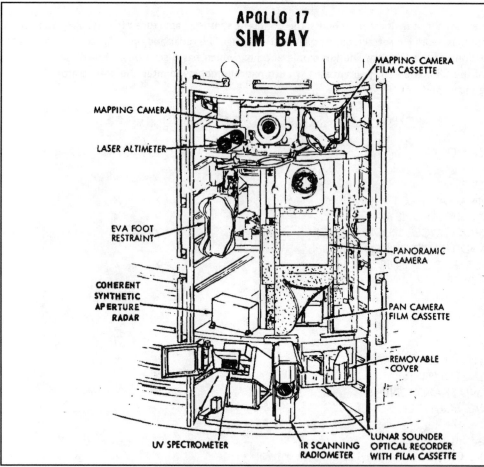

MAPPING CAMERA FILM CASSETTE

MAPPING CAMERA

LASER ALTIMETER

EVA FOOT RESTRAINT

PANORAMIC CAMERA

COHERENT SYNTHETIC APERTURE RADAR

PAN CAMERA FILM CASSETTE

REMOVABLE COVER

UV SPECTROMETER

IR SCANNING RADIOMETER

LUNAR SOUNDER OPTICAL RECORDER WITH FILM CASSETTE

APOLLO 17
LUNAR SOUNDER CONFIGURATION

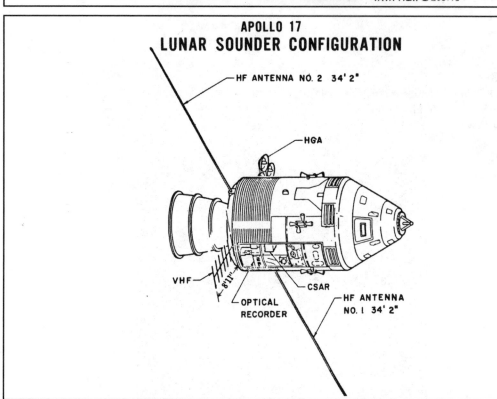

HF ANTENNA NO. 2   34' 2"

HGA

VHF

CSAR

OPTICAL RECORDER

HF ANTENNA NO. 1 34' 2"

he crew has controls for deploying and jettisoning (if required) the HF antenna and for selecting lunar ounder operating modes: VHF operate, HF operate and HF receive only. During the operating modes, an ectromagnetic pulse is transmitted toward the Moon and the return signal is recorded on film by the optical ecorder. In the HF "receive only" mode, electromagnetic background noises from galactic sources and effects f lunar occultation are received and recorded by the instrument. During lunar sounder "operate" modes he other SIM bay instruments are powered down. The sounder's data film cassette is retrieved by the ommand module pilot during the trans-Earth EVA at the same time the mapping and panoramic camera assettes are retrieved.

r. Stan Ward of the University of Utah and Dr. Walt Brown of the Jet Propulsion Laboratory, Pasadena, Calif., e lunar sounder principal investigators.

frared Scanning Radiometer (S-171):

he infrared scanning radiometer (ISR) experiment will provide a lunar surface temperature map with nproved temperature and spatial resolution over what has been possible before. Previous Earth-based bservations of the lunar surface thermal balance have been limited to the front side with a temperature solution of about 210°K (-80°F) and a surface resolution of about 15 km (9.3 miles). The ISR will permit easurements to be made on both the front and back side at a temperature resolution of better than 1°K .8°F) and a surface resolution of better than 2 km (1.2 miles). The ISR will locate and pinpoint anomalously old or hot regions on the lunar surface in addition to measuring the surface temperature of various areas s a function of Sun angle. These cooling curves (temperature vs Sun angle) will be used to calculate the termal properties of regions of varying geology, topography, and rock distribution.

Vhen correlated with orbital photography and lunar sounder data, ISR temperature measurements are xpected to aid in locating surface rock fields, crustal structural differences, volcanic activity and fissures mitting "hot" gases.

lounted on the bottom shelf of the SIM Bay, the infrared scanning radiometer is made up of an optical canning unit, a thermistor bolometer and associated processing electronics. The scanning unit consists of a olded cassegrain telescope with a rotating mirror which sweeps 162 degrees crosstrack. The thermal energy om about 2 to about 60m (7 to 200 feet) emitted from the lunar surface and reflected from the scanning nirror is focused by the telescope onto the thermistor bolometer. Changes in lunar surface temperature are etected by the bolometer. The output of the thermistor bolometer is processed by the electronics package hich splits the temperature readouts into three channels for telemetry: 0-160°K (-459 to -170°F), 0-250°K 459 to -10°F) and 0-400°K (-459 to 260°F). The lunar surface temperature at lunar noon is approximately 00°K (260°F) and drops to about 80°K (-315°F) just before lunar sunrise.

he CSM longitudinal axis will be aligned to the flight path during periods the experiment is turned on. The istrument will be operated over both the light side and darkside primarily during crew sleep periods.

r. Frank J. Low of the University of Arizona Lunar and Planetary Laboratory is the principal investigator.

ar-Ultraviolet Spectrometer (S-169):

tomic composition, density and scale height for several constituents of the lunar atmosphere will be easured by the far-ultraviolet spectrometer. Solar far-UV radiation reflected from the lunar surface as well s UV radiation emitted by galactic sources also will be detected by the instrument.

he far-UV spectrometer will gather data on the spectral emission in the range of 1175 to 1675 Å (angstrom) nd it is expected that among elements detected will be hydrogen (1216Å), carbon (1675Å), nitrogen 200Å), oxygen (1304Å), krypton (1236Å) and xenon (1470Å). The total lunar atmospheric density at the inar surface is estimated to be less than one trillionth of that at the Earth's surface.

The far-UV spectrometer, mounted on the bottom shelf of the SIM bay consists of an external baffle whic limits stray light, a 0.5 meter (20 inches) focal-length Ebert mirror spectrometer with .5x6 cm (.2x2.4 inch slits, a 10x10 cm (4x4 inch) reflection grating, a scan drive mechanism which provides the wavelength sca and a photomultiplier tube which measures the intensity of the incident ultraviolet radiation, and processing electronics. The spectrometer has a 12x12 degree field of view and is aligned 18 degrees to the right of th SIM bay centerline and 23 degrees forward of the CSM vertical axis.

Controls for activating and deactivating the experiment and for opening and closing a protective cover are located in the CM.

Dr. William G. Fastie of Johns Hopkins University. Baltimore, Md., is the principal investigator.

Gamma Ray Spectrometer (S-160):

As an adjunct to the gamma ray spectrometer experiment carried on Apollos 15 and 16, the aim of this tas is to gather data for a calibration baseline to support the overall S-160 experiment.

A SIM bay experiment on Apollo 16, the S-160 sensor was extended on a 7.5 meter (25 foot) boom t measure natural and cosmic rays, induced gamma radioactivity from the lunar surface and the radiation flu in cislunar space. A sodium iodide crystal was used as the detector's scintillator.

An identical sodium iodide crystal will be carried aboard the Apollo 17 command module to measur background galactic radiation and CM flux reaching it during the mission.

The measurements from this passive crystal will be "subtracted" from the Apollo 15 and Apollo 16 data t separate background noise from valid lunar measurements.

Immediately after splashdown, the 2.3-kilogram (five-pound) 7.5x7.5-cm (3x3-inch) cylinder containing th crystal will be removed and placed into a photomultiplier multichannel analyzer aboard the prime recover ship for measurements of short-period half-lives of several isotopes.

Panoramic Camera: 610mm (24-inch) SM orbital photo task: The camera gathers mono or stereo high resolution 2 meters (6.5ft.) photographs of the lunar surface from orbit. The camera produces an image siz of 28 x 334 kilometers (17 x 208 nm) with a field of view 11° along the track and 108° cross track. Th rotating lens system can be stowed face-inward to avoid contamination during effluent dumps and thruste firings.

The 33-kilogram (72-pound) film cassette of 1,650 frames will be retrieved by the command module pilo during a transearth coast EVA. The camera works in conjunction with the mapping camera and the lase altimeter to gain data to construct a *comprehensive* map of the lunar surface ground track flown by th mission —- about 2.97 million square meters (1.16 million square miles) or 8 percent of the lunar surface

Mapping Camera: 76mm (3-inch): Combines 20-meter (66ft) resolution terrain mapping photography o 12.5cm (5 in.) film with 76 mm (3-inch) focal length lens with stellar camera shooting the star field on 35mr film simultaneously at 96° from the surface camera optical axis. The stellar photos allow accurate orientatic of mapping photography postflight by comparing simultaneous star field photography with lunar surfac photos of the nadir (straight down). Additionally, the stellar camera provides pointing vectors for the lase altimeter during darkside passes. The mapping camera metric lens covers a 74° square field of view, or 17 x 170 km (92 x 92 nm) from 111.5 km (60 nm) in altitude. The stellar camera is fitted with a 76mm (3-inch f/2.8 lens covering a 24° field with cone flats. The 9-kg (20-lb) film cassette containing mapping camera fil (3,600 frames) and the stellar camera film will be retrieved during the same EVA described in the panoram camera discussion. The Apollo Orbital Science Photographic Team is headed by Frederick J. Doyle of the U. Geological Survey, McLean, Va.

Laser Altimeter: This altimeter measures spacecraft altitude above the lunar surface to within two meters (6.5ft). The instrument is boresighted with the mapping camera to provide altitude correlation data for the mapping camera as well as the panoramic camera. When the mapping camera is running, the laser altimeter automatically fires a laser pulse to the surface corresponding to mid-frame ranging for each frame. The laser light source is a pulsed ruby laser operating at 6,943 angstroms, and 200-millijoule pulses of 10 nanoseconds duration. The laser has a repetition rate up to 3.75 pulses per minute. On Apollo 15 and 16, this instrument revealed important new information on the shape of the Moon. A large depression on the backside was found and the separation of the center of figure and center of mass of the Moon by about 2 km (1.2 miles) was observed. The laser altimeter working group of the Apollo Orbital Science Photographic Team is headed by Dr. William M. Kaula of the UCLA Institute of Geophysics and Planetary Physics.

CSM/LM S-Band Transponder: The objective of this experiment is to detect variations in lunar gravity along the lunar surface track. These gravitational anomalies result in minute perturbations of the spacecraft motion and are indicative of magnitude and location of mass concentrations on the Moon. The Spaceflight Tracking and Data Network (STDN) and the Deep Space Network (DSN) will obtain and record S-band doppler tracking measurements from the docked CSM/LM and the undocked CSM while in lunar orbit; S-band doppler tracking measurements of the LM during non-powered portions of the lunar descent; and S-band doppler tracking measurements of the LM ascent stage during non-powered portions of the descent for lunar impact. The CSM and LM S-band transponders will be operated during the experiment period. The experiment was conducted on Apollo, 14, 15 and 16.

S-band doppler tracking data from the Lunar Orbiter missions were analyzed and definite gravity variations were detected. These results showed the existence of mass concentrations (mascons) in the ringed maria. Confirmation of these results has been obtained with Apollo tracking data, both from the CSM and in greater detail from the Apollo 15 and 16 subsatellites.

With appropriate spacecraft orbital geometry much more scientific information can be gathered on the lunar gravitational field. The CSM and/or LM in low-altitude orbits can provide new detailed information on local gravity anomalies. These data can also be used in conjunction with high-altitude data to possibly provide some description on the size and shape of the perturbing masses. Correlation of these data with photographic and other scientific records will give a more complete picture of the lunar environment and support future lunar activities. Inclusion of these results is pertinent to any theory of the origin of the Moon and the study of the lunar subsurface structure, since it implies a rigid outer crust for the Moon for at least the last three billion years. There is also the additional benefit of obtaining better navigational capabilities for future lunar missions in that an improved lunar gravity model will be known. William Sjogren, Jet Propulsion Laboratory, Pasadena, Calif. is principal investigator.

Apollo Window Meteoroid: This is a passive experiment in which command module windows are scanned under high magnification pre- and postflight for evidence of meteoroid cratering flux of one-trillionth gram or larger. Such particle flux may be a factor in degradation of surfaces exposed to space environment. Principal investigator is Burton Cour-Palais, NASA Manned Spacecraft Center.

APOLLO LUNAR ORBIT PHOTOGRAPHIC COVERAGE

	% LUNAR SURFACE		
	APOLLO 15	APOLLO 16	APOLLO 17
TOTAL AREA OVERFLOWN BETWEEN GROUNDTRACKS LOI TEI	17.0	7.2	13.5
MAPPING CAMERA -VERTICAL			
LUNAR SURFACE PHOTOGRAPHED	10.3	5.3	8.5
NEW AREA PHOTOGRAPHED	-	3.9	2.8
TOTAL NON-REDUNDANT PHOTOGRAPHY	10.3	14.2	17.0
PAN CAMERA			
UNRECTIFIED PHOTOGRAPHY	11.5	7.2	10.3
RECTIFIED PHOTOGRAPHY	6.7	4.2	6.0
NEW AREA - RECTIFIED PHOTOGRAPHY	-	3.1	2.0
TOTAL NON -REDUNDANT RECTIFIED PHOTOGRAPHY	6.7	9.8	11.8

MEDICAL TESTS AND EXPERIMENTS

Visual Light-Flash Phenomenon: Mysterious flashes of light penetrating closed eyelids have been reported by crewmen of every Apollo lunar mission since Apollo 11. Usually the light streaks and specks are observed in a darkened command module cabin while the crew is in a rest period. Averaging two flashes a minute, the phenomenon was observed in previous missions in translunar and transearth coast and in lunar orbit.

Two theories have been proposed on the origin of the flashes. One theory is that the flashes stem from visual phosphenes induced by cosmic rays. The other theory is that Cerenkov radiation by high-energy atomic particles either enter the eyeball or ionize upon collision with the retina or cerebral cortex.

The Apollo 17 crew will run a controlled experiment during translunar coast in an effort to correlate light flashes to incident primary cosmic rays. One crewman will wear an emulsion plate device on his head called the Apollo light flash moving emulsion detector (ALFMED), while his crewmates wear eyeshields. The ALFMED emulsion plates cover the front and sides of the wearer's head and will provide data on time, strength, and path of high-energy atomic particles penetrating the emulsion plates. This data will be correlated with the crewman's verbal reports on flash observations during the tests. The test will be repeated during transearth coast with all three crewman wearing eyeshields and without the ALFMED.

The experiment was also flown on Apollo 16.

Biocore (M-212: A completely passive experiment, Biocore is designed to determine whether ionizing heavy cosmic ray particles can injure non-regenerative (nerve) cells in the eye and brain. Cosmic Ray particles range from carbon particles to iron particles or even heavier.

Pocket mice, found in the California desert near Palm Springs, are used because they are hardy, small (weighing about one-third of an ounce), and drink no water (their water comes from the seeds they eat). Five pocket mice will have cosmic ray particle detectors, in sandwich form, implanted under their scalps. These detectors are made of the plastics lexan and cellulose nitrate. Under the microscope, the tracks that cosmic ray particles make in passing through these plastics can be seen. This enables physicists to determine the path of the particles passing into the brain.

Five mice, each in a perforated aluminum tube, are housed inside an aluminum canister which is 33.8 cm long and 17.8 cm wide (12 in. long, 7 in. wide). The tubes are small so that the mice cannot float free in the zero G environment and will have ample seeds for food. A central tube in the canister contains potassium superoxide. When the animals breathe, the moisture and carbon dioxide coming from their lungs activate the superoxide which gives off oxygen to sustain the mice. The canister is a self-sustained, closed unit, not requiring any attention by the astronauts during the flight.

Principal Investigator is Dr. Webb Haymaker, NASA Ames Research Center, Mountain View, Calif.

Biostack (M-211): The German Biostack experiment is a passive experiment requiring no crew action and is quite similar to the Biostack flown on Apollo 16. The experimental results obtained on Apollo 16 were considered quite good. Conducting the experiment again on Apollo 17 with six biological materials in lieu of the four flown on Apollo 16 should enlarge the data base obtained earlier.

Selected biological material will be exposed to high-energy heavy ions in cosmic radiation and the effects analyzed postflight. Heavy ion energy measurements cannot be gathered from ground-based radiation sources. The Biostack experiments will add to the knowledge of how these heavy ions may present a hazard for man during long space flights.

Alternate layers of biological materials and radiation track detectors are hermetically sealed in an aluminum cylinder measuring 12.5 cm in diameter and 9.8 cm high (4.8 x 3.4 inches) and weighing 2.4 kg (5.3 lbs.). The cylinder will be stowed aboard the command module preflight and removed postflight for analysis by the principal investigator, Dr. Horst Bucker of the University of Frankfurt am Main, Federal Republic of Germany.

he six biological materials in Biostack, none of which is harmful to man, are bacillus subtilis spores (hay acillus), arabiodopsis thaliana seeds (mouse-ear cress), vicia faba (broad bean roots), artemia saliva eggs rine shrimp), colpoda cucullus (protozoa cysts), and tribolium casteneum (beetle eggs).

ardiovascular Conditioning Garment: A counterpressure garment similar to a fighter pilot's "g-suit" which ill be donned by the command module pilot prior to entry and left on until completion of medical exams oard the prime recovery vessel. Lower body negative pressure tests will be run on the CMP pre-mission id again aboard ship as part of a program to evaluate the device as a potential tool for protecting the turned crewmen in the 1G environment of Earth from cardiovascular changes to the body resulting from ace flight.

ylab Mobile Laboratories (SML) Field Test: Apollo 17 recovery operations will serve as a practical field test r the Skylab Mobile Laboratory. The SML will be loaded aboard the prime recovery vessel and staffed with physicians and 4 para-medical professionals in a shakedown of the system in preparation for recovery perations for next year's Skylab mission.

he SML is made up of six basic U.S. Army Medical Unit Self-Contained Transportables (MUST) modified for xylab postflight crew medical examinations and processing of inflight and postflight medical experiment data. ach mini-lab is outfitted to meet a specific discipline: blood, cardiovascular, metabolic studies, microbiology, utrition and endocrinology, and operational medicine.

od Compatibility Assessment: This investigation will measure whole body metabolic gains or losses, gether with associated endocrinological and electrolytes controls. The purpose of the investigation is to sess food compatibility and to determine the effect of space flight upon overall body composition and upon e circulating and excretory levels of certain hormonal constituents which are responsible for maintaining omeostasis. This assessment is designed to acquire input and output information necessary not only to sess the metabolic consequences of the final lunar mission but also to provide a firmer basis for terpreting the results of the Skylab missions.

s part of this investigation, an improved urine collection system will be used by the Apollo 17 crew. Called e Biomedical Urine Sample System (BUSS), the device consists of a polyurethane film bag with 4,000 illiliters (120 ounces) capacity which contains 30 milligrams (.0012 ounces) of lithium for a tracer and 10 ams (.4 ounces) boric acid for preservation of organic constituents. One bag is furnished for each man/day the command module for a total of 34. Each bag is fitted with a roll-on cuff. At the end of each 24-hour mpling period, a small sample for postflight analysis is withdrawn from each of the three BUSS bags and beled and stowed for postflight analysis; the remainder is vented overboard through the command module aste management system. No samples will be collected in lunar module operations.

ENGINEERING/OPERATIONAL TESTS AND DEMONSTRATIONS

eat Flow and Convection

he flow of fluids on Earth are dependent upon gravity for motion, but in the absence of gravity in space ght fluids behave quite differently and depend more upon surface tension as their motive force. vestigations of fluid flow caused by surface tension gradients, interfacial tension of dissimilar fluids and xpansion are next to impossible in earthbound laboratories.

he Apollo 17 heat flow and convection demonstration will go beyond the demonstration carried out on pollo 14 by providing more exact data on the behavior of fluids in a low gravity field — data which will be aluable in the design of future science experiments and for manufacturing processes in space.

hree test cells are used in the Apollo 17 demonstration for measuring and observing fluid flow behavior. he tests will be recorded on motion picture film in addition to direct observation by the crew. Radial heat w will be induced in a circular cell which has an electrical heater in its center. A Liquid crystal material, which changes colors when heated, covers the argon-filled cell, thereby indicating heat flow through changin color patterns. Lineal heat flow will be demonstrated in a transparent cylinder filled with Krytox (heavy oi and color-indicating liquid crystal strips. Heat flow is induced by an electric heater in one end of the cylinde

The third device is a flow pattern test cell made up of a shallow aluminum dish into which layers of Kryto with suspended aluminum flakes are injected. A heater on the bottom of the dish causes flow patterns t form in the Krytox, with the aluminum flakes serving to make the flow more visible.

The demonstrations will be run first during translunar coast when the spacecraft rates are nulled in all thre axes, and again after passive thermal control (PTC) or "barbecue" roll has been set up. The 16-mm dat acquisition camera will be used to make sequence photos of the test cells — 10 minutes for the radial an lineal test cells, and 15 minutes for the flow pattern test cell.

Skylab Contamination Study

Since John Glenn reported seeing fireflies outside the tiny window of his Mercury spacecraft Friendship 7 decade ago, space crews have noted light-scattering particles that hinder visual observations as well a photographic tasks. These clouds of particles surrouonding spacecraft generally are from water dumps an escaping cabin gases changing into ice crystals. The phenomenon could be of concern in the Skylab missior during operation of the solar astronomy experiments. The light scattering from a 100-micron particle 1 kilometers (7.8 miles) away from the spacecraft, for example, is as bright as a third-magnitude star. A clou of particles with such a lightscattering effect would rule out any astronomical experiments being conducte on the sunlit portion of an orbit.

During translunar coast the Apollo 17 Infrared Scanning Radiometer and Far-Ultraviolet Spectrometer in th SIM Bay will provide data on optical contamination, adsorption of the contamination cloud, the scatterin effect of a particle cloud, and the duration of a cloud resulting from a specific waste water dump.

Light-scattering and contamination data gathered on Apollo 17 will aid in predicting contamination in th vicinity of the Skylab orbital workshop and its sensitive telescope mount, and in devising means of minimizir contaminate levels around the Skylab vehicle. Photography in support of the study was also conducted o Apollo 16.

LUNAR ROVING VEHICLE

The lunar roving vehicle (LRV), the third to be used on the Moon, will transport two astronauts on thre exploration traverses of the Moon's Taurus-Littrow region during the Apollo 17 mission. The LRV will als carry tools, scientific and communications equipment, and lunar samples.

The four-wheel, lightweight vehicle has greatly extended the lunar area that can be explored by man. It is th first manned surface transportation system designed to operate on the Moon, and it represents a solutio to challenging new problems without precedent in Earth-bound vehicle design and operation.

The LRV must be folded into a small package within a wedge-shaped storage bay of the lunar module descer stage for transport to the Moon. After landing, the vehicle must be unfolded from its stowed position an deployed on the surface. It must then operate in an almost total vacuum under extremes of surfac temperatures, low gravity, and on unfamiliar terrain.

The first lunar roving vehicle, used on the Apollo 15 lunar mission, was driven for three hours during it exploration traverses, covering a distance of 27.9 kilometers (17.3 statute miles) at an average speed of 9. kilometers an hour (5.8 miles an hour). The second lunar roving vehicle, used on Apollo 16, was driven thre hours and twenty-six minutes for a total distance of 26.9 kilometers (16.7 statute miles) at an average spee of 7.8 kilometers an hour (4.9 miles an hour).

General Description

The LRV is 3.1 meters long (10.2 feet); has a 1.8 - meter (six-foot) width; is 1.14 meters high (44.8 inches); and has a 2.3-meter wheel base (7.5 feet). Each wheel is powered by a small electric motor. The maximum speed reached on the Apollo 15 mission was 13 km/hr (eight mph), and 17 km/hr (11 mph) on Apollo 16.

Two 36-volt batteries provide vehicle power, and either battery can run all systems. The front and rear wheels have separate steering systems; if one fails it can be disconnected and the LRV will operate with the other system.

Weighing about 209 kilograms (461 pounds), Earth weight, when deployed on the moon the LRV can carry a total payload of about 490 kilograms (1,080 pounds), more than twice its own weight. The payload includes two astronauts and their portable life support systems (about 363 kilograms; 800 pounds), 68.0 kilograms (150 pounds) of communications equipment, 54.5 kilograms (150 pounds) of scientific equipment and photographic gear, and 40.8 kilograms (90 pounds) of lunar samples.

The LRV is designed to operate during a minimum period of 78 hours on the lunar surface. It can make several exploration sorties to a cumulative distance of 96 kilometers (57 miles). The maximum distance the LRV will be permitted to range from the lunar module will be approximately 9.4 kilometers (5.9 miles), the distance the crew could safely walk back to the LM in the unlikely event of a total LRV failure. This walkback distance limitation is based upon the quantity of oxygen and coolant available in the astronauts' portable life support systems.

This area contains about 292 square kilometers (113 square miles) available for investigation, 10 times the area that can be explored on foot.

The vehicle can negotiate obstacles 30.5 centimeters (one foot) high and cross crevasses 70 centimeters (28 inches). The fully loaded vehicle can climb and descend slopes as steep as 25 degrees, and park on slopes up to 35 degrees. Pitch and roll stability angles are at least 45 degrees, and the turn radius is three meters (10 feet).

Both crewmen sit so the front wheels are visible during normal driving. The driver uses an on-board dead reckoning navigation system to determine direction and distance from the lunar module, and total distance traveled at any point during a traverse.

The LRV has five major systems: mobility, crew station, navigation, power, and thermal control. Secondary systems include the deployment mechanism, LM attachment equipment, and ground support equipment.

The aluminum chassis is divided into three sections that support all equipment and systems. The forward and aft sections fold over the center one for stowage in the LM. The forward section holds both batteries, part of the navigation system, and electronics gear for the traction drive and steering systems. The center section holds the crew station with its two seats, control and display console, and hand controller. The floor of beaded aluminum panels can support the weight of both astronauts standing in lunar gravity. The aft section holds the scientific payload, television assembly (GCTA), scientific equipment, tools, and sample stowage bags.

Mobility System

The mobility system is the major LRV system, containing the wheels, traction drive, suspension, steering, and drive control electronics subsystems.

The vehicle is driven by a T-shaped hand controller located on the control and display console post between the crewmen. Using the controller, the astronaut maneuvers the LRV forward, reverse, left and right.

Each LRV wheel has a spun aluminum hub and a titanium bump stop (inner frame) inside the tire (outer frame). The tire is made of a woven mesh of zinc-coated piano wire to which titanium treads are riveted in a chevron pattern around the outer circumference. The bump stop prevents excessive *inflection* of the mesh tire during heavy impact. Each wheel weighs 5.4 kilograms (12 pounds) on Earth and is designed to be driven at least 180 kilometers (112 miles). The wheels are 81.3 centimeters (32 inches) in diameter and 22.9 centimeters (nine inches) wide.

A traction drive attached to each wheel has a motor harmonic drive gear unit, and a brake assembly. The harmonic drive reduces motor speed at an 80-to-1 rate for continuous operation at all speeds without gear shifting. The drive has an odometer pickup (measuring distance traveled) that sends data to the navigation system. Each motor develops 0.18 kilowatt (1/4-horsepower) and operates from a 36-volt input.

Each wheel has a mechanical brake connected to the hand controller. Moving the controller rearward de-energizes the drive motors and forces brake shoes against a drum, stopping wheel hub rotation. Full rear movement of the controller engages and locks a parking brake.

The chassis is suspended from each wheel by two parallel arms mounted on torsion bars and connected to each traction drive. Tire deflection allows a 35.6-centimeter (14-inch) ground clearance when the vehicle is fully loaded, and 43.2 centimeters (17 inches) when unloaded.

Both front and rear wheels have independent steering systems that allow a "wall-to-wall" turning radius of 3.1 meters (122 inches), exactly the vehicle length. If either set of wheels has a steering failure, its steering system can be disengaged and the traverse can continue with the active steering assembly. Each wheel can also be manually uncoupled from the traction drive and brake to allow "free wheeling" about the drive housing.

Pushing the hand controller forward increases forward speed; rear movement reduces speed. Forward and reverse are controlled by a knob on the controller's vertical stem. With the knob pushed down, the controller can only be pivoted forward; with it pushed up, the controller can be pivoted to the rear for reverse.

Crew Station
The crew station consists of the control and display console, seats, seat belts, an armrest, footrests, inboard and outboard handholds, toeholds, floor panels, and fenders.

The control and display console is separated into two main parts: The top portion holds navigation system displays; the lower portion contains monitors and controls. Attached to the upper left side of the console is an attitude indicator that shows vehicle pitch and roll.

At the console top left is a position indicator. Its outer circumference is a large dial that shows vehicle heading (direction) with respect to lunar north. Inside the dial are three digital indicators that show bearing and range to the LM and distance traveled by the LRV. In the middle of the console upper half is a Sun shadow device that is used to update the LRV's navigation system. Down the left side of the console lower half are control switches for power distribution, drive and steering, and monitors for power and temperature.

A warning flag atop the console pops up if a temperature goes above limits in either battery or in any drive motor.

The LRV seats are tubular aluminum frames spanned by nylon webbing. They are folded flat during launch and erected by crewmen after deployment. The seat backs support the astronaut portable life support systems. Nylon webbing seat belts, custom fitted to each crewman, snap over the outboard handholds with metal hooks.

The armrest, located directly behind the LRV hand controller, supports the arm of the driving crewman. The footrests, attached to the center floor section, are adjusted before launch to fit each crewman. Inboard handholds help crewmen get in and out of the LRV, and have receptacles for an accessory staff and the low gain antenna of the LCRU. The lightweight, fiberglass fenders keep lunar dust from being thrown on the astronauts, their equipment, sensitive vehicle parts, and from obstructing vision while driving. Front and rear fender sections are retracted during flight and extended by crewmen after LRV deployment on the lunar surface.

Navigation System

The navigation system is based on the principle of starting a sortie from a known point, recording speed, direction and distance traveled, and periodically calculating vehicle position.

The system has three major components: a directional gyroscope to provide vehicle headings; odometers on each wheel's traction drive unit to give speed and distance data; and a signal processing unit (a small, solid-state computer) to determine heading, bearing, range, distance traveled, and speed.

All navigation system readings are displayed on the control console. The system is reset at the beginning of each traverse by pressing a system reset button that moves all digital displays and internal registers to zero.

The directional gyroscope is aligned by measuring the inclination of the LRV (using the attitude indicator) and measuring vehicle orientation with respect to the Sun (using the shadow device). This information is relayed to ground controllers and the gyro is adjusted to match calculated values read back to the crew. Each LRV wheel revolution generates odometer magnetic pulses that are sent to the console displays.

Power System

The power system consists of two 36-volt, non-rechargeable batteries and equipment that controls and monitors electrical power. The batteries are in magnesium cases, use plexiglass monoblock (common cell walls) for internal construction, and have silver-zinc plates in potassium hydroxide electrolyte. Each battery has 23 cells and 121-ampere-hour capacity.

Both batteries are used simultaneously with an approximately equal load during LRV operation. Each battery can carry the entire electrical load; if one fails, its load can be switched to the other.

The batteries are activated when installed on the LRV at the launch pad about five days before launch. During LRV operation all mobility system power is turned off if a stop exceeds five minutes, but navigation system power remains on throughout each sortie. The batteries normally operate at temperatures of 4.4 to 51.7 degrees C. (40-125 degrees F.).

An auxiliary connector at the LRV's forward end supplies 150 watts of 36-volt power for the lunar communications relay unit.

Thermal Control

The basic concept of LRV thermal control is heat storage during vehicle operation and radiation cooling when it is parked between sorties. Heat is stored in several thermal control units and in the batteries. Space radiators are protected from dust during sorties by covers that are manually opened at the end of each sortie; when battery temperatures cool to about 7.2 degrees C. (45 degrees F.), the covers automatically close.

A multi-layer insulation blanket protects forward chassis components. Display console instruments are mounted to an aluminum plate isolated by radiation shields and fiberglass mounts. Console external surfaces are coated with thermal control paint and the face plate is anodized, as are handholds, footrests, tubular seat sections, and center and aft floor panels.

Stowage and Deployment

Space support equipment holds the folded LRV in the lunar module during transit and deployment at three attachment points with the vehicle's aft end pointing up.

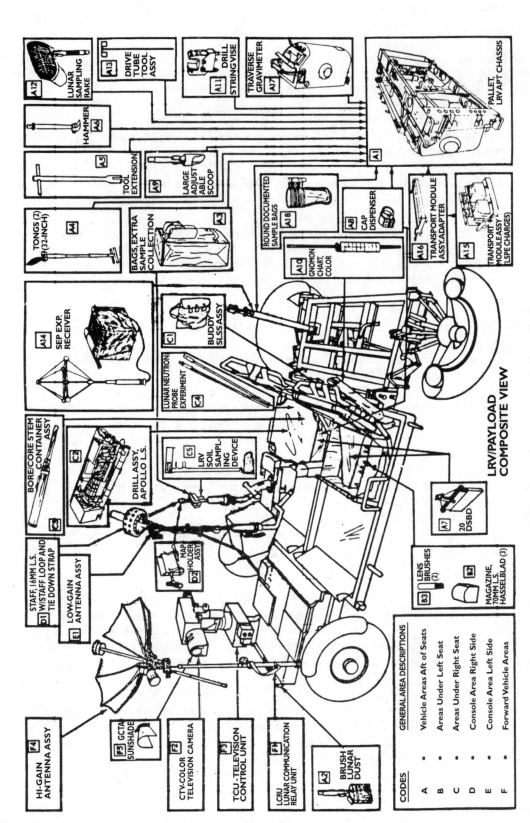

LRV/PAYLOAD COMPOSITE VIEW

FIGURE 3.6-5   LUNAR FIELD GEOLOGY EQUIPMENT STOWAGE ON LRV

GENERAL AREA DESCRIPTIONS

CODES		
A	-	Vehicle Areas Aft of Seats
B	-	Areas Under Left Seat
C	-	Areas Under Right Seat
D	-	Console Area Right Side
E	-	Console Area Left Side
F	-	Forward Vehicle Areas

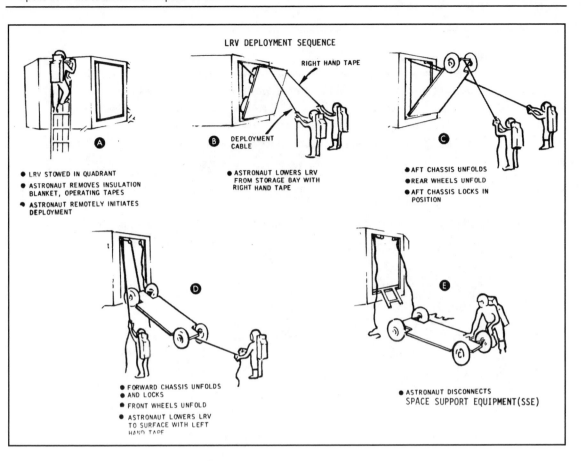

LRV DEPLOYMENT SEQUENCE

Deployment is essentially manual. One crewman releases a cable attached to the top (aft end) of the folded LRV as the first step in the deployment.

One of the crewmen then ascends the LM ladder part way and pulls a D-ring on the side of the descent stage. This releases the LRV, and lets the vehicle swing out at the top about 12.7 centimeters (five inches) until it is stopped by two steel cables. Descending the ladder, the crewman walks to the LRV's right side, takes the end of a deployment tape from a stowage bag, and pulls the tape hand-over-hand. This unreels two support cables that swivel the vehicle outward from the top. As the aft chassis is unfolded, the aft wheels automatically unfold and deploy, and all latches are engaged. The crewman continues to unwind the tape, lowering the LRV's aft end to the surface, and the forward chassis and wheels spring open and into place.

When the aft wheels are on the surface, the crewman removes the support cables and walks to the vehicle's left side. There he pulls a second tape that lowers the LRV's forward end to the surface and causes telescoping tubes to push the vehicle away from the LM. The two crewmen then deploy the fender extensions, set up the control and display console, unfold the seats, and deploy other equipment.

One crewman will board the LRV and make sure all controls are working. He will back the vehicle away slightly and drive it to the LM quadrant that holds the auxiliary equipment. The LRV will be powered down while the crewmen load auxiliary equipment aboard the vehicle.

HAND CONTROLLER OPERATION:

T-HANDLE PIVOT FORWARD - INCREASED DEFLECTION FROM NEUTRAL INCREASES FORWARD SPEED.

T-HANDLE PIVOT REARWARD - INCREASED DEFLECTION FROM NEUTRAL INCREASES REVERSE SPEED.

T-HANDLE PIVOT LEFT - INCREASED DEFLECTION FROM NEUTRAL INCREASES LEFT STEERING ANGLE.

T-HANDLE PIVOT RIGHT - INCREASED DEFLECTION FROM NEUTRAL INCREASES RIGHT STEERING ANGLE.

T-HANDLE DISPLACED REARWARD - REARWARD MOVEMENT INCREASES BRAKING FORCE. FULL 3 INCH REARWARD APPLIES PARKING BRAKE. MOVING INTO BRAKE POSITION DISABLES THROTTLE CONTROL AT 15° MOVEMENT REARWARD.

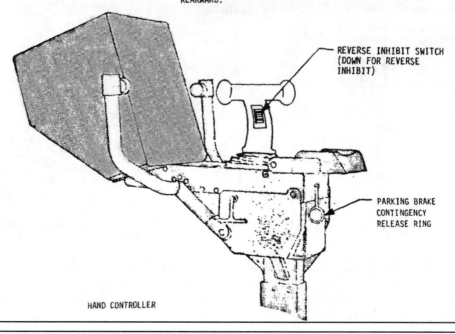

REVERSE INHIBIT SWITCH (DOWN FOR REVERSE INHIBIT)

PARKING BRAKE CONTINGENCY RELEASE RING

HAND CONTROLLER

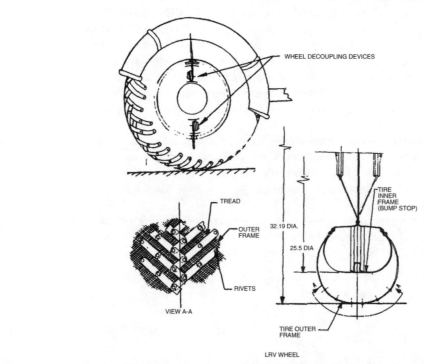

WHEEL DECOUPLING DEVICES

TREAD

OUTER FRAME

RIVETS

VIEW A-A

32.19 DIA.

25.5 DIA

TIRE INNER FRAME (BUMP STOP)

TIRE OUTER FRAME

LRV WHEEL

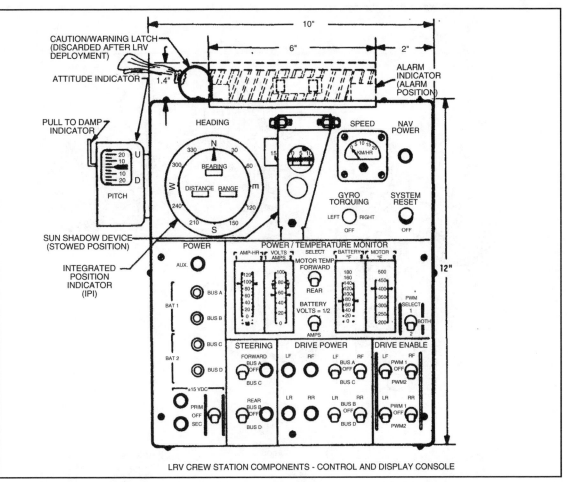

LRV CREW STATION COMPONENTS - CONTROL AND DISPLAY CONSOLE

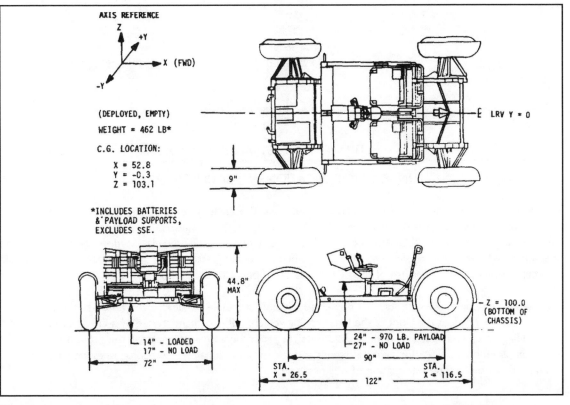

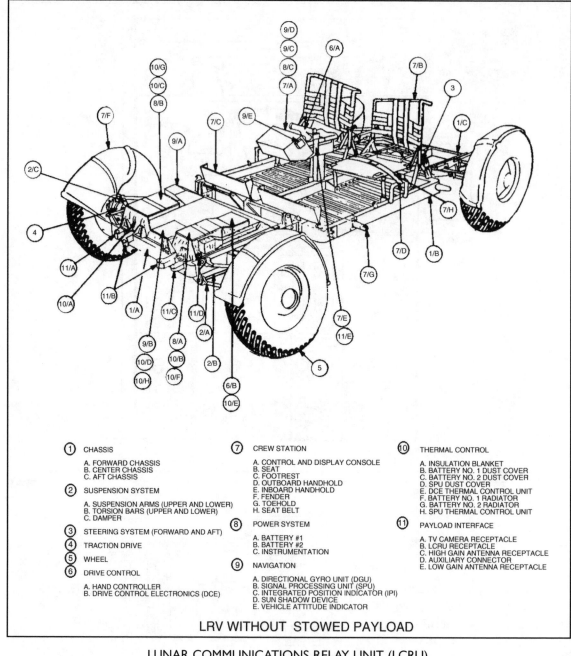

LRV WITHOUT STOWED PAYLOAD

① CHASSIS	⑦ CREW STATION	⑩ THERMAL CONTROL
A. FORWARD CHASSIS B. CENTER CHASSIS C. AFT CHASSIS	A. CONTROL AND DISPLAY CONSOLE B. SEAT C. FOOTREST D. OUTBOARD HANDHOLD E. INBOARD HANDHOLD	A. INSULATION BLANKET B. BATTERY NO. 1 DUST COVER C. BATTERY NO. 2 DUST COVER D. SPU DUST COVER E. DCE THERMAL CONTROL UNIT
② SUSPENSION SYSTEM	F. FENDER G. TOEHOLD H. SEAT BELT	F. BATTERY NO. 1 RADIATOR G. BATTERY NO. 2 RADIATOR H. SPU THERMAL CONTROL UNIT
A. SUSPENSION ARMS (UPPER AND LOWER) B. TORSION BARS (UPPER AND LOWER) C. DAMPER	⑧ POWER SYSTEM	⑪ PAYLOAD INTERFACE
③ STEERING SYSTEM (FORWARD AND AFT)	A. BATTERY #1 B. BATTERY #2 C. INSTRUMENTATION	A. TV CAMERA RECEPTACLE B. LCRU RECEPTACLE C. HIGH GAIN ANTENNA RECEPTACLE
④ TRACTION DRIVE		D. AUXILIARY CONNECTOR E. LOW GAIN ANTENNA RECEPTACLE
⑤ WHEEL	⑨ NAVIGATION	
⑥ DRIVE CONTROL	A. DIRECTIONAL GYRO UNIT (DGU) B. SIGNAL PROCESSING UNIT (SPU) C. INTEGRATED POSITION INDICATOR (IPI) D. SUN SHADOW DEVICE E. VEHICLE ATTITUDE INDICATOR	
A. HAND CONTROLLER B. DRIVE CONTROL ELECTRONICS (DCE)		

LUNAR COMMUNICATIONS RELAY UNIT (LCRU)

The range from which an Apollo crew can operate from the lunar module during EVAs while maintaining contact with the Earth is extended over the lunar horizon by a suitcase-size device called the lunar communications relay unit (LCRU). The LCRU acts as a portable direct relay station for voice, TV, and telemetry between the crew and Mission Control Center instead of through the lunar module communications system. First use of the LCRU was on Apollo 15.

Completely self-contained with its own power supply and erectable hi-gain S-Band antenna, the LCRU may be mounted on a rack at the front of the lunar roving vehicle (LRV) or handcarried by a crewman. In addition to providing communications relay, the LCRU receives ground-command signals for the ground commanded television assembly (GCTA) for remote aiming and focusing the lunar surface color television camera. The GCTA is described in another section of this press kit.

Between stops with the lunar roving vehicle, crew voice is beamed Earthward by a wide beam-width helical S-Band antenna. At each traverse stop, the crew must sight the highgain parabolic antenna toward Earth before television signals can be transmitted. VHF signals from the crew portable life support system (PLSS) transceivers are converted to S-Band by the LCRU for relay to the ground, and conversely, from S-Band to VHF on the uplink to the EVA crewmen.

The LCRU measures 55.9x40.6x15.2cm (22x16x6 inches) not including antennas, and weighs 25 Earth kilograms (55 Earth pounds) (9.2 lunar pounds). A protective thermal blanket around the LCRU can be peeled back to vary the amount of radiation surface which consists of 1.26 m^2 (196 square inches) of radiating mirrors to reflect solar heat. Additionally, wax packages on top of the LCRU enclosure stabilize the LCRU temperature by a melt-freeze cycle. The LCRU interior is pressurized to 7.5 psia differential (one-half atmosphere).

Internal power is provided to the LCRU by a 19-cell silver-zinc battery with a postassium hydroxide electrolyte. The battery weighs 4.1 kg (nine Earth lbs.) (1.5 lunar lbs.) and measures 11.8x23.9x11.8cm (4.7x9.4x4.65 inches). The battery is rated at 400 watt hours, and delivers 29 volts at a 3.1-ampere current load. The LCRU may also be operated from the LRV batteries.

The nominal plan is to operate the LCRU using LRV battery power during EVA-1. The LCRU battery will provide the power during EVA-2 and EVA-3.

Three types of antennas are fitted to the LCRU system: a low-gain helical antenna for relaying voice and data when the LRV is moving and in other instances when the high-gain antenna is not deployed; a .9 m (three-foot) diameter parabolic rib-mesh high-gain antenna for relaying a television signal; and a VHF omni-antenna for receiving crew voice and data from the PLSS transceivers. The high-gain antenna has an optical sight which allows the crewman to boresight on Earth for optimum signal strength. The Earth subtends a two degree angle when viewed from the lunar surface.

The LCRU can operate in several modes: mobile on the LRV, fixed base such as when the LRV is parked, or handcarried in contingency situations such as LRV failure. The LCRU is manufactured by RCA.

TELEVISION AND GROUND COMMANDED TELEVISION ASSEMBLY

Two different color television cameras will be used during the Apollo 17 mission. One, manufactured by Westinghouse, will be used in the command module. It will be fitted with a 5-centimeter (2-inch) black and white monitor to aid the crew in focus and exposure adjustment.

The other camera, manufactured by RCA, is for lunar surface use and will be operated from the lunar roving vehicle (LRV) with signal transmission through the lunar communication relay unit rather than through the LM communications system.

While on the LRV, the camera will be mounted on the ground commanded television assembly (GCTA). The camera can be aimed and controlled by astronauts or it can be remotely controlled by personnel located in the Mission Control Center. Remote command capability includes camera "on" and "off", pan, tilt, zoom, iris open/closed (f2.2 to f22) and peak or average automatic light control.

The GCTA is capable of tilting the TV camera upward 85 degrees, downward 45 degrees, and panning the camera 350 degrees between mechanical stops. Pan and tilt rates are approximately 3 degrees per second. The TV lens can be zoomed from a focal length of 12.5mm to 75mm corresponding to a field of view from 9 to 54 degrees.

At the end of the third EVA, the crew will park the LRV about 91.4 m (300 ft.) east of the LM so that the color TV camera can cover the LM ascent from the lunar surface. Because of a time delay in a signal going the quarter million miles out to the Moon, Mission Control must anticipate ascent engine ignition by about two seconds with the tilt command.

The GCTA and camera each weigh approximately 5.9 kg (13 lb.). The overall length of the camera is 46 cm (18.0 in.) its width is 17 cm (6.7 in.), and its height is 25 cm (10 in.), it is powered from the LCRU battery supply, or externally from the LRV batteries. The GCTA is built by RCA.

APOLLO 17 TELEVISION EVENTS

DATE (GET)	TIME (EST)	TIME (HRS:MIN)	DURATION	EVENT
7 DEC	4:12	0205	0:20	TD & E
11 DEC	117:55	1948	5:19	EVA-1
12 DEC	139:38	1731	6:21	EVA-2
13 DEC	163:05	1658	6:35	EVA-3
14 DEC	187:48	1741	0:25	LM LIFT-OFF
14 DEC	189:38	1931	0:06	RENDEZVOUS
14 DEC	190:01	1954	0:05	DOCKING
16 DEC	236:53	1846	0:32	POST TEI
17 DEC	257:26	1519	1:04	CMP EVA
18 DEC	284:07	1800	0:30	PRESS CONFERENCE

*LSPE CHARGES AND LM ASCENT STAGE IMPACT TELEVISION TIMES ARE NOT SHOWN

PHOTOGRAPHIC EQUIPMENT

Still and motion pictures will be made of most spacecraft maneuvers and crew lunar surface activities. During lunar surface operations, emphasis will be on documenting placement of lunar surface experiments, documenting lunar samples, and on recording in their natural state the lunar surface features.

Command module lunar orbit photographic tasks and experiments include high-resolution photography to aid exploration, photography of surface features of special scientific interest and astronomical phenomena such as solar corona, zodiacal light, and galactic poles.

Camera equipment stowed in the Apollo 17 command module consists of one 70mm Hasselblad electric camera, a 16mm Maurer motion picture camera, and a 35mm Nikon F single-lens reflex camera. The command module Hasselblad electric camera is normally fitted with an 80mm f/2.8 Zeiss Planar lens, but a bayonet-mount 250mm Zeiss Sonnar lens can be fitted for long-distance Earth/Moon photos.

The 35mm Nikon F is fitted with a 55mm f/1.2 Nikkor lens for the dim-light photographic experiments.

The Maurer 16mm motion picture camera in the command module has Kern-Switar lenses of 10,18 and 75mm focal length available. Accessories include a right-angle mirror, a power cable and a sextant adapter which allows the camera to film through the navigation sextant optical system.

Cameras stowed in the lunar module are two 70mm Hasselblad data cameras fitted with 60mm Zeiss Metric lenses, an electric Hasselblad camera with 500mm lens, and one 16mm Maurer motion picture camera with 10mm lenses.

The LM Hasselblads have crew chest mounts that fit dovetail brackets on the crewman's remote control unit, thereby leaving both hands free. The LM motion picture cameras will be mounted in the right-hand window to record descent, landing, ascent and rendezvous.

Descriptions of the 24-inch panoramic camera and the 3-inch mapping/stellar camera are in the orbital science section of this press kit.

ASTRONAUT EQUIPMENT

Space Suit

Apollo crewmen wear two versions of the Apollo space suit: the command module pilot version (A-7LB-CMP) for operations in the command module and for extravehicular operations during SIM bay film retrieval during transearth coast; and the extravehicular version (A-7LB-EV) worn by the commander and lunar module pilot for lunar surface EVAs.

The A-7LB-EV suit differs from Apollo suits flown prior to Apollo 15 by having a waist joint that allows greater mobility while the suit is pressurized — stooping down for setting up lunar surface experiments, gathering samples and for sitting on the lunar roving vehicle.

From the inside out, an integrated thermal meteroid suit cover layer worn by the commander and lunar module pilot starts with rubber-coated nylon and progresses outward with layers of nonwoven Dacron, aluminized Mylar film and Beta marquisette for thermal radiation protection and thermal spacers, and finally with a layer of nonflammable Teflon-coated Beta cloth and an abrasion-resistant layer of Teflon fabric — a total of 18 layers.

Both types of the A-7LB suit have a pressure retention portion called a torso limb suit assembly consisting of neoprene coated nylon and an outer structural restraint layer.

The space suit with gloves, and dipped rubber convolutes which serve as the pressure layer, liquid cooling garment, portable life support system (PLSS), oxygen purge system, lunar extravehicular visor assembly (LEVA), and lunar boots make up the extravehicular mobility unit (EMU). The EMU provides an extravehicular crewman with life support for a 7-hour period outside the lunar module without replenishing expendables.

Lunar extravehicular visor assembly - The assembly consists of polycarbonate shell and two visors with thermal control and optical coatings on them. The EVA visor is attached over the pressure helmet to provide impact, micrometeoroid, thermal and ultraviolet-infrared light protection to the EVA crewmen.

Extravehicular gloves - Built of an outer shell of Chromel-R fabric and thermal insulation the gloves provide protection when handling extremely hot and cold objects. The finger tips are made of silicone rubber to provide more sensitivity.

Constant-wear garment A one-piece constant-wear garment, similar to "long johns", is worn as an undergarment for the space suit in intravehicular and on CSM extravehicular operations and with the inflight coveralls. The garment is porous-knit cotton with a waist-to-neck zipper for donning. Biomedical harness attach points are provided.

Liquid-cooling garment - The knitted nylon-spandex garment includes a network of plastic tubing through which cooling water from the PLSS is circulated. It is worn next to the skin and replaces the constant-wear garment during lunar surface EVA.

Portable life support system (PLSS) - The backpack supplies oxygen at 3.7 psi and cooling water to the liquid cooling garment. Return oxygen is cleansed of solid and gas contaminants by a lithium hydroxide and activated charcoal canister. The PLSS includes communications and telemetry equipment, displays and controls, and a power supply. The PLSS is covered by a thermal insulation jacket, (two stowed in LM).

Oxygen, purge system (OPS)- Mounted atop the PLSS, the oxygen purge system provides a contingency 30-75 minute supply of gaseous oxygen in two bottles pressurized to 5,880 psia, (a minimum of 30 minutes in the maximum flow rate and 75 minutes in the low flow rate). The system may also be worn separately on the front of the pressure garment assembly torso for contingency EVA transfer from the LM to the CSM or behind the neck for CSM EVA. It serves as a mount for the VHF antenna for the PLSS, (two stowed in LM).

Coveralls - During periods out of the space suits, crewmen wear two-piece Teflon fabric inflight coveralls for warmth and for pocket stowage of personal items.

Communications carriers - "Snoopy hats" with redundant microphones and earphones are worn with the pressure helmet; a light weight headset is worn with the inflight coveralls, through a 1/8-inch-diameter tube within reach of his mouth. The bags are filled from the lunar module potable water dispenser.

Buddy Secondary Life Support System - A connecting hose system which permits a crewman with a failed PLSS to share cooling water in the other crewman's PLSS. The BSLSS lightens the load on the oxygen purge system in the event of a total PLSS failure in that the OPS would supply breathing and pressurizing oxygen while the metabolic heat would be removed by the shared cooling water from the good PLSS. The BSLSS will be stowed on the LRV.

Lunar Boots - The lunar boot is a thermal and abrasion protection device worn over the inner garment and boot assemblies. It is made up of layers of several different materials beginning with Teflon coated beta cloth for the boot liner to Chromel R metal fabric for the outer shell assembly. Aluminized Mylar, Nomex felt, Dacron, Beta cloth and Beta marquisette Kapton comprise the other layers. The lunar boot sole is made of high-strength silicone rubber.

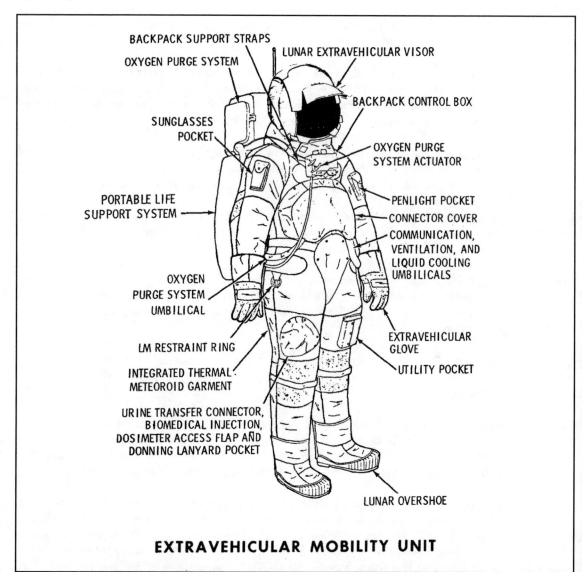

EXTRAVEHICULAR MOBILITY UNIT

Personal Hygiene

Crew personal hygiene equipment aboard Apollo 17 includes body cleanliness items, the waste management system, and one medical kit.

Packaged with the food are a toothbrush and a two-ounce tube of toothpaste for each crewman. Each man-meal package contains a 3.5-by-4-inch wet-wipe cleansing towel. Additionally, three packages of 12-by-12-inch dry towels are stowed beneath the command module pilot's couch. Each package contains seven towels. Also stowed under the command module pilot's couch are seven tissue dispensers containing 53 three-ply tissues each.

Solid body wastes are collected in plastic defecation bags which contain a germicide to prevent bacteria and gas formation. The bags are sealed after use, identified, and stowed for return to Earth for post-flight analysis.

Urine collection devices are provided for use while wearing either the pressure suit or the inflight coveralls. The urine is dumped overboard through the spacecraft urine dump valve in the CM and stored in the LM. On Apollo 16 urine specimens will be returned to Earth for analysis.

Survival Kit

The survival kit is stowed in two rucksacks in the righthand forward equipment bay of the CM above the lunar module pilot.

Contents of rucksack No. 1 are: two combination survival lights, one desalter kit, three pairs of sunglasses, one radio beacon, one spare radio beacon battery and spacecraft connector cable, one knife in sheath, three water containers, two containers of Sun lotion, two utility knives, three survival blankets and one utility netting.

Rucksack No. 2: one three-man life raft with CO_2 inflater, one sea anchor, two sea dye markers, three sunbonnets, one mooring lanyard, three manlines and two attach brackets.

The survival kit is designed to provide a 48-hour postlanding (water or land) survival capability for three crewmen between 40 degrees North and South latitudes.

Medical Kits

The command module crew medical supplies are contained in two kits. Included in the larger medical accessories kit are antibiotic ointment, skin cream, eye drops, nose drops, spare biomedical harnesses, oral thermometer and pills of the following types: 18 pain, 12 stimulant, 12 motion sickness, 48 diarrhea, 60 decongestant, 21 sleeping, 72 aspirin and 60 each of two types of antibiotic. A smaller command module auxiliary drug kit contains 80 and 12 of two types of pills for treatment of cardiac arrythymia and two injectors for the same symptom.

The lunar module medical kit contains eye drops, nose drops, antibiotic ointment, bandages and the following pills: 4 stimulant, 4 pain, 8 decongestant, 12 diarrhea, 12 aspirin and 6 sleeping. A smaller kit in the LM contains 8 and 4 units of injectable drugs for cardiac arrythymia and 2 units for pain suppression.

Crew Food System

The Apollo 17 crew selected menus for their flight from the largest variety of foods ever available for a U.S. manned mission. As on Apollo 16, the preflight, inflight, and postflight diets are being monitored to facilitate interpretation of the medical tests. Menus were designed upon individual crewmember physiological requirements in the unique conditions of weightlessness and one-sixth gravity on the lunar surface. Daily menus provide approximately 2500 calories per day for each crewmember.

Food items are assembled into meal units and identified as to crewmember and sequence of consumption. Foods stored in the "pantry" may be used as substitutions for nominal meal items so long as the nutrient intake for a 24-hour period is not altered significantly.

Apollo 17 Menu

There are various types of food used in the menus. These include freeze-dried rehydratables in spoon-bowl packages; thermostabilized foods (wet packs) in flexible packages and metal easy-open cans; intermediate moisture foods; dry bite-size cubes; and beverages.

Water for drinking and rehydrating food is obtained from two sources in the Command Module — a portable dispenser for drinking water and a water spigot at the food preparation station which supplies water at about 145 degrees and 55 degrees Fahrenheit. The potable water dispenser provides a continuous flow of water as long as the trigger is held down, while the food preparation spigot dispenses water in one-ounce increments.

A continuous-flow water dispenser similar to the one in the Command Module is used aboard the Lunar Module for cold water reconstitution of food stowed aboard the Lunar Module.

Water is injected into a food package and the package is kneaded and allowed to sit for several minutes. The bag top is then cut to open and the food eaten with a spoon. After a meal, germicide tablets are placed in each bag to prevent fermentation of any residual food and gas formation. The bags are then rolled and stowed in waste disposal areas in the spacecraft.

An improved Skylab beverage package design will be used by the crew to measure water consumption. Functional aspects of the package and the behavior of liquid during extended periods of weightlessness will be observed.

The in-suit drink device will contain water as on the Apollo 15 mission. As on Apollo 15 and 16, the crewmen on the lunar surface will have the option to snack on an insuit food bar.

The nutritionally complete fruitcake provides all the nutrients needed by man in their correct proportions. The fruitcake contains many ingredients such as: soy flour, wheat flour, sugar, eggs, salt, cherries, pineapple, nuts, raisins, and shortening. Vitamins have been added. The product is heat sterilized in an impermeable flexible pouch and is shelf-stable until opened. This fruitcake can provide a nutritious snack or meal. This food is planned for use in the future in the Space Shuttle program as a contingency food system.

The irradiated ham provides the crew with a shelf-stable slice of ham 12 mm thick. Each slice weighs about 100 grams and may be used for making sandwiches during flight. The radiation sterilization (radappertization) is performed while the ham is at -40°C. The absorbed irradiation dose is 3.7 to 4.3 million rads. This gives an excellent product with an expected shelf-life of 3 years.

The fruitcake and ham slices were specially developed and provided for Apollo 17 by the U.S. Army Natick Laboratories, Natick, Massachusetts. New foods for the Apollo 17 mission are irradiated sterilized ham, nutrient complete fruitcake, and rehydratable tea and lemonade beverages.

APOLLO 17 CSM MENU   Eugene A. Cernan, CDR (Red Velcro)

MEAL	Day 1*, 5, 9***, 13		Day 2,.6**, 10, 14**		Day 3, 11		Day 4, 12	
A	Bacon Squares (8)	IMB	Spiced Oat Cereal	RSB	Scrambled Eggs	RSB	Sausage Patties	R
	Scrambled Eggs	RSB	Sausage Patties	R	Bacon Squares (8)	IMB	Apricot Cereal Cubes (4)	DB
	Cornflakes	RSB	Mixed Fruit	WP	Peaches	WP	Fruit Cocktail	R
	Peaches	RSB	Cinnamon Toast Bread(4)	DB	Pineapple GF Drink	R	Pears	IMB
	Orange Beverage	R	Instant Breakfast	R	Cocoa w/K	R	Cocoa w/K	R
	Cocoa	R	Coffee w/K	R			Coffee	R
B	Chicken & Rice Soup	RSB	Corn Chowder	RSB	Lobster Bisque	RSB	Chicken Soup	RSB
	Meatballs and Sauce	WP	Frankfurters	WP	Peanut Butter	WP	Ham (Ir)	WP
	Fruitcake	WP	Bread, white (2)		Jelly	WP	Cheddar Cheese Spread	WP
	Lemon Pudding	WP	Catsup	WP	Bread, white (1)		Bread, Rye (1)	
	Orange P/A Drink	R	Apricots	IMB	Chocolate Bar	IMB	Cereal Bar	IMB
			Orange GF Drink	R	Orange GF Drink w/K	R	Orange Beverage	R

C	Potato Soup	RSB	Turkey and Gravy	WP	Shrimp Cocktail	RSB	Tomato Soup	RSB
	Beef and Gravy	WP	Pork & Potatoes	RSB	Beef Steak	WP	Hamburger	WP
	Chicken Stew	RSB	Brownies (4)	DB	Butterscotch Pudding	RSB	Mustard	WP
	Ambrosia, Peach	RSB	Orange Juice	R	Peaches	IMB	Vanilla Pudding	WP
	Gingerbread (4)	DB	Lemonade	R	Orange Drink w/K	R	Date fruitcake (4)	IMB
	Citrus Beverage	R					Orange P/A Drink w/K	R

* Meal C only ** Meal A only *** Meals B and C only
DB = Dry Bite IMB = Intermediate Moisture Bite R = Rehydratable RSB = Rehydratable Spoon Bowl WP = Wet Pack Ir = Irradiated

APOLLO 17 — LM MENU, Eugene A. Cernan, CDR (Red Velcro)

MEAL	Day 6		Day 7		Day 8		Day 9	
B	Corn Chowder	RSB A	Scrambled Eggs	RSB A	Sausage Patties	R A	Bacon squares {8}	IMB
	Franfurters	WP	Bacon Squares (8)	IMB	Apricot Cereal Cubes(6)	DB	Scrambled Eggs	RSB
	Bread, white (2)		Peaches	IMB	Fruit Cocktail	R	Cornflakes	RSB
	Catsup	WP	Peanut Butter	WP	Pears	IMB	Beef and Gravy	WP
	Apricots	IMB	Jelly	WP	Cereal Bar	IMB	Fruitcake	WP
	Orange GF Drink	R	Bread, white (1)		Cheese Cracker Cube(4)	DB	Peaches	RSB
	Tea	R	Chocolate Bar	IMB	Ham (Ir)	WP	Cocoa	R
	Lemonade	R	Pineapple GF Drink	R	Cocoa	R	Orange Beverage	R
			Orange GF Drink w/K	R	Tea	R	Tea	R
			Cocoa	R	Spiced Oat Cereal	RSB		
			Tea		Lemonade	R		
C	Spaghetti & Meat							
	Sauce	RSB B	Chicken and Rice	RSB B	Lobster Bisque	RSB		
	Turkey and Gravy	WP	Shrimp Cocktail	RSB	Hamburger	WP		
	Pork and Potatoes	RSB	Beef Steak	WP	Mustard	WP		
	Brownies (4)	DB	Beef Sandwiches (4)	DB	Cheddar Cheese Spread	WP		
	Orange Beverage	R	Butterscotch Pudding	RSB	Bread, rye (1)			
	Tea	R	Graham Cracker Cube (6)	DB	Date Fruitcake (4)	IMB		
			Orange Drink w/K	R	Orange PA Drink w/K	R		
			Tea		Orange Beverage	R		
					Tea	R		

In-Suit Food Bar Assembly	6	ea	P/N: SEB 13100318-301
In-Suit Drinking Device	4	ea	P/N: 14-0151-02
Spoon Assembly (2)	1	ea	P/N: 14-0144-01
Germicidal Tablets Pouch (42)	1	ea	P/N: 14-02166
Germicidal Tablets Pouch (20)	1	ea	P/N: 14-

DB = Dry Bite IMB = Intermediate Moisture Bite R = Rehydratable   RSB = Rehydratable Spoon Bowl WP = Wet Pack

APOLLO 17 CSM MENU, Harrison H. Schmitt, LMP (Blue Velcro)

MEAL	Day 1*, 5, 9***,13		Day 2, 6***, 10, 14**		Day 3, 11		Day 4, 12	
A	Bacon Squares (8)	IMB	Sausage Patties	R	Scrambled Eggs	RSB	Sausage Patties	R
	Scrambled Eggs	RSB	Cinnamon Toast Bread (4)	DB	Bacon Squares (8)	IMB	Grits	RSB
	Cornflakes	RSB	Mixed Fruit	WP	Peaches	WP	Peaches	RSB
	Apricots	IMB	Instant Breakfast	R	Orange P/A Drink w/K	R	Pears	IMB
	Cocoa	R	Coffee w/K	R	Cocoa	R	Pineapple GF Drink	R
							Coffee w/K	R
B	Chicken & Rice Soup	RSB	Corn Chowder	RSB	Potato Soup	RSB	Chicken Soup	RSB
	Meatballs w/Sauce	WP	Frankfurters	WP	Peanut Butter	WP	Ham (Ir)	WP
	Fruitcake	WP	Bread, White (2)		Jelly	WP	Cheddar Cheese Spread	WP
	Lemon Pudding	WP	Catsup	WP	Bread, White(1)		Bread, Rye (1)	
	Citrus Beverage	R	Chocolate Pudding.	RSB	Cherry Bar (1)	IMB	Cereal Bar	IMB
			Orange GF Drink w/K	R	Orange GF Drink w/K	R	Orange Drink w/K	R
C	Lemonade	R	Turkey & Gravy	WP	Shrimp Cocktail	RSB	Tomato Soup	RSB
	Beef & Gravy	WP	Pork and Potatoes	RSB	Beef Steak	WP	Hamburger	WP
	Chicken Stew	RSB	Carmel Candy	IMB	Butterscotch Pudding	RSB	Mustard	WP
	Ambrosia	RSB	Orange Juice	R	Peaches	IMB	Vanilla Pudding	WP
	Gingerbread (4)	DB			Orange Drink w/K	R	Chocolate Bar	IMB
	Grapefruit Drink	R					Grape Drink w/K	R

CALORIES
*Meal C only **Meal A only ***Meal B and C only
DB = Dry Bite IMB = Intermediate Moisture Bite R = Rehydratable WP = Wet Pack RSB = Rehydratable Spoon Bowl Ir = Irradiated

LM Menu Continued
APOLLO 17 - LM MENU, Harrison H. Schmitt, LMP (Blue Velcro)

MEAL	Day 6		Day 7		Day 8		Day 9	
B	Corn Chowder	RSB	Scrambled Eggs	RSB	Sausage Patties	R	Bacon Squares (8)	IMB
	Frankfurters	WP	Bacon Squares (8)	IMB	Spiced Oat Cereal	RSB	Scrambled Eggs	RSB
	Bread, White (2)		Peaches	IMB	Peaches	RSB	Cornflakes	RSB
	Catsup	WP	Peanut Butter	WP	Pears	IMB	Apricots	IMB
	Chocolate Pudding	RSB	Jelly	WP	Cereal Bar	IMB	Cocoa	R
	Orange GF Drink	R	Bread, White (1)		Gingerbread (6)	DB	Tea	R
	Tea	R	Orange GF Drink w/K	R	Ham (Ir)	WP	Beef and Gravy	WP
	Lemonade	R	Cocoa	R	Pineapple GF Drink	R	Fruitcake	WP
			Tea	R	Tea	R		
			Fruit Cocktail	R				
C	Turkey and Gravy	WPB	Chicken & Rice	RSB	Potato Soup	RSB		
	Pork and Potatoes	RSB	Shrimp Cocktail	RSB	Hamburger	WP		
	Carmel Candy	IMB	Beef Steak	WP	Mustard	WP		
	Orange Beverage	R	Beef Sandwiches (4)	DB	Cheddar Cheese Spread	WP		
	Tea	R	Butterscotch Pudding	RSB	Bread, Rye (1)			
			Graham Cracker Cube (6)	DB	Chocolate Bar	IMB		
			Orange Drink w/K	R	Banana Pudding	RSB		
			Orange P/A Drink	R	Orange Drink w/K	R		
			Tea	R	Grape Drink w/K	R		
					Tea	R		

APOLLO 17 CSM MENU, RONALD E. EVANS, CMP (White Velcro)

MEAL	Day 1*,5,9,13		Day 2,6,10,14**		Day 3,7,11		Day 4,8,12	
A	Bacon Squares (8)	IMB	Spiced Oat Cereal	RSB	Scrambled Eggs	RSB	Sausage	R
	Scrambled Eggs	RSB	Sausage Patties	R	Bacon Squares (8)	IMB	Grits	RSB
	Cornflakes	RSB	Mixed Fruit	WP	Peaches	WP	Fruit Cocktail	R
	Apricots	IMB	Instant Breakfast	R	Cinnamon Toast Bread(4)	DB	Orange Beverage	R
	Orange Juice	R	Coffee w/K	R	Orange Juice	R	Coffee w/K	R
					Cocoa w/K	R		
B	Chicken & Rice Soup	RSB	Frankfurters	WP	Lobster Bisque	RSB	Ham (Ir)	WP
	Meatballs w/Sauce	WP	Bread, white (2)		Peanut Butter	WP	Cheddar Cheese Spread	WP
	Fruitcake	WP	Catsup	WP	Jelly	WP	Bread, rye (1)	
	Butterscotch Pudding	WP	Pears	IMB	Bread, white (1)		Peaches	RSB
	Orange PA Drink	R	Chocolate Pudding	RSB	Cherry Bar (1)	IMB	Cereal Bar	IMB
			Grape Drink w/K	R	Citrus Beverage w/K	R	Orange PA Drink w/K	R
C	Potato Soup	RSB	Corn Chowder	RSB	Shrimp Cocktail	RSB	Tomato Soup	RSB
	Beef and Gravy	WP	Turkey & Gravy	WP	Beef Steak	WP	Hamburger	WP
	Chicken Stew	RSB	Chocolate Bar	IMB	Butterscotch Pudding	RSB	Mustard	WP
	Ambrosia	RSB	Orange Beverage	R	Orange Drink w/K	R	Vanilla Pudding	WP
	Brownies (4)	DB					Sugar Cookies (4)	DB
	Orange GF Drink	R					Carmel Candy	IMB
							Grape Drink w/K	R

* Meal C only ** Meal A only
DB = Dry Bite IMB = Intermediate Moisture Bite R = Rehydratable RSB = Rehydratable Spoon Bowl WP = Wet Pack Ir = Irradiated

APOLLO 17 PANTRY STOWAGE ITEMS

BEVERAGES	QTY.	ACCESSORIES	QTY.
Coffee (B)	20	Contingency Feeding System	1
Tea	20		
Grape Drink	10	Germicidal Tablets (42)	3
Grape Punch	10		
		Index Card	1

S/L Beverage Dispenser (empty) 3
Contingency Beverages   30
(For Contingency Use Only)
15 Instant Breakfast
5 Orange Drink
5 Pineapple Orange Drink
5 Lemonade

SNACK ITEMS

Bacon Squares (4)	9 Apricot Cereal Cubes (4)	6 Brownies (4)	3 Gingerbread (4)	3
Graham Crackers (4)	6 Jellied Candy	6 Peach Ambrosia	3 Pecans (6)	6
Fruitcake (WP)	3 Sugar Cookies (4)	6 Apricots (IMB)	3 Peaches (IMB)	3
Pears (IMB)	3 Chocolate Bar (IMB)	3 Tuna Salad Spread (WP)	2 (Small Cans)	
Catsup (WP)	3			

SATURN V LAUNCH VEHICLE

The Saturn V launch vehicle (SA-512) assigned to the Apollo 17 mission is similar to the vehicles used for the missions of Apollo 8 through Apollo 16.

First Stage

The five first stages (S-1C) F-1 engines develop about 34 million newtons (7.67 million pounds) of thrust at launch. Major stage components are the forward skirt, oxidizer tank, intertank structure, fuel tank, and thrust structure. Propellant to the five engines normally flows at a rate of about 13,200 kilograms (29,200 pounds; 3,370 gallons) a second. One engine is rigidly mounted on the stage's centerline; the outer four engines are mounted on a ring equally spaced around the center engine. These outer engines are gimbaled to control the vehicle's attitude during flight.

Second Stage

The five second stage (S-II) J-2 engines develop a total of about 5.13 million newtons (1.15 million pounds) of thrust during flight. Major components are the forward skirt, liquid hydrogen and liquid oxygen tanks (separated by an insulated common bulkhead) a thrust structure, and an interstage section that connects the first and second stages. The engines are mounted and used in the same arrangement as the first stage's F-1 engines: four outer engines can be gimbaled; the center one is fixed.

Third Stage

Major components of the third stage (S-IVB) are a single J-2 engine, aft interstage and skirt, thrust structure, two propellant tanks with a common bulkhead, and forward skirt.

The gimbaled engine has a maximum thrust of .93 million newtons (209, 000 pounds), and can be restarted in Earth orbit.

Instrument Unit

The instrument unit (IU) contains navigation, guidance and control equipment to steer the Saturn V into Earth orbit and translunar trajectory. The six major systems are structural, enviromental control, guidance and control, measuring and telemetry, communications, and electrical.

The IU's inertial guidance platform provides space-fixed reference coordinates and measures acceleration during flight. If the platform should fail during boost, systems in the Apollo spacecraft are programmed to provide launch vehicle guidance. After second stage ignition, the spacecraft commander can manually steer the vehicle if its guidance platform is lost.

Propulsion

The Saturn V has 31 propulsive units, with thrust ratings ranging from 311 newtons (70 pounds) to more than 6.8 million newtons (1.53 million pounds). The large main engines burn liquid propellants; the smaller units use solid or hypergolic (self-igniting) propellants.

The five F-1 engines give the first stage a thrust range of from 34,096,110 newtons, (7,665,111 pounds) at liftoff to 40,207,430 newtons (9,038,989 pounds) at center engine cutoff. Each F-1 engine weighs almost nine metric tons (10 short tons), is more than 5.5 meters long (18 feet), and has a nozzle exit diameter of nearly 4.6 meters (14 feet). Each engine uses almost 2.7 metric tons (3 short tons) of propellant a second.

The five J-2 engines on the second stage develop an average thrust of 5,131,968 newtons (1,153,712 pounds) during flight. The one J-2 engine of the third stage develops an average thrust of 926,307 newtons (208,242 pounds). The 1,590-kilogram (3,500-pound) J-2 engine uses high-energy, low-molecular-weight liquid hydrogen as fuel, and liquid oxygen as oxidizer.

The first stage has eight solid-fuel retro-rockets that fire to separate the first and second stages. Each rocket produces a thrust of 337,000 newtons (75,800 pounds) for 0.54 seconds.

Four retro-rockets, located in the third stage's aft interstage, separate the second and third stages. Two jettisonable ullage rockets settle propellants before engine ignition. Six smaller engines in two auxiliary propulsion system modules on the third stage provide three-axis attitude control.

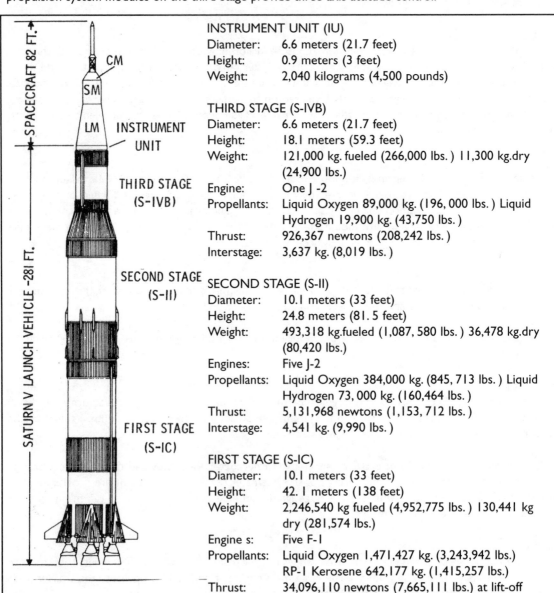

INSTRUMENT UNIT (IU)

Diameter:	6.6 meters (21.7 feet)
Height:	0.9 meters (3 feet)
Weight:	2,040 kilograms (4,500 pounds)

THIRD STAGE (S-IVB)

Diameter:	6.6 meters (21.7 feet)
Height:	18.1 meters (59.3 feet)
Weight:	121,000 kg. fueled (266,000 lbs.) 11,300 kg.dry (24,900 lbs.)
Engine:	One J -2
Propellants:	Liquid Oxygen 89,000 kg. (196, 000 lbs.) Liquid Hydrogen 19,900 kg. (43,750 lbs.)
Thrust:	926,367 newtons (208,242 lbs.)
Interstage:	3,637 kg. (8,019 lbs.)

SECOND STAGE (S-II)

Diameter:	10.1 meters (33 feet)
Height:	24.8 meters (81. 5 feet)
Weight:	493,318 kg.fueled (1,087, 580 lbs.) 36,478 kg.dry (80,420 lbs.)
Engines:	Five J-2
Propellants:	Liquid Oxygen 384,000 kg. (845, 713 lbs.) Liquid Hydrogen 73, 000 kg. (160,464 lbs.)
Thrust:	5,131,968 newtons (1,153, 712 lbs.)
Interstage:	4,541 kg. (9,990 lbs.)

FIRST STAGE (S-IC)

Diameter:	10.1 meters (33 feet)
Height:	42. 1 meters (138 feet)
Weight:	2,246,540 kg fueled (4,952,775 lbs.) 130,441 kg dry (281,574 lbs.)
Engine s:	Five F-1
Propellants:	Liquid Oxygen 1,471,427 kg. (3,243,942 lbs.) RP-1 Kerosene 642,177 kg. (1,415,257 lbs.)
Thrust:	34,096,110 newtons (7,665,111 lbs.) at lift-off

NOTE: Weights and measures given above are for the nominal vehicle configuration for Apollo 17. The figures may vary slightly due to changes before launch to meet changing conditions. Weights of dry stages and propellants do not equal total weight because frost and miscellaneous smaller items are not included in chart.

APOLLO SPACECRAFT

The Apollo spacecraft consists of the command module, service module, lunar module, a spacecraft lunar module adapter (SLA), and a launch escape system. The SLA houses the lunar module and serves as a mating structure between the Saturn V instrument unit and the SM.

Launch Escape System (LES) — The function of the LES is to propel the command module to safety in an aborted launch. It has three solid-propellant rocket motors: a 658,000 newton (147,000-pound)-thrust launch escape system motor, a 10,750-newton (2,400-pound)-thrust pitch control motor, and a 141,000 newton (31,500-pound)-thrust tower jettison motor. Two canard vanes deploy to turn the command module aerodynamically to an attitude with the heat-shield forward. The system is 10 meters (33 feet) tall and 1.2 meters (four feet) in diameter at the base, and weighs 4,158 kilograms (9,167 pounds).

Command Module (CM) — The command module is a pressure vessel encased in heat shields, cone-shaped, weighing 5,843.9 kg (12,874 lb.) at launch. The command module consists of a forward compartment which contains two reaction control engines and components of the Earth landing system; the crew compartment or inner pressure vessel containing crew accommodations, controls and displays, and many of the spacecraft systems; and the aft compartment housing ten reaction control engines, propellant tankage, helium tanks, water tanks, and the CSM umbilical cable. The crew compartment contains 6 cubic meters (210 cubic ft.) of habitable volume.

Heat-shields around the three compartments are made of brazed stainless steel honeycomb with an outer layer of phenolic epoxy resin as an ablative material.

The CSM and LM are equipped with the probe-and-drogue docking hardware. The probe assembly is a powered folding coupling and impact attentuating device mounted in the CM tunnel that mates with a conical drogue mounted in the LM docking tunnel. After the 12 automatic docking latches are checked following a docking maneuver, both the probe and drogue are removed to allow crew transfer between the CSM and LM.

Service Module (SM) — The Apollo 17 service module will weigh 24,514 kg (54,044 lb.) at launch, of which 18,415 kg (40,594 lb.) is propellant for the 91,840 newton (20,500 pound) - thrust service propulsion engine: (fuel: 50/50 hydrazine and unsymmetrical dimethyl-hydrazine; oxidizer: nitrogen textroxide). Aluminum honeycomb panels 2.54 centimeters (one inch) thick form the outer skin, and milled aluminum radial beams separate the interior into six sections around a central cylinder containing service propulsion system (SPS) helium pressurant tanks.

The six sectors of the service module house the following components: Sector I — oxygen tank 3 and hydrogen tank 3, J-mission Scientific Instrumentation Module (SIM) bay; Sector II — space radiator, +Y RCS package, SPS oxidizer storage tank; Sector III — space radiator, +Z RCS package, SPS oxidizer storage tank; Sector IV — three fuel cells, two oxygen tanks, two hydrogen tanks, auxiliary battery; Sector V — space radiator, SPS fuel sump tank, -Y RCS package; Sector VI — space radiator, SPS fuel storage tank, -Z RCS package.

Spacecraft-LM adapter (SLA) Structure — The spacecraft-LM adapter is a truncate cone 5 m (28 ft.) long tapering from 6.7 m (21.6 ft.) in diameter at the base to 3.9 m (12.8 ft.) at the forward end at the service module mating line. The SLA weighs 1,841 kg (4,059 lb.) and houses the LM during launch and the translunar injection manuever until CSM separation, transposition, and LM extraction. The SLA quarter panels are jettisoned at CSM separation.

Lunar Module (LM)

The lunar module is a two-stage vehicle designed for space operations near and on the Moon. The lunar module stands 7 m (22 ft. 11 in.) high and is 9.5 m (31 ft.) wide (diagonally across landing gear). The ascent and descent stages of the LM operate as a unit until staging, when the ascent stage functions as a single

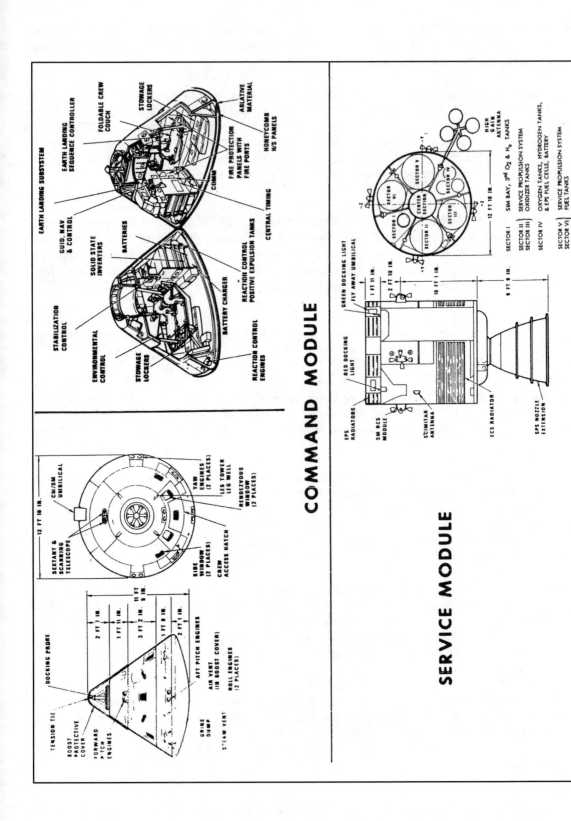

COMMAND MODULE

SERVICE MODULE

acecraft for rendezvous and docking with the CM.

scent Stage — Three main sections make up the ascent stage: the crew compartment, midsection, and aft quipment bay. Only the crew compartment and midsection are pressurized 337.5 grams per square entimeter (4.8 pounds per square inch gauge). The cabin volume is 6.7 cubic meters (235 cubic feet). The age measures 3.8 m (12 ft. 4 in.) high by 4.3 m (14 ft. 1 in.) in diameter. The ascent stage has six bstructural areas: crew compartment, midsection, aft equipment bay, thrust chamber assembly cluster upports, antenna supports, and thermal and micrometeoroid shield.

he cylindrical crew compartment is 2.35 m (7 ft. 10 in.) in diameter and 1.07 m (3 ft. 6 in.) deep. Two flight ations are equipped with control and display panels, armrests, body restraints, landing aids, two front indows, an overhead docking window, and an alignment optical telescope in the center between the two ght stations. The habitable volume is 4.5 cubic meters (160 cubic ft.)

 tunnel ring atop the ascent stage meshes with the command module docking latch assemblies. During ocking, the CM docking ring and latches are aligned by the LM drogue and the CSM probe.

he docking tunnel extends downward into the midsection 40 cm (16 in.). The tunnel is 81 cm (32 in.) in ameter and is used for crew transfer between the CSM and LM. The upper hatch on the inboard end of e docking tunnel opens inward and cannot be opened without equalizing pressure on both hatch surfaces.

 thermal and micrometeoroid shield of multiple layers of Mylar and a single thickness of thin aluminum skin ncases the entire ascent stage structure.

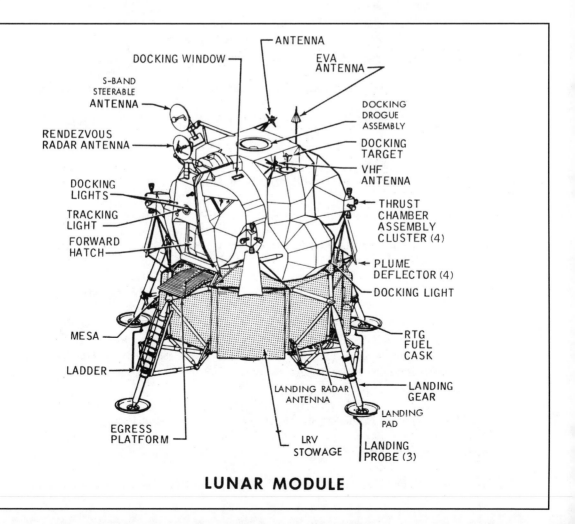

LUNAR MODULE

Descent Stage — The descent stage center compartment houses the descent engine, and descent propella▮ tanks are housed in the four bays around the engine. Quadrant II contains ALSEP. The radioisotop▮ thermoelectric generator (RTG) is externally mounted. Quadrant IV contains the MESA. The descent stag▮ measures 3.2 m (10 ft. 7 in.) high by 4.3 m (14 ft. 1 in.) in diameter and is encased in the Mylar and aluminu▮ alloy thermal and micrometeoroid shield. The LRV is stowed in Quadrant I.

The LM egress platform or "porch" is mounted on the forward outrigger just below the forward hatch. ▮ ladder extends down the forward landing gear strut from the porch for crew lunar surface operations.

The landing gear struts are released explosively and are extended by springs. They provide lunar surfac▮ landing impact attenuation. The main struts are filled with crushable aluminum honeycomb for absorbir▮ compression loads. Footpads 0.95 m (37 in.) in diameter at the end of each landing gear provide vehic▮ support on the lunar surface.

Each pad (except forward pad) is fitted with a 1.7-m (68-in.) long lunar surface sensing probe which upc▮ contact with the lunar surface signals the crew to shut down the descent engine.

The Apollo LM has a launch weight of 16,429 kg (36,244 lb.). The weight breakdown is as follows:

			kilograms	pounds	
1.	Ascent stage,	dry*	2,059	4,729	
2.	APS propellants	(loaded)	2,378	5,243	
3.	Descent stage,	dry	2,791	6,155	
4.	DPS propellants	(loaded)	8,838	19,486	
5.	RCS propellants	(loaded)	286	631	
			16,352 kg	36,244 lbs	

* Includes water and oxygen; no crew.

NATIONAL AERONAUTICS AND SPACE ADMINISTRATION
WASHINGTON, D.C. 20546

BIOGRAPHICAL DATA

NAME: Eugene A. Cernan (Captain, USN) NASA Astronaut - Apollo 17 Commander

BIRTHPLACE AND DATE: Born in Chicago, Illinois, on March 14, 1934. His mother, Mrs. Andrew G. Cerna▮ resides in Bellwood, Illinois.

PHYSICAL DESCRIPTION: Brown hair; blue eyes; height: 6 feet; weight: 175 pounds.

EDUCATION: Graduated from Proviso Township High School in Maywood, Illinois; received a Bachelor ▮ Science degree in Electrical Engineering from Purdue University and a Master of Science degree ▮ Aeronautical Engineering from the U.S. Naval Postgraduate School; recipient of an Honorary Doctorate ▮ Laws from Western State University College of Law in 1969 and an Honorary Doctorate of Engineerir▮ from Purdue University in 1970.

MARITAL STATUS: Married to the former Barbara J. Atchley of Houston, Texas.

CHILDREN: Teresa Dawn, March 4, 1963.

RECREATIONAL INTERESTS: His hobbies include horses, motorcycling, and all sports activities.

ORGANIZATIONS: Member of the Society of Experimental Test Pilots; Tau Beta Pi, national engineering society; Sigma Xi, national science research society; and Phi Gamma Delta, national social fraternity.

SPECIAL HONORS: Awarded the NASA Distinguished Service Medal, the NASA Exceptional Service Medal, the MSC Superior Achievement Award, the Navy Distinguished Service Medal, the Navy Astronaut Wings, the Navy Distinguished Flying Cross, the National Academy of Television Arts and Sciences Special Trustees Award (1969), and an Honorary Lifetime Membership in the American Federation of Radio and Television Artists.

EXPERIENCE: Cernan, a United States Navy Captain, received his commission through the Navy ROTC program at Purdue. He entered flight training upon graduation. He was assigned to Attack Squadrons 126 and 113 at the Miramar, California, Naval Air Station and subsequently attended the Naval Postgraduate School. He has logged more than 3,800 hours flying time, with more than 3,600 hours in jet aircraft.

CURRENT ASSIGNMENT: Captain Cernan was one of the third group of astronauts selected by NASA in October 1963. He occupied the pilot seat along side of command pilot Tom Stafford on the Gemini 9 mission. During this 3-day flight which began on June 3, 1966, the spacecraft attained a circular orbit of 161 statute miles; the crew used three different techniques to effect rendezvous with the previously launched Augmented Target Docking Adapter; and Cernan logged two hours and ten minutes outside the spacecraft in extravehicular activity. The flight ended after 72 hours and 20 minutes with a perfect reentry and recovery as Gemini 9 landed within 1 1/2 miles of the prime recovery ship USS WASP and 3/8 of a mile from the predetermined target Cernan subsequently served as backup pilot for Gemini 12 and as backup lunar module pilot for Apollo VII.

He was lunar module pilot on Apollo X, May 18-26, 1969, the first comprehensive lunar-orbital qualification and verification flight test of an Apollo lunar module. He was accompanied on the 248,000 nautical mile sojourn to the Moon by Thomas P. Stafford (spacecraft commander) and John W. Young (command module pilot). In accomplishing all of the assigned objectives of this mission, Apollo X confirmed the operational performance, stability, and reliability of the command/service module/lunar module configuration during translunar coast, lunar orbit insertion, and lunar module separation and descent to within 8 nautical miles of the lunar surface. The latter maneuver involved employing all but the final minutes of the technique prescribed for use in an actual lunar landing, and completing critical evaluations of the lunar module propulsion systems and rendezvous and landing radar devices in subsequent rendezvous and re-docking maneuvers. In addition to demonstrating that man could navigate safely and accurately in the Moon's gravitational fields, Apollo X photographed and mapped tentative landing sites for future missions. This was Captain Cernan's second space flight giving him more than 264 hours and 24 minutes in space. Captain Cernan has since served as backup spacecraft commander for the Apollo XIV flight.

AME: Ronald E. Evans (Commander, USN) NASA Astronaut - Apollo 17 command module pilot

BIRTHPLACE AND DATE: Born November 10, 1933, in St. Francis, Kansas. His father, Mr. Clarence E. Evans, lives in Bird City, Kansas, and his mother, Mrs. Marie A. Evans, resides in Topeka, Kansas.

PHYSICAL DESCRIPTION: Brown hair; brown eyes; height: 5 feet 11 1/2 inches; weight: 160 pounds.

EDUCATION: Graduated for Highland Park High School in Topeka, Kansas; received a Bachelor of Science degree in Electrical Engineering from the University of Kansas in 1956 and a Master of Science degree in Aeronautical Engineering from the U.S. Naval Postgraduate School in 1964.

MARITAL STATUS: Married to the former Jan Pollom of Topeka, Kansas; her parents, Mr. & Mrs. Harry M. Pollom, reside in Salina, Kansas.

CHILDREN: Jaime D. (daughter), August 21, 1959; Jon P. (son), October 9, 1961

RECREATIONAL INTERESTS: Hobbies include golfing, boating, swimming, fishing, and hunting.
ORGANIZATIONS: Member of Tau Beta Pi, Society of Sigma Xi., and Sigma Nu.

SPECIAL HONORS: Presented the MSC Superior Achievement Award (1970), and winner of eight Ai
Medals, the Viet Nam Service Medal, and the Navy Commendation Medal with combat distinguishing device

EXPERIENCE: When notified of his selection to the astronaut program, Evans was on sea duty in the Pacifi
— assigned to VF-51 and flying F8 aircraft from the carrier USS TICONDEROGA during a period of seve
months in Viet Nam combat operations. He was a Combat Flight Instructor (F8 aircraft) with VF-124 fror
January 1961 to June 1962 and, prior to this assignment, participated in two WESTPAC aircraft carrie
cruises while a pilot with VF-142. In June 1957, he completed flight training after receiving his commissio
as an Ensign through the Navy ROTC program at the University of Kansas. Total flight time accrued durin
his military career is 4,041 hours.

CURRENT ASSIGNMENT: Commander Evans is one of the 19 astronauts selected by NASA in April 196
He served as a member of the astronaut support crews for the Apollo VII and XI flights and as backu
command module for Apollo XIV.

NAME: Harrison H. Schmitt (PhD) NASA Astronaut - Apollo 17 lunar module pilot

BIRTHPLACE AND DATE: Born July 3, 1935, in Santa Rita, New Mexico. His mother, Mrs. Harrison A
Schmitt, resides in Silver City, New Mexico.

PHYSICAL DESCRIPTION: Black hair; brown eyes; height: 5 feet 9 inches; weight: 165 pounds.

EDUCATION: Graduated from Western High School, Silver City, New Mexico; received a Bachelor c
Science degree in science from the California Institute of Technology in 1957; studied at the University c
Oslo in Norway during 1957-58; received Doctorate in Geology from Harvard University in 1964.

MARITAL STATUS: Single

RECREATIONAL INTERESTS: His hobbies include skiing, hunting, fishing, carpentry and hiking.

ORGANIZATIONS: Member of the Geological Society of America, the American Geophysical Union, th
American Association for the Advancement of Science, the American Association of Petroleum Geologist
the American Institute of Aeronautics and Astronautics, and Sigma Xi.

SPECIAL HONORS: Winner of a Fulbright Fellowship (1957-58); a Kennecott Fellowship in Geology (1958
59); a Harvard Fellowship (1959-60); a Harvard Traveling Fellowship (1960); a Parker Traveling Fellowshi
(1961-62); a National Science Foundation Post-Doctoral Fellowship, Department of Geological Science
Harvard University (1963-64); and presented the MSC Superior Achievement Award (1970).

EXPERIENCE: Schmitt was a teaching fellow at Harvard in 1961; he assisted in the teaching of a course i
ore deposits there. Prior to his teaching assignment, he did geological work for the Norwegian Geologic
Survey in Oslo, Norway, and for the U.S. Geological Survey in New Mexico and Montana. He also worke
as a geologist for two summers in southeastern Alaska. Before coming to the Manned Spacecraft Cente
he served with the U.S. Geological Survey's Astrogeology Branch at Flagstaff, Arizona. He was project chi
for lunar field geological methods and participated in photo and telescopic mapping of the Moon; he wa
among the USGS astrogeologists instructing NASA astronauts during their geological field trips, He ha
logged more than 1,665 hours flying time.

CURRENT ASSIGNMENT; Dr. Schmitt was selected as a scientist-astronaut by NASA in June 1965. H
completed a 53-week course in flight training at Williams Air Force Base, Arizona, and, in addition to trainir
for future manned space flights, has been instrumental in providing Apollo flight crews with detaile

instruction in lunar navigation, geology, and feature recognition. He has also assisted in the integration of scientific activities into the Apollo lunar missions and participated in research activities requiring the conduct of geologic, petrographic, and stratigraphic analysis of samples returned from the Moon by Apollo missions.

NAME: John W. Young (Captain, USN) NASA Astronaut - Apollo 17 backup commander

BIRTHPLACE AND DATE: Born in San Francisco, California, on September 24, 1930. His parents, Mr. and Mrs. William H. Young, reside in Orlando, Florida.

PHYSICAL DESCRIPTION: Brown hair; green eyes; height: 5 feet 9 inches; weight: 165 pounds.

EDUCATION: Graduated from Orlando High School, Orlando, Florida; received a Bachelor of Science degree in Aeronautical Engineering from the Georgia Institute of Technology in 1952; recipient of an Honorary Doctorate of Laws degree from Western State University College of Law in 1969, and an Honorary Doctorate of Applied Science from Florida Technological University in 1970.

MARITAL STATUS: Married to the former Susy Feldman of St. Louis, Missouri.

CHILDREN: Sandy, April 30, 1957; John, January 17, 1959, by a previous marriage.

RECREATIONAL INTERESTS: He plays handball, runs and works out in the full pressure suit to stay in shape.

ORGANIZATIONS: Fellow of the American Astronautical Society, Associate Fellow of the Society of Experimental Test Pilots, and a member of the American Institute of Aeronautics and Astronautics.

SPECIAL HONORS: Awarded the NASA Distinguished Service Medal, two NASA Exceptional Service Medals, the MSC Certificate of Commendation (1970), the Navy Astronaut Wings, the Navy Distinguished Service Medals, and three Navy Distinguished Flying Crosses.

EXPERIENCE: Upon graduation from Georgia Tech, Young entered the U.S. Navy in 1952; he holds the rank of Captain in that service. He completed test pilot training at the U.S. Naval Test Pilot School in 1959, and was then assigned as a test pilot at the Naval Air Test Center until 1962. Test projects in which he participated include evaluations of the F8D "Crusader" and the F4B "Phantom" fighter weapons systems, and in 1962, he set world time-to-climb records to 3,000 and 25,000 meter altitudes in the Phantom. Prior to his assignment to NASA, he was maintenance officer of All-Weather Fighter Squadron 143 at the Naval Air Station, Miramar, California. He has logged more than 6,380 hours flying time, and completed three space flights totaling 267 hours and 42 minutes.

CURRENT ASSIGNMENT: Captain Young was selected as an astronaut by NASA in September 1962. He served as pilot with command pilot Gus Grissom on the first manned Gemini flight — a 3-orbit mission, launched on March 23, 1965, during which the crew accomplished the first manned spacecraft orbital trajectory modifications and lifting reentry, and flight tested all systems in Gemini 3. After this flight, he was backup pilot for Gemini 6. On July 18, 1966, Young occupied the command pilot seat for the Gemini 10 mission and, with Michael Collins as pilot, effected a successful rendezvous and docking with the Agena target vehicle. He was then assigned as the backup command module pilot for Apollo VII. Young was command module pilot for Apollo X, May 18-26, 1969, the comprehensive lunar-orbital qualification test of the Apollo lunar module. He was accompanied on the 248,000 nautical mile lunar mission by Thomas P. Stafford (spacecraft commander) and Eugene A. Cernan (lunar module pilot). Captain Young then served as backup spacecraft commander for Apollo XIII. Young was commander of the Apollo 16 mission to the Descartes highlands of the Moon in April 1972.

NAME: Stuart Allen Roosa (Lieutenant Colonel, USAF) NASA Astronaut - Apollo 17 backup comman module pilot

BIRTHPLACE AND DATE: Born August 16, 1933, in Durango, Colorado. His parents, Mr. and Mrs. Dewe Roosa, now reside in Tucson, Arizona.

PHYSICAL DESCRIPTION: Red hair; blue eyes; height: 5 feet 10 inches; weight: 155 pounds.

EDUCATION: Attended Justice Grade School and Claremore High School in Claremore, Oklahom studied at Oklahoma State University of Arizona and was graduated with honors and a Bachelor of Scienc degree in Aeronautical Engineering from the University of Colorado; presented an Honorary Doctorate c Letters from the University of St. Thomas (Houston, Texas) in 1971.

MARITAL STATUS: His wife is the former Joan C. Barrett of Tupelo, Mississippi; and her mother, Mrs, Joh T. Barrett, resides in Sessums, Mississippi.

CHILDREN: Christopher A., June 29, 1959; John D., January 2, 1961; Stuart A., Jr., March 12, 1962; Rosemar D., July 23, 1963.

RECREATIONAL INTERESTS: His hobbies are hunting, boating, and fishing.

ORGANIZATIONS: Associate Member of the Society of Experimental Test Pilots.

SPECIAL HONORS: Presented the NASA Distinguished Service Medal, the MSC Superior Achievemen Award (1970), the Air Force Command Pilot Astronaut Wings, the Air Force Distinguished Service Meda the Arnold Air Society's John F. Kennedy Award (1971), and the City of New York Gold Medal in 1971.

EXPERIENCE: Roosa, a Lt. Colonel in the Air Force, has been on active duty since 1953. Prior to joinin NASA, he was an experimental test pilot at Edwards Air Force Base, California — an assignment he hel from September 1965 to May 1966, following graduation from the Aerospace Research Pilots School. H was a maintenance flight test pilot at Olmsted Air Force Base, Pennsylvania, from July 1962 to August 196 flying F-101 aircraft. He served as Chief of Service Engineering (AFLC) at Tachikawa Air Base for two year following graduation from the University of Colorado under the Air Force Institute of Technology Progran Prior to this tour of duty, he was assigned as a fighter pilot at Langley Air Force Base, Virginia, where h flew the F-84F and F-100 aircraft. He attended Gunnery School at Del Rio and Luke Air Force Bases an is a graduate of the Aviation Cadet Program at Williams Air Force Base, Arizona, where he received hi flight training and commission in the Air Force. Since 1953, he has acquired 4,797 flying hours.

CURRENT ASSIGNMENT: Lt. Colonel Roosa is one of the 19 astronauts selected by NASA in April 196 He was a member of the astronaut support crew for the Apollo IX flight. He completed his first space fligh as command module pilot on Apollo XIV, January 31 - February 9, 1971. With him on man's third luna landing mission were Alan B. Shepard (spacecraft commander) and Edgar D. Mitchell (lunar module pilot In completing his first space flight, Roosa logged a total of 216 hours and 42 minutes. He was subsequentl designated to serve as backup command module for Apollo XVI.

NAME: Charles Moss Duke, Jr. (Colonel, USAF) NASA Astronaut - Apollo 17 backup lunar module pilo

BIRTHPLACE AND DATE: Born in Charlotte, North Carolina, on October 3, 1935. His parents, Mr. an Mrs. Charles M. Duke, make their home in Lancaster, South Carolina.

PHYSICAL DESCRIPTION: Brown hair; brown eyes; height: 5 feet 11 1/2 inches; weight: 155 pounds.

EDUCATION: Attended Lancaster High School in Lancaster, South Carolina, and was graduate valedictorian from the Admiral Farragut Academy in St. Petersburg, Florida; received a Bachelor of Scienc degree in Naval Sciences from the U.S. Naval Academy in 1957 and a Master of Science degree in Aeronautics from the Massachusetts Institute of Technology in 1964.

MARITAL STATUS: Married to the former Dorothy Meade Claiborne of Atlanta, Georgia; her parents are Dr. and Mrs. T. Sterling Claiborne of Atlanta.

CHILDREN: Charles M., March 8, 1965; Thomas C., May 1, 1967.

RECREATIONAL INTERESTS: Hobbies include hunting, fishing, reading, and playing golf.

ORGANIZATIONS: Member of the Air Force Association, the Society of Experimental Test Pilots, the Rotary Club, the American Legion, and the American Fighter Pilots Association.

SPECIAL HONORS: Awarded the MSC Certificate of Commendation (1970)

EXPERIENCE: When notified of his selection as an astronaut, Duke was at the Air Force Aerospace Research Pilot School as an instructor teaching control systems and flying in the F-104, F-101, and T-33 aircraft. He was graduated from the Aerospace Research Pilot School in September 1965 and stayed on there as an instructor. He is an Air Force Colonel and was commissioned in 1957 upon graduation from the Naval Academy. Upon entering the Air Force, he went to Spence Air Base, Georgia, for primary flight training and then to Webb Air Force Base, Texas, for basic flying training, where in 1958 he became a distinguished graduate. He was again a distinguished graduate at Moody Air Force Base, Georgia, where he completed advanced training in F-86L aircraft. Upon completion of this training he was assigned to the 526th Fighter Interceptor Squadron at Ramstein Air Base, Germany, where he served three years as a fighter interceptor pilot. He has logged 3,862 hours flying time.

CURRENT ASSIGNMENT: Colonel Duke is one of the 19 astronauts selected by NASA in April 1966. He served as a member of the astronaut support crew for the Apollo X flight and as backup lunar module pilot for the Apollo XIII flight. He served as lunar module pilot for the Apollo 16 mission.

SPACEFLIGHT TRACKING AND DATA SUPPORT NETWORK

NASA's worldwide Spaceflight Tracking and Data Network (STDN) will provide communication with the Apollo astronauts, their launch vehicle and spacecraft. It will also maintain the communications link between Earth and the Apollo experiments left on the lunar surface by earlier Apollo crews.

The STDN is linked together by the NASA Communication Network (NASCOM) which provides for all information and data flow.

In support of Apollo 17, the STDN will employ 11 ground tracking stations equipped with 9.1-meter (30-foot) and 25.9 m (85-ft) antennas, and instrumented tracking ship, and four instrumented aircraft. This portion of the STDN was known formerly as the Manned Space Flight Network. For Apollo 17, the network will be augmented by the 64-m (210-ft.) antenna system at Goldstone, Calif. (a unit of NASA's Deep Space Network), and if required the 64-m (210-ft.) radio antenna of the National Radio Astronomy Observatory Parkes, Australia.

The STDN is maintained and operated by the NASA Goddard Space Flight Center, Greenbelt, Md., under the direction of NASA's Office of Tracking and Data Acquisition. Goddard will become an emergency control center if the Houston Mission Control Center is impaired for an extended time.

NASA Communications Network (NASCOM).

The tracking network is linked together by the NASA Communications Network. All information flows to and from Mission Control Center, (MCC), Houston, and the Apollo spacecraft over this communication system.

The NASCOM consists of more than 3.2 million circuit kilometers (1.7 million nautical miles), using satellite submarine cables, land lines, microwave systems, and high frequency radio facilities. NASCOM control center is located at Goddard. Regional communication switching centers are in Madrid; Canberra, Australia; Honolulu; and Guam.

Intelsat communications satellites will be used for Apollo 17. One satellite over the Atlantic will link Goddard with Ascension Island and the Vanguard tracking ship. Another Atlantic satellite will provide a direct link between Madrid and Goddard for TV signals received from the spacecraft. One satellite positioned over the mid-Pacific will link Carnarvon, Australia; Canberra, Guam and Hawaii with Goddard through the Jamesburg, California ground station. An alternate route of communications between Spain and Australia is available through another Intelsat satellite positioned over the Indian Ocean if required.

Mission Operations: Prelaunch tests, liftoff, and Earth orbital flight of the Apollo 17 are supported by the Apollo subnet station at Merritt Island, Fla., 6.4 km (3.5 nm) from the launch pad.

During the critical period of launch and insertion of the Apollo 17 into Earth orbit, the USNS Vanguard provides tracking, telemetry, and communications functions. This single sea-going station of the Apollo subnet will be stationed about 1,610 km (870 nm) southeast of Bermuda.

When the Apollo 17 conducts the translunar injection (TLI) Earth orbit for the Moon, two Apollo range instrumentation aircraft (ARIA) will record telemetry data from Apollo and relay voice communication between the astronauts and the MCC at Houston. These aircraft will be airborne between South America and the west coast of Africa. ARIA 1 will cover TLI ignition and ARIA 2 will monitor TLI burn completion.

Approximately 1 hour after the spacecraft has been injected into a translunar trajectory, three prime MSFN stations will take over tracking and communication with Apollo. These stations are equipped with 25.9 m (85 ft.) antennas.

Each of the prime stations, located at Goldstone, Madrid, and Honeysuckle, Australia is equipped with dual systems for tracking the command module in lunar orbit and the lunar module in separate flight paths or at rest on the Moon.

For reentry, two ARIA (Apollo Range Instrumented Aircraft) will be deployed to the landing area to relay communications between Apollo and Mission Control at Houston. These aircraft also will provide position information on the Apollo after the blackout phase of reentry has passed.

An applications technology satellite (ATS) terminal has been placed aboard the recovery ship USS Ticonderoga to relay command control communications of the recovery forces, via NASA's ATS satellite. Communications will be relayed from the deck-mounted terminal to the NASA tracking stations at Mojave, Calif. and Rosman, N.C., through Goddard to the recovery control centers located in Hawaii and Houston.

Prior to recovery, the astronauts aeromedical records are transmitted via the ATS satellite to the recovery ship for comparison with the physical data obtained in the postflight examination performed aboard the recovery ship.

Television Transmissions: Television from the Apollo spacecraft during the journey to and from the Moon and on the lunar surface will be received by the three prime stations, augmented by the 64-m (210-ft.) antennas at Goldstone and Parkes. The color TV signal must be converted at MSC, Houston. A black and white version of the color signal can be released locally from the stations in Spain and Australia.

While the camera is mounted on the lunar roving vehicle (LRV), the TV signals will be transmitted directly to tracking stations as the astronauts explore the Moon.

Once the LRV has been parked near the lunar module, its batteries will have about 80 hours of operating life. This will allow ground controllers to position the camera for viewing the lunar module liftoff, post liftoff geology, and other scenes.

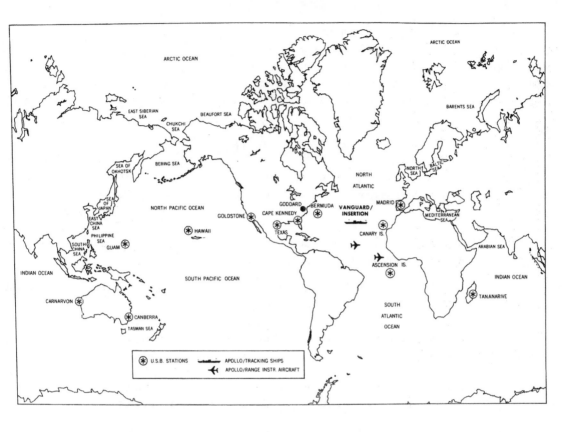

ENVIRONMENTAL IMPACT OF APOLLO/SATURN V MISSION

Studies of NASA space mission operations have concluded that Apollo does not significantly effect the human environment in the areas of air, water, noise or nuclear radiation.

During the launch of the Apollo/Saturn V space vehicle, products exhausted from Saturn first stage engines in all cases are within an ample margin of safety. At lower altitudes, where toxicity is of concern, the carbon monoxide is oxidized to carbon dioxide upon exposure at its high temperature to the surrounding air. The quantities released are two or more orders of magnitude below the recognized levels for concern in regard to significant modification of the environment. The second and third stage main propulsion systems generate only water and a small amount of hydrogen. Solid propellant ullage and retro rocket products are released and rapidly dispersed in the upper atmosphere at altitudes above 70 kilometers (43.5 miles). This material will effectively never reach sea level and, consequently, poses no toxicity hazard.

Should an abort after launch be necessary, some RP-1 fuel (kerosene) could reach the ocean. However, toxicity of RP-1 is slight and impact on marine life and waterfowl are considered negligible due to its dispersive characteristics. Calculations of dumping an aborted S-IC stage into the ocean showed that spreading and evaporating of the fuel occurred in one to four hours.

There are only two times during a nominal Apollo mission when above normal overall sound pressure levels are encountered. These two times are during vehicle boost from the launch pad and the sonic boom experienced when the spacecraft enters the Earth's atmosphere. Sonic boom is not a significant nuisance since it occurs over the mid-Pacific Ocean.

NASA and the Department of Defense have made a comprehensive study of noise levels and other hazards to be encountered for launching vehicles of the Saturn V magnitude. For uncontrolled areas the overall sound pressure levels are well below those which cause damage or discomfort. Saturn launches have had no deleterious effects on wildlife which has actually increased in the NASA-protected areas of Merritt Island.

A source of potential radiation hazard but highly unlikely, is the fuel capsule of the radioisotope thermoelectric generator supplied by the Atomic Energy Commission which provides electric power for Apollo lunar surface experiments. The fuel cask is designed to contain the nuclear fuel during normal operations and in the event of aborts so that the possibility of radiation contamination is negligible. Extensive safety analyses and tests have been conducted which demonstrated that the fuel would be safely contained under almost all credible accident conditions.

PROGRAM MANAGEMENT

The Apollo program is the responsibility of the office of Manned Space Flight (OMSF), National Aeronautics and Space Administration, Washington, D. C. Dale D. Myers is Associate Administrator for Manned Space Flight.

NASA Manned Spacecraft Center (MSC), Houston, is responsible for development of the Apollo spacecraft, flight crew training, and flight control. Dr. Christopher C. Kraft, Jr. is Center Director.

NASA Marshall Space Flight Center (MSFC), Huntsville, Ala., is responsible for development of the Saturn launch vehicles. Dr. Eberhard F. M. Rees is Center Director.

NASA John F. Kennedy Space Center (KSC), Fla., is responsible for Apollo/Saturn launch operations. Dr. Kurt H. Debus is Center Director.

The NASA Office of Tracking and Data Acquisition (OTDA) directs the program of tracking and data flow on Apollo. Gerald M. Truszynski is Associate Administrator for Tracking and Data Acquisition.

NASA Goddard Space Flight Center (GSFC), Greenbelt, Md., manages the Manned Space Flight Network and Communications Network. Dr. John F. Clark is Center Director.

The Department of Defense is supporting NASA during launch, tracking, and recovery operations. The Air Force Eastern Test Range is responsible for range activities during launch and down-range tracking. Recovery operations include the use of recovery ships and Navy and Air Force aircraft.

APOLLO/SATURN OFFICIALS

NASA Headquarters
Dr. Rocco A. Petrone	Apollo Program Director, OMSF
Chester M. Lee (Capt., USN, Ret.)	Apollo Mission Director, OMSF
John K. Holcomb (Capt., USN, Ret.)	Director of Apollo Operations, OMSF
William T. O'Bryant	Director of Apollo Lunar
(Capt., USN, Ret.)	Exploration, OMSF
Charles A. Berry, M.D.	Director for Life Sciences

Kennedy Space Center
Miles J. Ross	Deputy Center Director
Walter J. Kapryan	Director of Launch Operations
Peter A. Minderman	Director of Technical Support
Robert C. Hock	Apollo/Skylab Program Manager
Dr. Robert H. Gray	Deputy Director, Launch Operations
Dr. Hans F. Gruene	Director, Launch Vehicle Operations

ohn J. Williams   Director, Spacecraft Operations
aul C. Donnelly   Associate Director Launch Operations
om A. Rigell   Deputy Director Launch Vehicle Operations

Manned Spacecraft Center
igurd A. Sjoberg   Deputy Center Director
Ioward W. Tindall   Director, Flight Operations
Owen G. Morris   Manager, Apollo Spacecraft Program
Donald K. Slayton   Director, Flight Crew Operations
ete Frank   Flight Director
Jeil Hutchinson   Flight Director
ierald D. Griffin   Flight Director
ugene F. Kranz   Flight Director
Charles Lewis   Flight Director
ichard S. Johnston   Director, Life Sciences

Marshall Space Flight Center
Or. William R. Lucas   Deputy Center Director, Technical
ichard W. Cook   Deputy Center Director, Management
ichard G. Smith   Manager, Saturn Program Office (SPO)
ohn C. Rains   Manager, S-IC Stage Project, SPO
Villiam F. LaHatte   Manager, S-II, S-IVB Stage Projects SPO
ames B. Bramlet   Manager, Instrument Unit, GSE Projects SPO
ames M. Sisson   Manager, LRV Project, SPO
.P. Smith   Manager, Engines Project, SPO
Ierman F. Kurtz   Manager, Mission Operations Office

Goddard Space Flight Center
ecwyn Roberts   Director, Networks
Villiam P. Varson   Chief, Network Computing & Analysis Division
Valter Lafleur   Chief, Network Operations Division
obert Owen   Chief, Network Engineering Division
.R. Stelter   Chief, NASA Communications Division

Department of Defense
Maj. Gen. David M. Jones, USAF   DOD Manager for Manned Space Flight Support
                                       Operations
Col. Alan R. Vette, USAF   Deputy DOD Manager for Manned Space Flight Support
                                       Operations, and Director, DOD Manned Space Flight

upport
                                       Office
ear Adm. J.L. Bulls   Commander, Task Force 130, USN Pacific Recovery Area
ear Adm. Roy G. Anderson, USN   Commander Task Force 140, Atlantic Recovery Area
Capt. Norman K. Green, USN   Commanding Officer, USS Ticonderoga, CVS-14 Primary
                                       Recovery Ship
rig. Gen. Frank K. Everest, Jr.   Commander Aerospace Rescue USAF and Recovery
                                       Service

CONVERSION TABLE

Multiply	By	To Obtain
Distance:		
inches	2.54	centimeters
feet	0.3048	meters
meters	3.281	feet
kilometers	3281	feet
kilometers	0.6214	statute miles
statute miles	1.609	kilometers
nautical miles	1.852	kilometers
nautical miles	1.1508	statute miles
statute miles	0.8689	nautical miles
statute miles	1760	yards
Velocity:		
feet/sec	0.3048	meters/sec
meters/sec	3.281	feet/sec
meters/sec	2.237	statute mph
feet/sec	0.6818	statute miles/hr
feet/sec	0.5925	nautical miles/hr
statute miles/hr	1.609	km/hr
nautical miles/hr	1.852	km/hr
(knots)		
km/hr	0.6214	statute miles/hr
Liquid measure:		
gallons	3.785	liters
liters	0.2642	gallons
Weight:		
pounds	0.4536	kilograms
kilograms	2.205	pounds
metric ton	1000	kilograms
short ton	907.2	kilograms
Volume:		
cubic feet	0.02832	cubic meters
Pressure:		
pounds/sq. inch	70.31	grams/sq. cm
Thrust:		
pounds	4.448	newtons
newtons	0.225	pounds
Temperature:		
Centigrade	1.8 add 32	Fahrenheit

Report No. M-933-72-17

MISSION OPERATION REPORT

APOLLO 17 MISSION

OFFICE OF MANNED SPACE FLIGHT

Prelaunch Mission Operation Report

No. M-933-72-17
28 November 1972
MEMORANDUM
TO: A/Administrator FROM: M/Apollo Program Director
SUBJECT: Apollo 17 Mission (AS-512)

We plan to launch Apollo 17 from Pad A of Launch Complex 39 at the Kennedy Space Center no earlier than 6 December 1972. This will be the Apollo Program's sixth and last manned lunar landing and the third consecutive mission to carry the Lunar Roving Vehicle for surface mobility, added Lunar Module consumables for a longer surface stay time, and the Scientific Instrument Module for extensive lunar orbital science investigations.

Primary objectives of this mission are selenological inspection, survey, and sampling of materials and surface features in a preselected area of the Taurus-Littrow region of the moon; emplacement and activation of surface experiments; and the conduct of in-flight experiments and photographic tasks. In addition to the standard photographic documentation of operational and scientific activities, television coverage is planned for selected periods in the spacecraft and on the lunar surface. The lunar surface TV coverage will include remote controlled viewing of astronaut activities at each major science station on the three EVA traverses.

The 12.7-day mission will be terminated with the Command Module landing in the mid-Pacific Ocean about 650 km (350 NM) southeast of Samoa Islands.

Rocco A. Petrone APPROVAL: Dale D. Myers Associate Administrator for Manned Space Flight

Report No. M-933-72- 17 MISSION OPERATION REPORT APOLLO 17 MISSION
OFFICE OF MANNED SPACE FLIGHT

FOREWORD

MISSION OPERATION REPORTS are published expressly for the use of NASA Senior Management, as required by the Administrator in NASA Management Instruction HQMI 8610. I, effective 30 April 1971. The purpose of these reports is to provide NASA Senior Management with timely, complete, and definitive information on flight mission plans, and to establish official Mission Objectives which provide the basis for assessment of mission accomplishment.

Prelaunch reports are prepared and issued for each flight project just prior to launch. Following launch, updating (Post Launch) reports for each mission are issued to keep General Management currently informed of definitive mission results as provided in NASA Management Instruction HQMI 8610. I.

Primary distribution of these reports is intended for personnel having program/project management responsibilities which sometimes results in a highly technical orientation. The Office of Public Affairs publishes a comprehensive series of reports on NASA flight missions which are available for dissemination to the Press.

APOLLO MISSION OPERATION REPORTS are published in two volumes: the MISSION OPERATION REPORT (MOR); and the MISSION OPERATION REPORT, APOLLO SUPPLEMENT. This format was designed to provide a mission-oriented document in the MOR, with supporting equipment and facility description in the MOR, APOLLO SUPPLEMENT. The MOR, APOLLO SUPPLEMENT is a program-oriented reference document with a broad technical description of the space vehicle and associated equipment, the launch complex, and mission control and support facilities.

Published and Distributed by PROGRAM and SPECIAL REPORTS DIVISION (AXP) EXECUTIVE SECRETARIAT - NASA HEADQUARTERS

SUMMARY OF APOLLO/SATURN FLIGHTS

Mission	Launch Date	Launch Vehicle	Payload	Description
AS-201	2/26/66	SA-201	CSM-009	Launch vehicle and CSM development. Test of CSM subsystems and of the space vehicle. Demonstration of reentry adequacy of the CM at earth orbital conditions.
AS-203	7/5/66	SA-203	LH2 in S-IVB	Launch vehicle development. Demonstration of control of LH2 by continuous venting in orbit.
AS-202	8/25/66	SA-202	CSM-011	Launch vehicle and CSM development. Test of CSM subsystems and of the structural integrity and compatibility of the space vehicle. Demonstration of propulsion and entry control by G&N system. Demonstration of entry at 8689 meters per second.
Apollo 4	11/9/67	SA-501	CSM-017 LTA-10R	Launch vehicle and spacecraft development. Demonstration of Saturn V Launch Vehicle performance and of CM entry at lunar return velocity.
Apollo 5	1/22/68	SA-204	LM-1 SLA-7	LM development. Verified operation of LM subsystems: ascent and descent propulsion systems (including restart) and structures. Evaluation of LM staging. Evaluation of S-IVB/IU orbital performance.
Apollo 6	4/4/68	SA-502	CM-020 SM-014 LTA-2R SLA-9	Launch vehicle and spacecraft development. Demonstration of Saturn V Launch Vehicle performance.
Apollo 7	10/11/68	SA-205	CM-101 SM-101 SLA-5	Manned CSM operations. Duration 10 days 20 hours.
Apollo 8	12/21/68	SA-503	CM-103 SM-103 LTA- B SLA-11	Lunar orbital mission. Ten lunar orbits. Mission duration 6 days 3 hours. Manned CSM operations.
Apollo 9	3/3/69	SA-504	CM-104 SM-104 LM-3 SLA-12	Earth orbital mission. Manned CSM/LM operations. Duration 10 days 1 hour.
Apollo 10	5/18/69	SA-505	CM-106 SM-106 LM-4 SLA-13	Lunar orbital mission. Manned CSM/LM operations. Evaluation of LM performance in cislunar and lunar environment, following lunar landing profile. Mission duration 8 days.
Apollo 11	7/16/69	SA-506	CM-107 SM-107 LM-5 SLA-14 EASEP	First manned lunar landing mission. Lunar surface stay time 21.6 hours. One dual EVA (5 man hours). Mission duration 8 days 3.3 hours.
Apollo 12	11/14/69	SA-507	CM-108 SM-108 LM-6 SLA-15 ALSEP	Second manned lunar landing mission. Demonstration of point landing capability. Deployment of ALSEP I. Surveyor III investigation. Lunar surface stay time 31.5 hours. Two dual EVAs (15.5 man hours). Mission duration 10 days 4.6 hours.
Apollo 13	4/11/70	SA-508	CM- 109 SM-109 LM-7 SLA-16 ALSEP	Planned third lunar landing. Mission aborted at approximately 56 hours due to loss of SM cryogenic oxygen and consequent loss of capability to generate electrical power and water. Mission duration 5 days 22.9 hours.
Apollo 14	1/31/71	SA-509	CM-110 SM-110 LM-8 SLA-17 ALSEP	Third manned lunar landing mission. Selenological inspection, survey and sampling of materials of Fra Mauro Formation. Deployment of ALSEP. Lunar surface stay time 33.5 hours. Two dual EVAs (18.8 man hours). Mission duration 9 days.
Apollo 15	7/26/71	SA-510	CM-112 SM-112 LM-10 SLA-19 LRV-1 ALSEP Subsatellite	Fourth manned lunar landing mission. Selenological inspection, survey and sampling of materials of the Hadley-Apennine Formation. Deployment of ALSEP. Increased lunar stay time to 66.9 hours. First use of Lunar Roving Vehicle and direct TV and voice communications to earth during EVAs. Total distance traversed on lunar surface 27.9 km. Three dual EVAs (37.1 man hours). Mission duration 12 days 7.2 hours.

Apollo 16 4/16/72   SA-511   CM-113 SM-113 LM-11 SLA-20 ALSEP Fifth manned lunar landing mission. Selenological inspection, survey and sampling of materials of the Descartes Formation. Deployment of ALSEP. Lunar surface stay time 71.2 hours. Use of second Lunar Roving Vehicle and direct TV and voice communications to earth during EVAs. Total distance traversed on lunar surface 26.7 km. Three dual EVAs (40.5 man hours). Use of Far UV camera/spectroscope on lunar surface. Mission duration 11 days 1. 8 hours.

NASA OMSF MISSION OBJECTIVES FOR APOLLO 17 PRIMARY OBJECTIVES

Perform selenological inspection, survey, and sampling of materials and surface features in a preselected area of the Taurus-Littrow region.

Emplace and activate surface experiments.

Conduct in-flight experiments and photographic tasks.

Rocco A.Petrone   Dale D. Myers
Apollo Program Director   Associate Administrator for Manned Space Flight
Date: 22 November 1972   Date: Nov 22 1972

MISSION OPERATIONS

The following paragraphs contain a brief description of the nominal launch, flight, recovery, and post-recovery operations. For launch opportunities which may involve a T-24 or T+24 hour launch, there will be a revised flight plan. Overall mission profile is shown in Figure 1.

LAUNCH WINDOWS

The mission planning considerations for the launch phase of a lunar mission are, to a major extent, related to launch windows. Launch windows are defined for two different time periods: a "daily window" has a duration of a few hours during a given 24-hour period; a "monthly window" consists of a day or days which meet the mission operational constraints during a given month or lunar cycle.

Launch windows are based on flight azimuth limits of 72° to 100° (earth-fixed heading of the launch vehicle at end of the roll program), on booster and spacecraft performance, on insertion tracking, and on lighting constraints for the lunar landing sites.

The Apollo 17 launch windows and associated lunar landing sun elevation angles are presented in Table 1.

TABLE 1
LAUNCH WINDOWS

WINDOWS	(EST)		SUN ELEVATION
LAUNCH DATE	OPEN	CLOSE	ANGLE (degrees)
6 December	2153	0131	13.3
7 December	2153	0131	16.9- 19.1
4 January *	2150	2352	6.8
5 January	2021	2351	10.2 - 11.1
6 January	2028	2356	20.3 - 22.4
3 February	1847	2213	13.3 - 15.5
4 February	1858	2220	13.3 - 15.5

* Launch azimuth limits for 4 January are 84 to 100 degrees

LAUNCH THROUGH TRANSLUNAR INJECTION

The space vehicle will be launched from Pad A of launch complex 39 at the Kennedy Space Center. The boost into a 167 km (90 NM) earth parking orbit (EPO) will be accomplished by sequential burns and staging of the S-IC and S-II launch vehicle stages and a partial burn of the S-IVB stage. The S-IVB/instrument unit (IU) and spacecraft will coast in a circular EPO for approximately 2 revolutions while preparing for the first opportunity S-IVB translunar injection (TLI) burn, or 3 revolutions if the second opportunity TLI burn is required. Both injection opportunities are to occur over the Atlantic Ocean. TLI targeting will permit an acceptable earth return to be achieved using the service propulsion system (SPS) or LM descent propulsion system (DPS) until at least pericynthion plus 2 hours if lunar orbit insertion (LOI) is not performed. A reaction control system (RCS) capability to return the command service module/lunar module (CSM/LM) combination to an acceptable earth return exists to as great as 57 hours; for a CSM only case the RCS capability exists to about 69 hours. In the unlikely event of no separation from the S-IVB a combination RCS burn, LOX dump, and auxiliary propulsion system (APS) burn permits an acceptable earth return as late as 46 hours GET.

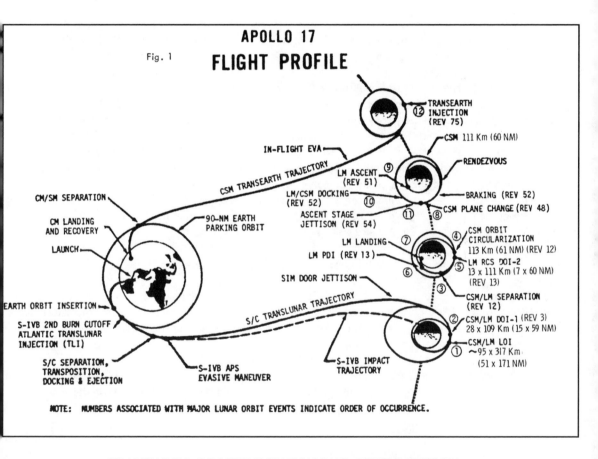

Fig. 1

APOLLO 17 FLIGHT PROFILE

NOTE: NUMBERS ASSOCIATED WITH MAJOR LUNAR ORBIT EVENTS INDICATE ORDER OF OCCURRENCE.

TRANSLUNAR COAST THROUGH LUNAR ORBIT INSERTION

Within 2 hours after injection the CSM will separate from the S-IVB/IU and spacecraft-LM adapter (SLA) and will transpose, dock with the LM, and eject the LM/CSM from the S-IVB/IU. Subsequently, the S-IVB/IU will perform an evasive maneuver to alter its circumlunar coast trajectory clear of the spacecraft trajectory.

The spent S-IVB/IU will be impacted on the lunar surface at 7°00'S and 8°00'W providing a stimulus for the Apollo 12, 14, 15, and 16 emplaced seismology experiments. The necessary delta velocity required to alter the S-IVB/IU circumlunar trajectory to the desired impact trajectory will be derived from dumping of residual liquid oxygen (LOX) and burn(s) of the S-IVB/auxiliary propulsion system (APS) and ullage motors. The final maneuver will occur within about 10 hours of liftoff. The IU will have an S-band transponder for trajectory tracking. A frequency bias will be incorporated to insure against interference between the S-IVB/IU and LM communications during translunar coast.

Spacecraft passive thermal control will be initiated after the first midcourse correction (MCC) opportunity and will be maintained throughout the translunar-coast phase unless interrupted by subsequent MCCs and/or navigational activities. The scientific instrument module (SIM) bay door will be jettisoned shortly after the MCC-4 point, about 4.5 hours before LOI.

Multiple-operation covers over the SIM bay experiments and cameras will provide thermal and contamination protection whenever they are not in use.

A LOI retrograde SPS burn will be used to place the docked spacecraft into a 95 x 317 km (51 x 171 NM) orbit, where they will remain for approximately two revolutions.

DESCENT ORBIT INSERTION THROUGH LANDING

The descent orbit insertion (DOI-1) maneuver, a SPS second retrograde burn, will place the CSM/LM combination into a 28 x 109 km (15 x 59 NM) orbit (Figure 2). A "soft" undocking will be made during the twelfth revolution, using the docking probe capture latches to reduce the imparted delta-V. Spacecraft separation will be executed by the SM RCS. Following separation, the CSM will maneuver into a 100 x 130 km (54 x 70 NM) orbit during the twelfth revolution to achieve an approximate 113 km (61 NM) circular orbit at the time of CSM/LM rendezvous. After the CSM circularization maneuver, DOI-2 will be performed with the LM RCS to lower the pericynthian to about 13,170 meters (43,200 feet) which will be about 100 west of the landing site with PDI nominally commencing at about 17,200 meters (56,400 feet). During the thirteenth revolution the LM DIPS will be used for powered descent which will begin approximately 26° east of pericynthian. A terrain profile model will be available in the LM guidance computer (LGC) program to minimize unnecessary LM pitching or thrusting maneuvers. A LM yaw maneuver may be performed to ensure good LM steerable antenna communications coverage with the Spaceflight Tracking and Data Network (STDN). A descent path of 25° will be used from high gate to about 61 meters (200 feet) altitude, or to crew manual takeover, to enhance landing site visibility.

LANDING SITE (LITTROW REGION)

Taurus-Littrow is designated as the landing site for the Apollo 17 Mission. This site is located on the southeastern rim of Mare Serenitatis in a dark deposit between massif units of the southwestern Taurus Mountains, south of the crater Littrow. It is about 750 km (405 NM) east of the Apollo 15 site and about the same distance north of the Apollo 11 site.

The massif units of the Taurus Mountains are believed to be ancient highland crustal blocks (pre-Imbrian in geologic age) which were emplaced by faulting and uplifting both during and after formation of the Serenitatis basin.

However, a thin (because of the base distances) ejecta blanket from the younger impacts of Crisium and Imbrium may have mantled some of the massif units.

Fresh and blocky slopes in excess of 25° are common, which indicates that later debris movements have exposed the origin massif surfaces.

One large landslide or debris flow southwest of the landing site offers an excellent opportunity to sample both the massif materials and any later ejecta materials.

The dark deposit, which occupies the low-lands between, and in a few cases the tops of the massif units, is believed to be among the youngest lunar surface units.

It is characterized by a smooth appearance and lack of large blocks as indicated by photogeologic analysis and earth-based radar studies. This deposit is associated with numerous dark halo craters and is believed to be a volcanic (pyroclastic) mantle which originated deep within the moon. It offers, therefore, an opportunity to sample a relatively young volcanic material which may shed light on the composition as well as thermal history of the lunar interior.

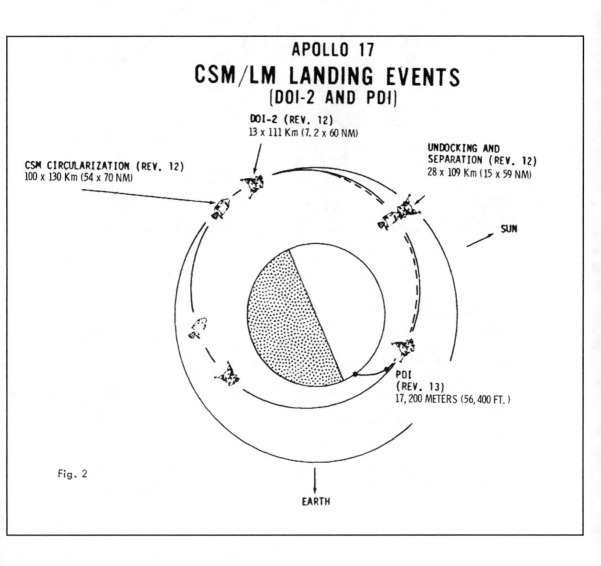

Fig. 2

The Taurus-Littrow site is geologically complex: it offers a number of rock types which apparently vary in age, albedo, composition and probable origin; it also portrays numerous stratigraphic-structural problems which will be investigated by the traverse geophysics to be carried for the first time on Apollo 17.

The location of the site makes the inclination of the orbital tracks suitable for the SIM bay experiments, especially those to be carried for the first time on Apollo 17.

The planned landing coordinates for the Taurus-Littrow area are 20°09'50"N latitude and 30°44'58"E longitude (Figure 3).

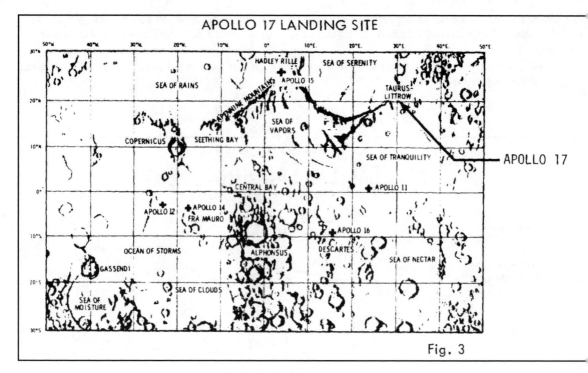

Fig. 3

LUNAR SURFACE OPERATIONS

The nominal stay time on the lunar surface is planned for about 75 hours, with the overall objective of optimizing effective surface science time relative to hardware margins, crew duty cycles, and other operational constraints. Photographs of the lunar surface will be taken through the LM cabin window after landing. The nominal extravehicular activity (EVA) is planned for three periods of up to 7 hours each.

The duration of each EVA period will be based upon real time assessment of the remaining consumables. As in Apollo 15 and 16, this mission will employ the lunar roving vehicle (LRV) which will carry both astronauts, experiment equipment, and independent communications systems for direct contact with the earth when out of the line-of-sight of the LM relay system. Voice communication will be continuous and color TV coverage will be provided at each major science stop (Figure 4) where the crew will align the high gain antenna. The ground controllers will then assume control of the TV through the ground controlled television assembly (GCTA) mounted on the LRV. A TV panorama is planned at each major science stop, as well as coverage of the astronauts' scientific activities.

The radius of crew operations will be constrained by the portable life support system (PLSS) walkback capability, or the buddy secondary life support system (BSLSS) rideback capability, whichever is less.

If a walking traverse must be performed, the radius of operations will be constrained by the BSLSS capability to return the crew to the LM in the event of a PLSS failure.

EVA PERIODS

The activities to be performed during each EVA period are described below. Rest periods are scheduled prior to the second and third EVAs and prior to LM liftoff. The three traverses planned for Apollo 17 are designed for flexibility in selection of science stops as indicated by the enclosed areas shown along traverses II and III (Figure 4).

First EVA Period

The first EVA will include: LM inspection, LRV deployment and checkout, and deployment and activation of the Apollo lunar surface experiments package (ALSEP). Television will be deployed on the LRV as soon as possible in this period for observation of crew activities near the LM (Figure 5).

ALSEP deployment will be approximately 90 meters (300 feet) west of the LM (Figure 6). After ALSEP activation the crew will perform a geology traverse (see Figure 4).

Lunar samples collected will be verbally and photographically documented. Sample return must be assured; therefore, a contingency sample of lunar soil will be collected in the event of a contingency during the EVA, but only if no other soil sample has been collected and is available for return to earth. Experiment activities other than ALSEP include the deployment and activation of the surface electrical properties transmitter at least 70 meters (230 feet) east of the LM, deployment of some of the seismic profiling charges, obtaining some traverse gravimeter readings, obtaining a measurement of surface electric properties, and emplacement of the lunar neutron probe. The planned timeline for EVA-1 activities is presented in Figure 7.

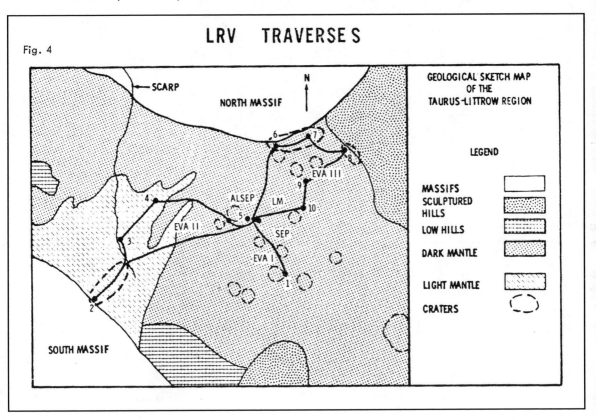

Fig. 4

LRV TRAVERSES

Second and Third EVA Periods

Traverses in the second and third EVA periods (Figures 8 and 9) are planned to maximize the scientific return in support of the primary objectives.

LRV sorties will be planned for flexibility in selecting stops and conducting experiments. Consumables usage will be monitored at Mission Control Center (MCC) to assist in real time traverse planning.

The major portion of the lunar geology investigation (S-059) will be conducted during the second and third EVAs and will include voice and photographic documentation of sample material as it is collected and descriptions of lurain features. If time does not permit filling the sample containers with documented samples, the crew may fill the containers with samples selected for scientific interest.

Deployment of the lunar seismic profiling (S-203) explosive charges will be completed on EVA-2 and 3. Readings from the traverse gravimeter experiment (S-199) will be taken at specified traverse stations. The

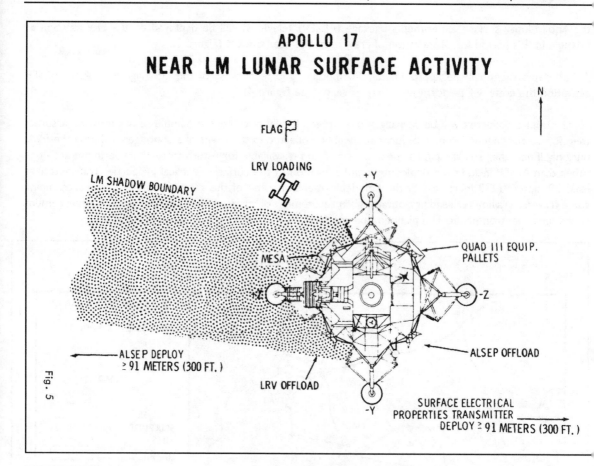

APOLLO 17
NEAR LM LUNAR SURFACE ACTIVITY

Fig. 5

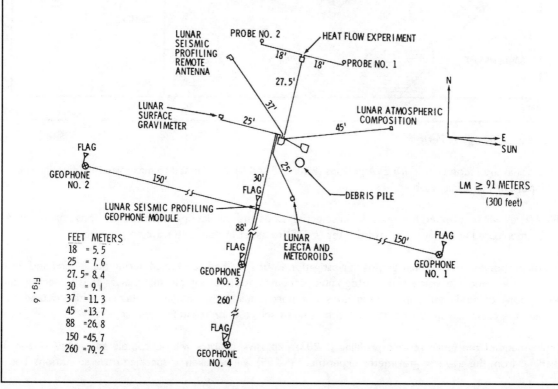

APOLLO 17
ALSEP DEPLOYMENT

FEET	METERS
18	= 5.5
25	= 7.6
27.5	= 8.4
30	= 9.1
37	= 11.3
45	= 13.7
88	= 26.8
150	= 45.7
260	= 79.2

Fig. 6

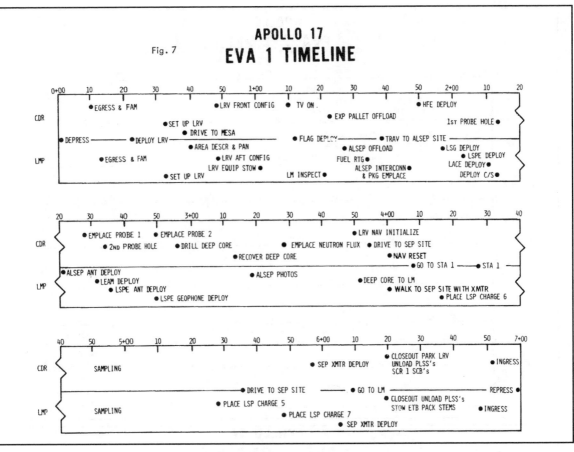

Fig. 7

APOLLO 17
EVA 1 TIMELINE

surface electrical properties experiment (S-204) will be continued during the second and third EVAs.

The LRV will be positioned at the end of the EVA-3 traverse to enable GCTA-monitored ascent and other TV observations of scientific interest.

LUNAR ORBIT OPERATIONS

The Apollo 17 Mission is the third with the modified Block II CSM configuration. An increase in cryogenic storage provides increased mission duration for the performance of both an extended lunar surface stay time and a lunar orbit science period. The SIM in the SM provides for the mounting of scientific experiments and for their operation in flight.

After the SIM door is jettisoned by pyrotechnic charges and until completion of lunar orbital science tasks, selected RCS thrusters will be inhibited or experiment protective covers will be closed to minimize contamination of experiment sensors during necessary RCS burns. Attitude changes for thermal control and experiment alignment with the lunar surface and deep space (and away from direct sunlight) will be made with the active RCS thrusters.

Orbital science activities have been planned at appropriate times throughout the lunar phase of the mission and consist of the operation of five cameras (35mm Nikon, 16mm data acquisition, 70mm Hasselblad, 24-inch panoramic and a 3-inch mapping), a color TV camera, a laser altimeter, a gamma ray spectrometer, a lunar sounder, a far ultraviolet spectrometer, and infrared scanning radiometer. Pre-Rendezvous Lunar Orbit Science Orbital science operations will be conducted during the 109 x 28 km (59 x 14 NM) orbits after DOI-I, while in the docked configuration. Orbital science operations will be stopped for the separation and circularization maneuvers performed during the 12th revolution, then restarted after CSM circularization.

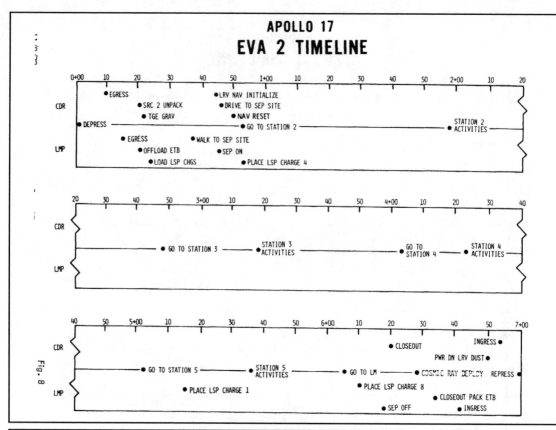

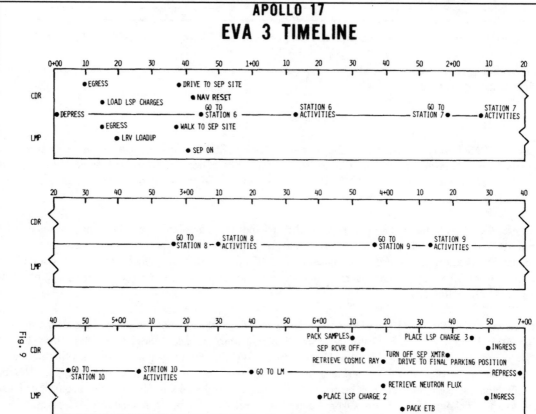

In the event of a T-24 launch, the additional day in the 111 x 26 km (60 x 14 NM) orbit prior to lunar landing will also be used for orbital science.

The experiments timeline has been developed in conjunction with the surface timeline to provide, as nearly as possible, 16-hour work days and concurrent 8-hour CSM and LM crew sleep periods. Experiment activation cycles are designed to have minimum impact on crew work-rest cycles.

Conduct of orbital experiments and photographic tasks have been planned in consideration of: outgassing, standby, and warmup periods; experiments fields-of-view limitations; and STDN data collection requirements. Water and urine dumps and fuel cell purges have been planned to avoid conflict with operation cycles. Prior to LM liftoff, the CSM will perform a plane change maneuver to provide the desired coplanar orbit at the time of the LM rendezvous.

LM Ascent, Rendezvous and Jettison After completion of lunar surface activities and ascent preparations, the LM ascent stage propulsion system (APS) and RCS will be used to launch and rendezvous with the CSM. The direct ascent rendezvous technique initiated on Apollo 14 and subsequently used on Apollo 15 and 16 will be performed.

The LM ascent stage liftoff window duration is about 30 seconds and is constrained to keep the perilune above 15 km (8 NM). The ascent stage will be inserted into a 89 x 17 km (48 x 9 NM) orbit so that an APS terminal phase initiation (TPI) burn can be performed approximately 38 minutes after insertion. The final braking maneuver will occur about 45 minutes later. The total time from ascent stage liftoff to the final braking maneuver will be about 90 minutes.

Docking will be accomplished by the CSM with SM RCS maneuvers. Once docked, the two LM crewmen will transfer to the CSM with lunar sample material, exposed films, and designated equipment.

The LM ascent stage will be jettisoned and subsequently deorbited to impact on the lunar surface to provide a known stimulus for the emplaced seismic experiment. The impact will be targeted for 30°32.4'E and 19°55.8'N about 9 km (5 NM) from the Apollo 17 ALSEP.

Post-LM Jettison Lunar Orbit Science

After LM ascent stage jettison, additional scientific data will be obtained by the CSM over a 2-day period. Conduct of the SIM experiments and both SM and CM photographic tasks will take advantage of the extended ground track coverage during this period.

TRANSEARTH INJECTION THROUGH LANDING

After completion of the post-rendezvous CSM orbital activities, the SPS will perform a posigrade burn to inject the CSM onto the transearth trajectory. The nominal return time will not exceed 110 hours and the return inclination will not exceed 70° with relation to the earth's equator.

During the transearth phase there will be continuous communications coverage from the time the spacecraft appears from behind the moon until shortly prior to entry. MCCs will be made, if required.

A 6-hour period, including pre- and post-EVA activities, will be planned to perform an in-flight EVA to retrieve film cassettes and the lunar sounder tape from the SIM bay in the SM TV and photographic tasks will be conducted during transearth coast. The CM will separate from the SM 15 minutes before the entry interface. Earth touchdown will be in the mid-Pacific and will nominally occur approximately 12.7 days after launch. Targeted landing coordinates are 17°54'S, 166°00'W.

POST-LANDING OPERATIONS

Flight Crew Recovery

Following splashdown, the recovery helicopter will drop swimmers and life rafts near the CM. The swimmers will install the flotation collar on the CM, attach the life raft, and pass fresh flight suits in through the hatch for the flight crew to don before leaving the CM. The crew will be transferred from the spacecraft to the recovery ship via life raft and helicopter and will return to Houston, Texas, for debriefing. Quarantine procedures were eliminated prior to Apollo 15; therefore, the mobile quarantine facility will not be used.

However, biological isolation garments will be available for use in the event of unexplained crew illness. The Skylab medical laboratory will be aboard the prime recovery ship and will be used in lieu of the ship's post-flight medical facilities. However, the ship's facilities will be used for X-rays.

CM and Data Retrieval Operations

An attempt will be made to recover the earth landing system main parachutes on this mission.

In addition, the CM RCS propellants will not be vented during the landing in order to preclude possible damage to the parachutes. After flight crew pickup by helicopter, the CM will be retrieved and placed on a dolly aboard the recovery ship, USS TICONDEROGA. The CM RCS helium pressure will be vented and the CM will be stowed near the ship's elevator to insure adequate ventilation. Lunar samples, film, flight logs, etc., will be retrieved for shipment to the Lunar Receiving Laboratory (LRL). The spacecraft will be offloaded from the ship in San Diego and transported to an area where deactivation of the propellant system will be accomplished. The CM will then be returned to contractor facilities.

ALTERNATE MISSIONS

If an anomaly occurs after liftoff that would prevent the space vehicle from following its nominal flight plan, an abort or an alternate mission will be initiated. An abort will provide for acceptable flight crew and CM recovery in the Atlantic or Pacific Ocean.

An alternate mission is a modified flight plan that results from a launch vehicle, spacecraft, or support equipment anomaly that precludes accomplishment of the primary mission objectives. The purpose of the alternate mission is to provide the flight crew and flight controllers with a plan by which the greatest benefit can be gained from the flight using the remaining systems capabilities.

The two general categories of alternate missions that can be performed during the Apollo 17 Mission are (1) earth orbital and (2) lunar orbital.

Both of these categories have several variations which depend upon the nature of the anomaly leading to the alternate mission and the resulting systems status of the LM and CSM. A brief description of these alternate missions is contained in the following paragraphs.

Earth Orbit

In case of no TLI burn, an earth orbital mission of up to about 6-1/2 days will be conducted to obtain maximum benefit from the scientific equipment onboard. Subsequent to transfer of necessary equipment to the CM, the LM will be deorbited into the Pacific Ocean. The SPS will be used to attain the optimum orbit for conducting orbital science and to increase the orbital inclination up to 45°. A backup RCS deorbit capability will be maintained at all times. The lunar sounder, far UV Spectrometer, and IR scanning radiometer will be used to obtain data on the earth's surface and atmosphere and for galactic observations. The mapping camera and pan camera will be used to photograph selected earth targets. The lunar sounder optical recorder film, mapping camera film, and pan camera film will be retrieved by EVA on the last day of the mission.

Lunar Orbit
Lunar orbit missions of the following types will be planned if spacecraft systems will enable accomplishment of orbital science objectives in the event a lunar landing is not possible.

If the SIM bay cameras and/or the lunar sounder are used, film cassettes will be retrieved by EVA during transearth coast. An attempt should be made to follow the nominal timeline in order to minimize real time flight planning activities. An SPS capability to perform TEI on any revolution will be maintained.

Generally, when the LM is available for a lunar orbit alternate mission it will be retained. The decision on when to jettison the LM will be made in real time.

CSM/LM (Operable DPS)
The translunar trajectory will be maintained to be within the DPS capability of an acceptable earth return until LOI plus 2 hours in the event LOI is not performed. If it is determined during translunar coast that a lunar landing mission cannot be performed, either the SPS or the LM DPS may be used to perform the LOI-1 maneuver to put the CSM/LM into an appropriate orbit.

In the event the SIM bay door is not jettisoned prior to LOI, the LOI-1 maneuver will be performed with the SPS. The LOI-2 maneuver will be performed with the SPS. The LOI-2 maneuver will place the CSM/LM into a 111 km (60 NM) orbit. Orbital science and photographic tasks will be performed for up to approximately 6 days in lunar orbit.

CSM/LM Inoperable DPS
If, following a nominal LOI maneuver it is determined that the DPS is inoperable, an SPS circularization maneuver will be performed to obtain near nominal orbital inclination. The CSM will generally remain in a 111 km (60 NM) orbit. Orbital science and photographic tasks will be performed for up to approximately 6 days in the lunar orbit per the nominal flight plan.

CSM Alone
In the event the LM is not available following a nominal TLI burn, an SPS MCC-1 maneuver will place the CSM on a trajectory such that an acceptable return to earth can be achieved within the SM RCS capability. LOI will not be performed if the SIM bay door cannot be jettisoned.

If the SIM bay door has been jettisoned, orbital science and photographic tasks will be performed in an orbit of near nominal (approximately 20°) inclination but with an easterly node shift of from 40° to 60°. The CSM will remain in a 111 km (60 NM) orbit. The duration in lunar orbit will be up to approximately 6 days.

CSM Alone (From Landing Abort)
In the event the lunar landing is aborted, an orbital science mission will be accomplished by the CSM alone after rendezvous, docking, and LM jettison. The total lunar orbit time will be approximately 6 days.

EXPERIMENTS, DETAILED OBJECTIVES, IN-FLIGHT DEMONSTRATIONS, AND OPERATIONAL TESTS

The technical investigations to be performed on the Apollo 17 Mission are classified as experiments, detailed objectives, in-flight demonstrations, or operational tests:

Experiment - A technical investigation that supports science in general or provides engineering, technological, medical or other data and experience for application to Apollo lunar exploration or other programs and is recommended by the Manned Space Flight Experiments Board (MSFEB) and assigned by the Associate Administrator for Manned Space Flight to the Apollo Program for flight.

Detailed Objective - A scientific, engineering, medical or operational investigation that provides important data and experience for use in development of hardware and/or procedures for application to Apollo missions. Orbital photographic tasks, though reviewed by the MSFEB, are not assigned as formal experiments and will be processed as CM and SM detailed objectives.

In-flight Demonstration - A technical demonstration of the capability of an apparatus and/or process to illustrate or utilize the unique conditions of space flight environment.

In-flight demonstration will be performed only on a noninterference basis with all other mission and mission-related activities. Utilization performance, or completion of these demonstrations will in no way relate to mission success.

Operational Test - A technical investigation that provides for the acquisition of technical data or evaluates operational techniques, equipment, or facilities but is not required by the objectives of the Apollo flight mission. An operational test does not affect the nominal mission timeline, adds no payload weight, and does not jeopardize the accomplishment of primary objectives, experiments, or detailed objectives.

EXPERIMENTS

The Apollo 17 Mission includes the following experiments:

Lunar Surface Experiments
Lunar surface experiments are deployed and activated or conducted by the Commander and the Lunar Module Pilot during EVA periods. Those experiments which are part of the ALSEP are so noted.

Lunar Heat Flow (S-037) (ALSEP)
The heat flow experiment is designed to measure the subsurface vertical temperature gradients and brightness temperature in the lunar surface layer, and the absolute temperature and thermal conductivity of the lunar subsurface material to a depth of approximately 2 meters (7 feet). The experiment includes two sensor probes which are placed in bore holes drilled with the Apollo lunar surface drill (ALSD).

Lunar Ejecta and Meteorites (S-202) (ALSEP)
The lunar ejecta and meteorites experiment is designed to measure physical parameters of primary cosmic dust particle impacts on sensors in cislunar space, and of lunar ejecta emanating from the sites of meteorite impacts on the lunar surface. It will measure the direction, mass distribution, and speed of both the primary and secondary particles.

Lunar Seismic Profiling (S-203) (ALSEP)
The lunar seismic profiling experiment is designed to obtain data on physical properties of the lunar surface and subsurface by generating and monitoring artificial seismic waves in the surface and near subsurface in the active mode, and by detecting moonquakes and meteorite impacts in the passive listening mode. The experiment electronics package and the four-geophone array will be deployed in the ALSEP area during EVA-1. Transport modules with explosive charges will be mounted on the LRV for deployment during the EVA traverses within 2.5 km (1.3 NM) of ALSEP. The eight charges will be detonated remotely from earth subsequent to lunar liftoff.

Lunar Atmospheric Composition (S-205) (ALSEP)
The lunar atmospheric composition experiment is designed to obtain data on composition of the lunar atmosphere in mass range 1-110 AMU at the lunar surface. A secondary goal is the detection of transient changes in composition due to emission of gasses from the surface or from man-made sources. The instrument is a magnetic sector field mass spectrometer with a Nier-type thermionic electronic bombardment ion source.

Lunar Surface Gravimeter (S-207) (ALSEP)
The lunar surface gravimeter is designed to gather the following information: absolute lunar gravity, tidal changes in local gravity due to the change in relative position of celestial bodies, low level lunar gravity changes with periods between 10 seconds and 20 minutes due to natural oscillations of the moon excited by gravity waves, and vertical axis seismic activity.

From this information conclusions may be drawn about the internal constitution of the moon, about associated seismic activity, and about the existence of gravity waves. The vertical component of gravity is measured in three different frequency ranges.

Lunar Geology Investigation (S-059)

The lunar geology experiment is designed to provide data for use in the interpretation of the geologic history of the moon in the vicinity of the landing site. The investigation will be carried out during the planned lunar surface traverses and will utilize astronaut descriptions, camera systems, hand tools, drive tubes, the ALSD, and sample containers. The battery powered ALSD will be used to obtain core samples to a maximum depth of approximately 3 meters (11 feet). There are two major aspects of the experiment:

Documented Samples - Rock and soil samples representing different morphological and petrological features will be described, photographed, and collected in individual pre-numbered bags for return to earth.

Geological Description and Special Samples - Descriptions and photographs of the field relationships of all accessible types of lunar features will be obtained. Special samples, such as drive tube samples, will be collected and documented for return to earth.

Cosmic Ray Detector (S-152)

The Apollo 17 cosmic ray detector is designed to measure the elemental composition, abundance and energy spectrum of the solar and galactic cosmic rays, with emphasis on quiet sun conditions, and to detect radon emissions from the lunar surface. All elements heavier than hydrogen in the energy range from 1 kev/AMU to 25 Mev/AMU will be detected.

The experiment will be in two sections, each about 5 x 10 cm (2 x 4 inches). One section holding mica, glass, aluminum foil and platinum foil detector elements will be placed in the sun in order to detect solar emissions. A second section, containing in addition Lexan sheets, will be placed in the shade in order to detect galactic cosmic rays and radon emission.

Traverse Gravimeter (S-199)

The traverse gravimeter experiment is designed to measure and map the gravity in the area traversed by the LRV. The data can provide resolution of surface and subsurface gravity anomalies, and their relationships to observed geological features. The experiment will also provide surface verification of Lunar Orbiter gravity measurements and information on higher harmonic contents of the lunar gravitational field. It will also provide a comparison of terrestrial and lunar gravity. The instrument mounted on the LRV uses a vibrating string accelerometer to measure gravity fields at the traverse stations. Data will be read to the ground by the crew. Stations will be accurately located by orbital stereo photo coverage and surface photos.

Surface Electrical Properties (S-204)

The surface electrical properties experiment is designed to determine layering in the lunar surface, to search for the presence of water below the surface, and to measure the electrical properties of the lunar material in situ. Instrumentation includes a solar panel powered transmitter and multiple frequency antenna deployed at least 70 meters (230 feet) from the LM and 70 meters (230 feet) from the ALSEP, and a receiver with tri-loop receiving antenna and data recorder mounted on the LRV. It is utilized on the EVAs while the LRV is in motion. Six frequencies ranging from 1 to 32 MHz allow probing of the subsurface from a few meters to several kilometers. The data recorder is to be returned to earth for analysis.

Lunar Neutron Probe (S-229)

The neutron probe experiment is designed to measure neutron capture rates as a function of depth in the lunar regolith. The neutron probe data will permit an unambiguous interpretation of neutron dosage measurements on the lunar samples. The lunar dosage enables one to calculate regolith erosion and accretion rates, regolith mixing depths, and rock irradiation depths. The experiment probe is a two-section, 2 meter (7 feet rod which is activated assembled, and inserted into a core stem hole in the lunar surface during the early stages of extravehicular activity and retrieved, deactivated, and disassembled at the end of the final extravehicular activity. The two sections of the rod consist of concentric tubes. One half of the inner diameter of the outer tube is lined with a plastic track detector. Half of the outer diameter of the center tube is lined with a boron film which emits alpha particles when struck by neutrons. These alpha particles leave traces in the plastic film when the two films are in registration.

In-Flight Experiments
In-flight experiments may be conducted during any phase of the mission. They are performed within the command module (CM), and from the scientific instrument module (SIM) located in sector I of the service module (SM).

S-band Transponder (CSM/LM) (S-164)
The S-band transponder experiment is designed to detect variations in the lunar gravity field caused by mass concentrations and deficiencies and to establish gravitational profiles of the ground tracks to the spacecraft. The experiment data are obtained by analysis of the S-band Doppler tracking data for the CSM and LM in lunar orbit. Minute pertubations of the spacecraft motion are correlated to mass anomalies in the lunar structure.

Far UV Spectrometer (S-169)
The far UV spectrometer experiment is designed to determine the atomic composition, density, and scale height for each constituent in the lunar atmosphere, and to repeatedly scan the spectral region of 1175 to 1675Å with primary emphasis on hydrogen Lyman Alpha (1216Å) and xenon (1470Å).

The instrument will detect spatial and temporal variations in the lunar atmosphere, measure the temporary atmosphere created by the LM descent and ascent engines, measure the UV albedo and its graphic variations, study the fluorescence on the lunar darkside, measure the UV galactic emission, and study the atomic hydrogen distribution between the earth and the moon.

IR Scanning Radiometer (S-171)
The IR scanning radiometer experiment is designed to obtain data to construct a high resolution temperature map of the lunar surface. It will measure lunar surface thermal emissions along the spacecraft orbital track, and will be used to determine thermal conductivity, bulk density, and specific heat on the lunar surface. Data from this experiment will also be correlated with data obtained from other lunar orbiting spacecraft.

Lunar Sounder (S-209)
The lunar sounder experiment is designed to obtain lunar subsurface and near subsurface stratigraphic structural, and tectonic data for development of a geological model of the lunar interior to a depth of approximately 1.3 km (.7 NM). The instrument utilizes three tranceivers (frequencies are 150 MHz, 15 MHz and 5 MHz), a VHF antenna, an HF antenna, and an optical recorder. Return echos will be linearly detected displayed in the recorder's cathode ray tube, and photographed. The film cassette will be retrieved during in-flight EVA and returned to earth.

Passive Experiments
Additional experiments assigned to the Apollo 17 Mission which are completely passive are discussed in this section only. Completely passive means no crew activities are required during the mission to perform these experiments.

Gamma Ray Spectrometer (S-160)
This extension of the basic S-160 experiment is designed to obtain measurements of background caused by activation products produced with the Sodium Iodide crystal by cosmic ray interaction, and is in support of the analysis of astronomical gamma ray data collected on the Apollo 15 and 16 Missions. The experiment package is completely passive, consisting of a NaI crystal and plastic scintillator, and is stowed in the CM for the duration of the mission. Immediately upon return to earth, the detector crystal count rate will be measured to determine the background counts produced by the cosmic ray flux interaction with the crystal

Apollo Window Meteoroid (S-176) (CM)

The Apollo window meteoroid experiment is designed to obtain data on the cislunar meteoroid flux of mass range 10-12 grams. The returned CM window will be analyzed for meteoroid impacts by comparison with a preflight photomicroscopic window map. The photomicroscopic analysis will be compared with laboratory calibration velocity data to define the mass of impacting meteoroids.

Soil Mechanics Experiment (S-200)

The soil mechanics experiment is designed to obtain data on the mechanical properties of the lunar soil from the surface to depths of several meters. Data are derived from LM landing, flight crew observations and debriefings, examination of photographs, analysis of lunar samples, and astronaut activities using the Apollo hand tools.

Biostack IIA Experiment (M-211)

The Biostack experiment is designed to investigate the biologic effects of cosmic radiation during space flight. The Biostack consists of layers of several selected kinds of biological objects (Bacillus subtilis spores, Colpoda cucullus cysts, Arabidopsis thaliana seeds, Vicia faba radiculae, Artemia salina eggs, Tribolium castaneum eggs) stacked alternatively with different physical track detectors (nuclear emulsions, plastics, AgCl-crystals). The biologic affects of galactic cosmic particles under consideration are: molecular and cellular inactivation; damage to nuclei and other sub-cellular systems; induction of mutations leading to genetic changes; and modification in growth and development of tissues. The experiment is stowed in the CM for the duration of the mission. The research is of special interest because of its possible relationship to the biologic effects of space flight on man.

Biocore Experiment (M-212)

The Biocore experiment is designed to ascertain whether heavy particles of cosmic radiation of known trajectory and terminating in the brain and eyes, will produce morphologically demonstrable damage. Five pocket mice will be used to establish occurrence or nonoccurrence of brain lesions caused by the particles. The experiment is stowed in the CM for the duration of the mission. For a T-24 launch opportunity this experiment will not be flown.

DETAILED OBJECTIVES

Following is a brief description of each of the launch vehicle and spacecraft detailed objectives planned for this mission.

Launch Vehicle Detailed Objectives

Impact the expended S-IVB/IU in a preselected zone on the lunar surface under nominal flight profile conditions to stimulate the ALSEP passive seismometers. Post-flight determination of actual S-IVB/IU point of impact within 5 km (2.7 NM), and time of impact within 1 second.

Spacecraft Detailed Objectives

Obtain SM high resolution panoramic and high quality metric lunar surface photographs and altitude data from lunar orbit to aid in the overall exploration of the moon. Obtain CM photographs of lunar surface features of scientific interest and of low brightness astronomical and terrestrial sources. Record visual observations of farside and nearside lunar surface features and processes to complement photographs and other remote-sensed data. Obtain data on whole body metabolic gains or losses, together with associated endocrinological controls for food compatibility assessment. Obtain data on the effectiveness of the protective pressure garment in counteracting orthostatic intolerance. Obtain more definitive information on the characteristics and causes of visual light flashes. Obtain data on Apollo spacecraft induced contamination.

IN-FLIGHT DEMONSTRATION

Heat Flow and Convection

This demonstration will be performed to show the convective instability existing in fluids containing temperature gradients at low acceleration levels. The demonstration is designed to obtain further information on the effects detected in the Apollo 14 demonstration.

OPERATIONAL TESTS

The operational tests listed below have been approved for conduct in conjunction with the Apollo 17 Mission, with the qualification stated in connection with the ETR 0.13 Radar Skin Track Test.

The ETR 0.13 Radar Skin Track Test and Chapel Bell actively radiate, whereas the remaining tests are passive. The operational tests using the Air Force Eastern Test Range (AFETR) instrumentation or facilities are to be scheduled by the AFETR.

Chapel Bell (TEPEE)
The DOD operational test Chapel Bell has been coordinated between centers and has received OMSF approval. This test has been conducted on previous Apollo launches and has received OMSF approval for the Apollo 17 Mission.

ETR 0.13 Radar Skin Track Test
Subject to certification by KSC, in coordination with other OMSF Centers, the AFETR will operate a research and development FPQ-13, C-band radar during the launch and orbital phases of the mission. Certification will be based upon a satisfactory radio frequency interference test. The AFETR Superintendent of Range Operations will respond to the MSC Network Controller and turn the radar off if radio frequency interference should occur.

Ionospheric Disturbances from Missiles
This test studies the long-range propagation of acoustic waves of continuous sources, such as rocket vehicles, in the upper atmosphere and in the lower ionosphere at Grand Bahama Island.

Observations will be made of electromagnetic radiation at audiofrequencies expected to result from vehicle flight.

Acoustic Measurement of Missile Exhaust Noise
The acoustic noise of the rocket's exhaust is recorded at Cape Kennedy Air Force Station by a cross-shaped array of nine microphones, which are sheltered from wind noise by heavy vegetation. The Air Force performs analyses of the recorded data to determine wind speed and direction from sea level to 80 km (43 NM).

Army Acoustic Test
This DOD operational test has been conducted on previous Apollo launches and has received OMSF approval for Apollo 17.

Long Focal Length Optical System
Photographic information on selected launch and orbital operations is collected by the AFETR, using a long focal length optical system.

Sonic Boom Measurement
The sonic boom overpressure levels of the Apollo 17 space vehicle during launch will be measured in Atlantic launch abort area. The data will be used to assist in developing high-altitude, high-Mach number, accelerated flight sonic boom prediction techniques.

Skylab Medical Mobile Laboratory
The Skylab Medical Mobile Laboratory will be on board the prime recovery ship when the crew is recovered. The Skylab medical laboratory will be used to replace the ship's post-flight medical facilities, except for shipboard X-ray. No new medical protocol changes will be implemented above the normal Apollo Program requirements.

MISSION CONFIGURATION AND DIFFERENCES MISSION HARDWARE AND SOFTWARE CONFIGURATION

The Saturn V Launch Vehicle and the Apollo Spacecraft for the Apollo 17 Mission will be operational configurations.

CONFIGURATION	DESIGNATION NUMBERS
Space Vehicle	AS-512
Launch Vehicle	SA-512
First Stage	S-IC-12
Second Stage	S-II-12
Third Stage	S-IVB-512
Instrument Unit	S-IU-512
Spacecraft-LM Adapter	SLA-21
Lunar Module	LM-12
Lunar Roving Vehicle	LRV-3
Service Module	SM-114
Command Module	CM-114

Onboard Programs	
Command Module	Colossus 3
Lunar Module	Luminary 1G
Experiments Package	Apollo 17 ALSEP
Launch Complex	LC-39A

CONFIGURATION DIFFERENCES

The following summarizes the significant configuration differences associated with the AS-512 Space Vehicle and the Apollo 17 Mission

Spacecraft
Command/Service Module

Added 5 micron filter in ECS	To prevent particulate contamination from
Suit P gauge	causing erroneous pressure readings.
Added battery manifold relief valve	To permit overboard dumping of battery gasses in event manual vent valve is closed during a pressure buildup.
Added lunar sounder booms	To provide equipment for lunar sounder experiment.
Added fuel and oxidizer pressure	To provide redundant fuel and oxidizer transducers system pressure measurements.

Crew Systems

Added third EMU maintenance kit	To provide additional lubricant due to amount of dust encountered on previous flights.
Enlarged PGA bag	To preclude dust from contaminating wrist connects.
Modified OPS purge valve	To preclude inadvertant pulling of locking pin.
Added spare OPS antenna	To provide spare in event of breakage.
Eliminated EVA Visor Stop in helmet	To prevent difficulty in retracting center eye shade.

LRV

Added fender extension fasteners	To preclude loss of fender extensions.

Lunar Module

Modified S-band steerable antenna latch mechanism	To improve unstow capability.
Removed thermal paint from RCS	To preclude peeling during space flight
Improved structural attachment panels between base heatshield and support tube for aft equipment rack panels	To preclude shearing during LM lunar liftoff

Launch Vehicle
S-IVB Redesigned S-IVB APS Helium   To reduce possibility of internal system
pressure transducer mounting   leakage block

Replaced APS Helium bulkhead   To reduce possibility of external system leakage
fittings and seals

Interconnected stage Helium   To provide capability to charge APS Helium spheres
repressurization system with APS   from stage Helium spheres in event of APS leak
Helium system

IU
Modified EDS distributor to   To provide two out of three voting logic
provide three independent liftoff   for start of Time Base one to eliminate single failure point
signals and change Time Base one
and start logic

Added redundant path across   To reduce possibility of S-IC engine shutdown after ignition
IU/ESE interface to assure power
to IU measuring bus prior to liftoff

Removed modulating flow control   Not required. In line production change.
valve and associated circuitry

Added lightning detection devices   To determine magnitude of lightning
to Saturn launch vehicle   strike to provide ample data for retest requirements

J-2 Engine
Replaced engine control assembly   To eliminate single failure points by
(ECA) with ECA incorporating   providing redundant circuits redundant timers

Modified LOX dome and gas   To prevent excessive Helium loss due to
generator purge control valve   slow engine purge control valve closure

TV AND PHOTOGRAPHIC EQUIPMENT

Standard and special-purpose cameras, lenses, and film will be carried to support the objectives, experiments
and operational requirements. Table 2 lists the TV and camera equipments and shows their stowage locations

Table 2

Nomenclature	CSM at Launch	LM at Launch	CM to LM	LM to CM	CM at Entry
TV, Color, Zoom Lens Monitor with CM System)	I	I			I
Camera, Data Acquisition, 16mm	I	I			I
Lens - 10mm	I	I			I
- 18mm	I				I
- 75mm	I				I
Film Magazines	13		3	3	13
Camera, 35mm Nikon	I				I
Lens - 55mm	I				I
Cassette, 35mm	8				8
Camera, Hasselblad, 70mm Electric	I				I
Lens - 80mm	I				I
- 250mm	I				I
Film Magazines	8				8

Camera, Hasselblad				
Electric Data (Lunar Surface)		2		
Lens - 60mm		2		
Film Magazines	14		14	14	14
Polarizing Filter		1		
Camera, 24-in Panoramic (In SIM)	1			
Film Magazine (EVA Transfer)	1			1
Camera, 3-in Mapping Stellar (SIM)	1			
Film Magazine Containing 5-in Mapping and 35mm Stellar Film (EVA Transfer)	1			1
Lunar L Sounder Film Magazine (SIM) (EVA Transfer)	1			1

FLIGHT CREW DATA
PRIME CREW (Figure 10)

Commander: Eugene A. Cernan (Captain, USN) Space Flight Experience: Captain Cernan was selected as an astronaut by NASA in October 1963. Captain Cernan was the LM Pilot for the Apollo 10 Mission, which included all phases of a lunar mission except the final minutes of an actual landing. He also served as pilot for the Gemini 9 Mission. Captain Cernan has logged more than 264 hours in space.

Command Module Pilot: Ronald E. Evans (Commander, USN) Space Flight Experience: Commander Evans was selected as a NASA astronaut in April 1966. He served as a member of the astronaut support crews for the Apollo 7 and 11 flights and the backup command pilot for Apollo 14. Commander Evans has flown 3400 hours in jet aircraft.

Lunar Module Pilot: Harrison H. Schmitt (Dr. Phd) Space Flight Experience: Dr. Schmitt was selected as a scientist astronaut by NASA in June 1965. He served as backup lunar module pilot for Apollo 15. He has logged 1100 hours in jet aircraft.

APOLLO 17 PRIME CREW   Fig 10

BACKUP CREW

Commander: John W. Young (Captain, USN) Space Flight Experience: Captain Young was selected as a astronaut by NASA in September 1962. Captain Young was the commander for the lunar landing Apollo 1 Mission and command module pilot for the Apollo 10 Mission. He also served as pilot for the Gemini Mission and commander of the Gemini 10 Mission. Captain Young has logged more than 533 hours in space

Command Module Pilot: Stuart A. Roosa (Lieutenant Colonel, USAF) Space Flight Experience: Lt. Colone Roosa was selected as an astronaut by NASA in April 1966. He was the command module pilot for the Apoll 14 lunar landing mission, backup command module pilot for the Apollo 16 Mission, and a member of th Apollo 9 Mission. He has logged more than 216 hours of space flight.

Lunar Module Pilot: Charles M. Duke, Jr. (Colonel, USAF) Space Flight Experience: Colonel Duke was selecte as an astronaut by NASA in April 1966. He served as backup lunar module pilot for the Apollo 15 Missio and was the lunar module pilot for the Apollo 16 lunar landing. Colonel Duke has been on active duty sinc graduating from the U.S. Naval Academy in 1957. He has logged more than 267 hours in space.

MISSION MANAGEMENT RESPONSIBILITY

Title	Name	Organization
Director, Apollo Program	Dr. Rocco A. Petrone	OMSF
Mission Director	Capt. Chester M. Lee, USN (Ret)	OMSF
Saturn Program Manager	Mr. Richard G. Smith	MSFC
Apollo Spacecraft	Mr. Owen G. Morris	MSC Program Manager
Apollo Program Manager,	Mr. Robert C. Hock	KSC KSC
Director of Launch Operations	Mr. Walter J. Kapryan	KSC
Director of Flight Operations	Mr. Howard W. Tindall, Jr.	MSC
Launch Operations Manager	Mr. Paul C. Donnelly	KSC
Flight Directors	Mr. M. P. Frank	MSC
	Mr. Eugene F. Kranz	MSC
	Mr. Gerald Griffin	MSC

ABBREVIATIONS AND ACRONYMS

AGS	Abort Guidance System	EI	Entry Interface
ALSEP	Apollo Lunar Surface Experiments Package	EMU	Extravehicular Mobility Unit
AOS	Acquisition of Signal	EPO	Earth Parking Orbit
APS	Ascent Propulsion System (LM)	EST	Eastern Standard Time
APS	Auxiliary Propulsion System (S-IVB)	ETA	Equipment Transfer Bag
ARIA	Apollo Range Instrumentation Aircraft	EVA	Extravehicular Activity
AS	Apollo/Saturn	FM	Frequency Modulation
BIG	Biological Isolation Garment	fps	Feet Per Second
BSLSS	Buddy Secondary Life Support System	FDAI	Flight Director Attitude Indicator
CCATS	Communications, Command, and Telemetry System	FTP	FUJI Throttle Position
CCGE	Cold Cathode Gauge Experiment	GCTA	Ground Commanded Television
CDR	Commander	GET	Ground Elapsed Time
CPLEE	Charged Particle Lunar Environment Experiment	GNCS	Guidance, Navigation, and Control System (CSM)
CM	Command Module	GSFC	Goddard Space Flight Center
CMP	Command Module Pilot	HBR	High Bit Rate
CSI	Concentric Sequence Initiation	HFE	Heat Flow Experiment
CSM	Command/Service Module	HTC	Hand Tool Carrier
DAC	Data Acquisition Camera	IMU	Inertial Measurement Unit
DDAS	Digital Data Acquisition System	IU	Instrument Unit
DOD	Department of Defense	IVT	Intravehicular Transfer
DOI	Descent Orbit Insertion	KSC	Kennedy Space Center
DPS	Descent Propulsion System	LBR	Low Bit Rate
DSKY	Display and Keyboard Assembly	LCC	Launch Control Center
ECS	E-ironmentol Control System	LCRU	Lunar Communications Relay Unit

LMK	Landmark	PTC	Passive Thermal Control
LEC	Lunar Equipment Conveyor	QUAD	Quadrant
LES	Launch Escape System	RCS	Reaction Control System
LET	Launch Escape Tower	RLS	Radius Landing Site
LGC	LM Guidance Computer	RTCC	Real-Time Computer Complex
LH2	Liquid Hydrogen	RTG	Radioisotope Thermoelectric Generator
LiOH	Lithium Hydroxide	S/C	Spacecraft
LM	Lunar Module	SEA	Sun Elevation Angle
LMP	Lunar Module Pilot	STDN	Spaceflight Tracking and Data Network
LOI	Lunar Orbit Insertion	S-IC	Saturn V First Stage
LOS	Loss of Signal	S-II	Saturn V Second Stage
LOX	Liquid Oxygen	S-IVB	Saturn V Third Stage
LPO	Lunar Parking Orbit	SIDE	Suprathermal Ion Detector Experiment
LR	Landing Radar	SIM	Scientific Instrument Module
LRL	Lunar Receiving Laboratory	SLA	Spacecraft-LM Adapter
LRRR	Laser Ranging Retro-Reflector	SM	Service Module
LSM	Lunar Surface Magnetometer	S PS	Service Propulsion System
LV	Launch Vehicle	SRC	Sample Return Container
MCC	Midcourse Correction	SSB	Single Side Band
MCC	Mission Control Center	SSR	Staff Support Room
MESA	Modularized Equipment Stowage Assembly	SV	Space Vehicle
MHz	Megahertz	SWC	Solar Wind Composition Experiment
MOCR	Mission Operations Control Room	TD&E	Transposition, Docking and LM Ejection
MOR	Mission Operations Report	TEC	Transearth Coast
MPL	Mid-Pacific Line	TEI	Transearth Injection
MSC	Manned Spacecraft Center	TFI	Time From Ignition
MS FC	Marshall Space Flight Center	TLC	Translunar Coast
MSFEB	Manned Space Flight Experiment Board	TLI	Translunar Injection
MSFN	Manned Space Flight Network	TLM	Telemetry
NASCOM	NASA Communications Network	TPF	Terminal Phase Finalization
NM	Nautical Mile	TPI	Terminal Phase Initiation
OMSF	Office of Manned Space Flight	T-time	Countdown Time (referenced to liftoff time)
OPS	Oxygen Purge System	TV	Television
ORDEAL	Orbital Rate Display Earth and Lunar	USB	Unified S-Band
PCM	Pulse Code Modulation	USN	United States Navy
PDI	Powered Descent Initiation	USAF	United States Air Force
PGA	Pressure Garment Assembly	VAN	Vanguard
PGNCS	Primary Guidance Navigation & Control System (LM)	VHF	Very High Frequency
PLSS	Portable Life Support System	Delta-V	Differential Velocity
PSE	Passive Seismic Experiment		

Post Launch Mission Operations Report

No, M-933-72-17
December 19, 1972
TO: A/Administrator
FROM: MA/Apollo Program Director
SUBJECT: Apollo 17 Mission (AS-512)
Post Mission Operation Report No. 1

The Apollo 17 mission was successfully launched from the Kennedy Space Center on Thursday, December 7, 1972. The mission was completed successfully, with recovery on December 19, 1972. Initial review of the mission events indicates that all mission objectives were accomplished. Detailed analysis of all data is continuing and appropriate refined results of the mission will be reported in the Manned Space Flight Centers' technical reports.

Attached is the Mission Director's Summary Report for Apollo 17, which is submitted as Post Launch Operations Report No. 1. Also attached are (he NASA OMSF Primary Objectives for Apollo 17, The Apollo 17 mission has achieved all the assigned primary objectives and I judge it to be a success.

Rocco A. Petrone
Approval: Dale D. Myers - Associate Administrator for Manned Space Flight

---

NASA OMSF MISSION OBJECTIVES FOR APOLLO 17 PRIMARY OBJECTIVES

Perform selenological inspection, survey, and sampling of materials and surface features in a preselected area of the Taurus-Littrow region.

Emplace and activate surface experiments.

Conduct in-flight experiments and photographic tasks.

Rocco A. Petrone
Apollo Program Director
Date: 22 November 1972

Dale D. Myers
Associate Administrator For Manned Space Flight
Date: Nov 22 1972

ASSESSMENT OF THE APOLLO 17 MISSION

Based upon a review of the assessed performance of Apollo 17, launched 7 December 1972 and completed 19 December 1972, this mission is adjudged a success in accordance with the objectives stated above.

Rocco A. Petrone
Apollo Program Director
Date: 20 December 1972

Dale D. Myers
Associate Administrator for Manned Space Flight
Date: Jan 10, 1973

NATIONAL AERONAUTICS AND
SPACE ADMINISTRATION
WASHINGTON, D.C. 20546

IN. REPLY REFER TO.

MAO   December 19, 1972
TO: Distribution
FROM: MA/Apollo Mission Director
SUBJECT: Mission Director's Summary Report, Apollo 17

INTRODUCTION

The Apollo 17 Mission was planned as a lunar landing mission to accomplish selenological inspection, survey, and sampling of materals and surface features in a preselected area of the Taurus-Littrow region of the moon; emplace and activate surface experiments; and in-flight experiments and photographic tasks. Flight crew members were Commander (CDR) Captain Eugene A. Cernan (USN), Command Module Pilot (CMP) Commander Ronald E, Evans (USN), Lunar Module Pilot (LMP) Dr. Harrison H. Schmitt (PhD). Significant detailed information is contained in Tables 1 through 15. Initial review indicates that all primary mission objectives were accomplished (reference Table 1). Table 2 lists the Apollo 17 achievements.

PRELAUNCH

The launch countdown proceeded smoothly until T-30 seconds, at which time an automatic cutoff occurred. Subsequent to a recycle and hold at T-22 minutes an additional hold was called at T-8 minutes. The duration of the holds delayed the launch by 2 hours 40 minutes.

The hold was caused when the Terminal Countdown Sequencer (TCS) failed to command pressurization of the S-IVB LOX tank. This command: (1) Closes the LOX tank vent; (2) Opens the LOX tank pressurization valve; and (3) Arms the S-IVB LOX tank pressurized interlock. The tank was pressurized manually satisfying (1) and (2) above, but the absence of (3) prevented actuation of the interlock in the S-IVB ready to launch logic train. The result was automatic cutoff at T-30 seconds. The launch was accomplished with the interlock bypassed by a jumper. Investigation indicates cause of failure to be a defective diode on a printed circuit card in the TCS.

The workaround to jump the single failed function was thoroughly analyzed and satisfactorily checked out on the breadboard at MSFC and, subsequently a decision was made to proceed with the countdown. The countdown then proceeded smoothly. Launch day weather conditions were clear, visibility 13 kilometers (km) (7 nautical miles (nm)), winds 4 meters per second (mps) (8 knots), and scattered cloud cover at 1200 meters (m) (4000 feet).

LAUNCH, EARTH PARKING ORBIT, AND TRANSLUNAR INJECTION

The Apollo 17 space vehicle was successfully launched from Kennedy Space Center, Florida, 2 hours 40 minutes late, at 00:33 EST, December 7, 1972. The S-IVB/IU/LM/CSM was inserted into an earth orbit of 170 x 168 km (92.5 x 91.2 nm) at 00:11:52 GET (hrs:min:sec). Following nominal checkout of the space vehicle, the S-IVB performed the Translunar Injection Maneuver over the Atlantic Ocean as planned. The maneuver occurred at 3:12:36 GET and was nominal in all respects.

TRANSLUNAR COAST

The CSM separated from the S-IVB/IU/LM at 3:42:29 GET, transposed, and then docked with the LM at 3:57:10 GET. However, during docking, a talkback barberpole indicated a possible ring latch malfunction. Subsequent LM pressurization, hatch removal, and troubleshooting revealed that latches 7, 9, and 10 handles were not locked. Latch 10 handle was locked by pushing on the latch handle. Latches 7 and 9 were locked and manually fired to lock the handles and the system talkback indicated normal. Following hatch replacement, the CSM/LM combination was successfully ejected 4:45:20 GET.

The S-IVB lunar impact maneuver targeted the stage for impact at lunar coordinates 7.0°S and 8.0°W at about 87:05 GET.

The spacecraft trajectory was such that midcourse correction (MCC) 1 was not performed. MCC-2 maneuver was performed on time at 35:29:59 GET. The Service Propulsion System (SPS) was fired for 1.7 seconds, resulting in a velocity change of 3 mps (9.9 feet per second (fps)). This was 0.2 mps (0.7 fps) less than planned because tank pressures at the time of the maneuver were slightly lower than those used to predict the firing duration. This residual was trimmed out manually to near zero using the service module Reaction Control System (RCS).

The Commander (CDR) and Lunar Module Pilot (LMP) began IVT to the LM at approximately 40:10 GET. At ingress, it was discovered that #4 docking latch was not properly latched. The CMP moved the latch handle about 30°-45°, disengaging the hook from the docking ring. After discussion between the ground controllers and the flight crew, it was decided to curtail further action on the latch until the second IVT/LM Activation at 59:59 GET. The remainder of the LM housekeeping was nominal and the LM was closed out at 42:11 GET.

The Heat Flow and Convection demonstrations were conducted as planned. The first demonstration was performed with the spacecraft in attitude hold while the second run was accomplished with the spacecraft in the passive thermal control (PTC) mode. The radial, lineal and flow pattern demonstrations produced satisfactory results.

Since the spacecraft trajectory was near nominal, MCC-3 was not required. IVT/LM housekeeping commenced about 59:59 GET and completed about 62:16 GET. All LM systems checks were nominal. During the LM housekeeping period, the Command Module Pilot (CMP) performed troubleshooting on the docking latch #4 problem experienced during the first IVT/LM. Following instructions from the ground controllers the CMP stroked latch #4 handle and succeeded in cocking the latch. The latch was left in the cocked position for CSM/LM rendezvous.

As the delay in liftoff was being experienced, the planned trajectory was continually being modified to speed up the translunar coast so that the spacecraft would arrive at lunar orbit insertion (LOI) at the same GMT time. Subsequently, in order to adjust the GET to allow for the delay in liftoff, a 2-hour and 40-minute GET clock update was performed at 65:00 GET, placing all events back on schedule with the flight plan.

At 68:19 GET, a 1-hour Visual Light Flash Phenomenon observation was conducted by the crew. The crew reported seeing light flashes ranging from bright to dull.

The spacecraft entered the moon's sphere of influence at about 73:18 GET. MCC- 4 was not performed since the spacecraft trajectory was near normal.

The SIM door was successfully jettisoned at 84:12 GET. The crew stated that the SIM bay looked good.

LUNAR ORBIT INSERTION AND S-IVB IMPACT

LOI was performed with the service propulsion system (SPS) at 88:54:22 GET. The 398-second maneuver produced a velocity change (delta V) of -910 mps (-2988 fps) and inserted the CSM/LM into a 315 x 97 km (170.0 x 52.6 nm) lunar orbit. The resultant orbit was very close to the prelaunch planned orbit of 317 x 95 km (171 x 61 nm).

S-IVB impact on the lunar surface occurred at 89:39 GET about 18 minutes later than the prelaunch prediction. Impact coordinates were 4°12'S and 12°18'W, about 160 km (86 nm) northwest of the planned arget point. The event was recorded by the Apollo 12, 14, 15, and 16 Apollo Lunar Surface Experiment 'ackages (ALSEPs).

DESCENT ORBIT UNDOCKING POWERED DESCENT AND LANDING

.2-second burn at 93:11 GET was nominal and produced a delta-V of -60 mps (-197 fps) and a resultant orbit »f 109 x 27 km (59 x 14.6 nm).

√T/LM activation occurred at about 107:42 GET (hr:min). The LM was powered up and all systems were ιominal.

'he CSM and LM performed the undocking and separation maneuver on schedule at 110:27:55 GET. The ¦SM then performed the circularization maneuver at 111:57:28 GET which placed the CSM into a 129 x 100 .m (70 x 54 nm) orbit.

'he Descent Orbit Insertion-2 maneuver occurred at 112:02:41 GET and inserted the LM into a 111 x 11 .m (59.6 x 6.2 nm) orbit. Powered Descent initiation performed at 112:49:52 GET and landing at Taurus-.ittrow occurred at 113:01:53 GET. The landing coordinates were 20°12.6'N and 30°45.0'E.

LUNAR SURFACE

xtra-Vehicular Activity (EVA)-1 commenced at 117:01.36 GET and terminated at 124:13:47 for a total luration of 7 hrs. 12 min. 11 secs. After deploying the Lunar Roving Vehicle (LRV) and prior to traversing to he ALSEP site, the CDR inadvertently knocked the right rear fender extension off of the LRV fender. The ender extension was subsequently secured to the fender with tape. The ALSEP and the Cosmic Ray xperiment were deployed. Steno Crater was used as Station 1A in lieu of the preplanned station (Emory ¦rater), The new station was selected because of the accumulated delay time in the EVA by completion of \LSEP deployment. During the traverse to Station 1A, the fender extension came off and as a result, the crew ιnd LRV experienced a great deal of dust. The Surface Electrical Properties (SEP) Transmitter was deployed .ear the end of the EVA.

ince the crew did not get far enough out to deploy the 1.4 Kilogram (Kg) (3 pound (lb)) Explosive Package EP), only the 0.23 Kg (lb) and 0.45 Kg (1 lb) EP's were deployed on EVA-1, EVA-2 started at 140:34:48 GET, pproximately 1 hour, 20 minutes late and ended at 148:12:10 GET. Total EVA time was 7 hours, 37 minutes, 2 seconds.

'rior to starting the EVA traverse, the crew received instructions from the ground controllers for nprovising a replacement for the lost fender extension. A rig of 4 chronopaque maps, taped together and ιeld in position by two clamps from portable utility lights made an excellent substitute for the extension and he crew did not experience the dust problem as on EVA-1.

¦tations 2 (Nansen), 3 (Lara), 4 (Shorty) , and 5 (Camelot) were visited according to premission plan although ιtation times were modified. An additional brief stop at Station 2A, 0.7 km at 71° from Station 2, was made ι order to obtain an additional traverse Gravimeter reading and additional samples. During the traverse, the .rew deployed the 0.06 Kg (1/8 lb), 2.7 kg (6 lb) and 0.11Kg (¼lb) EPs, obtained photographs, and locumented samples.

\n orange colored material, believed to be of volcanic origin, was found at Station 4. The LMP revisited the \LSEP site at the end of the EVA in order to verify that the Lunar Surface Gravimeter (LSG) was properly leployed and leveled. Total distance covered was approximately 19 km (12 nm).

:VA 3 was initiated at 163:32:35 GET about 50 minutes late, and was terminated 7 hours and 15 minutes 31 econds later at 170:48:06 GET. Exploration of the stations was modified during the traverse. In lieu of raversing Stations 6 (North Massif) through 10B (Sherlock) as planned, the crew was instructed to spend less time at Station 7 (North Massif) due to a longer stay time at Station 6. Station 9 (Van Serg) was explored as planned. Station 10B was deleted and Station 8A (Sculptured Hills) was added. Photographs, and documented samples were obtained at all stations. About 66 Kg (145 pounds) of samples were retrieved, and the LRV traversed a total of 11.6 km (6 nm).

The 1.4 Kg (3lb) EP, left over from EVA-1, was deployed in addition to 0.11 Kg (¼ lb) and 0.06 Kg (1/8 lb) EPs. (See Table)

The total time for the three EVAs was 22 hours, 5 minutes, 4 seconds. The total distance traveled in the lunar rover was about 35 km. The combined weight of samples was about 115 kg (250 lbs), plus double cores and 1 deep drill core. Surface photographs taken during the three EVAs total at least 2120.

Good quality television transmission was received throughout the three EVAs.

Equipment jettison #1 was completed at 171:59 GET

Since the CSM orbit did not decay to the planned orbit for CSM/LM rendezvous, a trim maneuver was initiated at 181:34:01 GET. The 2.8 mps (9.2 fps) burn for 30 seconds produced a 124 x 115 Km (67.3 x 62.. nm)orbit. The Lunar Orbit Plane Change to orient the CSM orbit for rendezvous occurred at 182:33:53 GET. The maneuver produced a change of 3.2° in inclination with a 6.1° shift in the line of nodes.

Equipment jettison #2 was completed at 186:04 GET.

ASCENT, RENDEZVOUS, DOCKING, AND LM IMPACT

LM Ascent Stage lift-off occurred at 188:01:36 GET. A 3 mps (10 fps) tweak burn for 10 seconds at 188:12:1 GET set up the LM orbit for a nominal rendezvous. The Terminal Phase Initiate burn of 3.2 seconds was executed at 188:55:57 GET and resulted in a velocity change of 16.4 mps (53.8 fps) and an orbit of 119 x 9 Km (64.7 x 48.5 nm). CSM/LM docking was completed at 190:17.03 GET. Following equipment and crew transfer the CSM, the LM Ascent Stage Jettison and CSM Separation were completed as planned. Ascent Stage Deorbit was initiated at 195:38:13 GET and lunar surface impact occurred at 195:57:18 GET. The event was observed by the four Apollo 17 Geophones and the Apollo 12, 14, 15, and 16 ALSEPs. Initial data indicate the impact point location to be at 19°54'N and 30°30'E.

The crew did not obtain photographs of Solar Corona at 208:17 GET due to extension of the rest period. The task was not rescheduled. Other photographic tasks were performed as planned.

Explosive Package (EP #6 0.45 Kg (1 lb) was detonated at 212:55:35 GET and EP #7 0.23 Kg (½lb) at 215:25:01 GET. Both events were picked up by the Lunar Seismic (LSP) geophones. The resulting flash and dust from the EP #7 explosion was seen on Television. (See Table 10).

The Ground Controlled Television Assembly (GCTA) and Lunar Communications Relay Unit (LCRU) failed to operate when attempts were made to command the camera on at 221:20 GET. Additional attempts between 237:44 and 237:53 GET to command the systems on were unsuccessful. It was later determined the LCRU experienced an over-temperature failure.

Explosive Package (EP) #4 0.06Kg (1/8 lb) detonated at 232:15:46 GET.

TRANSEARTH INJECTION AND COAST

The Transearth Injection (TEI) maneuver was performed at 236:42:08 GET. The Service Propulsion System (SPS) 144.9-second burn produced a change in velocity of 928 mps (3046.3 fps). EP#1 2.7Kg (6 lb) detonated at 237:49:52 GET and EP#8 0.11 Kg (¼lb) at 240:52:50 GET; see Table. The Lunar Seismic Profiling geophones received strong signals from EP's 4; 1, and 8.

Midcourse Correction (MCC)-5 was not performed since the spacecraft trajectory was near nominal.

The spacecraft left the moon's sphere of influence at 250:39:50 GET, traveling at a velocity of 1173 mps (3851fps), CMP in-flight EVA commenced at 257:34:24 GET. The CMP retrieved the Lunar Sounder film, Panoramic Camera, and Mapping Camera Cassettes in three trips to the SIM Bay. The CDR reported the SIM Bay was in good condition. EVA termination occurred at 258:41:42 GET for a total of 1 hour, 7 minutes, 18 seconds.

Explosive Packages #5 1.4Kg (3 lb) , #2 0.11 Kg (¼lb) and #3 0.06Kg (1/8 lb) were detonated at 260:23:56, 261:52:02, and 264:14:29 GET, respectively. The detonations were received by Lunar Surface Profiling Geophones. The spacecraft trajectory was such that MCC-6 was not performed.

MCC- 7 at 301:18:00 was performed with the RCS firing of 9 seconds. The burn produced a change in velocity of .63 mps (2.1 fps).

ENTRY AND LANDING

The CM separated from the SM at about 304:03:50 GET, 15 minutes before entry interface (EI) at 121,920 m (400,000 feet). Drogue and main parachutes deployed normally; landing occurred in the mid-Pacific Ocean at 304:31:58 GET at approximately 166°07'W longitude and 17°53'S latitude. The CM landed in a stable one position, about 6.4 km (3.5 nm) from the prime recovery ship, USS Ticonderoga, and about 2.4 km (1.3 nm) from the planned landing point.

Weather in the prime recovery area was as follows: Visibility 18 km (10 miles), wind 130° at 5 mps (10 knots), scattered cloud cover 914 m (3,000 feet) and wave height of .6 - .9 m (2 - 3 feet).

ASTRONAUT RECOVERY OPERATIONS

Following CM landing, the recovery helicopter dropped swimmers who installed the flotation collar and attached the life raft. Fresh flight suits were passed through the hatch for the flight crew. The host ventilation fan was turned off, the CM was powered down, the crew egressed, and the CM hatch was secured.

The helicopter recovered the astronauts and transferred them to the recovery ship. After landing on the recovery ship, the astronauts proceeded to the Biomed area for a series of examinations. Following the examinations the astronauts departed the USS Ticonderoga the next day for Samoa, were flown to Norton Air Force Base, California, and then to Ellington Air Force Base, Texas.

COMMAND MODULE RETRIEVAL OPERATIONS

After astronaut pickup by the helicopter, the CM was retrieved and placed on a dolly aboard the recovery ship. Since the CM had propellants onboard, it was stowed near the elevator shaft to insure adequate ventilation. All lunar samples, data, and equipment will be removed from the CM and subsequently returned to the Manned Spacecraft Center via Ellington Air Force Base, Texas. The CM will be offloaded at San Diego, California, where deactivation of the CM propellant system will take place.

SYSTEMS PERFORMANCE

Systems performance on Apollo 17 was very near nominal throughout the entire mission. Only minor discrepancies occurred which had no affect on safety or mission objectives.

FLIGHT CREW PERFORMANCE

The crew's condition was good throughout the mission with the exception of some occasional minor discomfort due to gas.

All information and data in this report are preliminary and subject to revision by the normal Manned Spaceflight Center's technical reports. C. M. Lee

SURFACE SCIENCE

As in previous missions, the first surface science event was the S-IVB impact. S-IVB impact occurred at 89:39:40 GET, about 18 minutes later than the prelaunch prediction. The event was recorded by the Apollo 12, 14, 15, and 16 passive seismometers. The best estimate of the impact point location is 4° 12'S and 12° 18'W. The Apollo 14 Charged Particle Lunar Environment Experiment (CPLEE) and Suprathermal Ion Detector Experiment (SIDE), located approximately 85 nm from the impact point, recorded small events. The Apollo 15 SIDE approximately 549 nm away did not record the impact. As a result of initial analysis by the principal investigator for the passive seismometer some changes may be in order for the interpretation of the moon's internal structure. There is a suggestion that the thickness of the lunar crust should be reduced by about one half of the previously held thickness, to approximately 25 KM. Seismic velocities in the moon's mantle may also be less than previously estimated with the new velocity approximately 7.5 KM/Sec.

The LM touched down in the valley at Taurus-Littrow at 113:01:58 GET. Landing site landmarks were clearly visible during descent and the crew reported that they believed they were abeam of the western-most Trident crater and about 100M north of Poppy. Final estimates of the landing point placed the LM at 20°10'N 30° 46'E. Very little dust was noted during descent. The surface at the landing site was described as undulating and with a much higher abundance of blocks than anticipated. Many large blocks could be seen from the LM windows and it was estimated that LRV traversing would not be difficult.

EVA-1 commenced at 117:01:36. The Cosmic Ray Detector was deployed by the LMP prior to removing the ALSEP from the SEQ Bay. During LRV deployment, the Surface Electrical Properties (SEP) Receiver and the Traverse Gravimeter Experiment (TGE) were mounted on the rear of the LRV. ALSEP removal was nominal but some difficulty was experienced removing the dome from the RTG fuel cask. This was finally accomplished by prying up the dome and then removing the fuel element. After carrying the ALSEP to the general deployment area, LMP found it difficult to locate a large enough area, free of boulders, to carry out the deployment. The final area chosen included some boulders but it is not believed that they will affect any of the experiments. ALSEP deployment was nominal. A deep core (2.8m) was successfully drilled. Retraction from the hole was difficult but eventually completed. The Neutron Probe was inserted into the hole and covered with its thermal blanket.

With various small delays encountered up to the end of ALSEP, deployment and Neutron Probe insertion, a new Station 1A was selected at the crater Steno. LMP returned to the LM and carried the Surface Electrical Properties (SEP) transmitter to its deployment site east of the LM. The crew then rode to Station 1A. Dark mantle material and subfloor samples, described by the crew as a vesicular, gabbroic, basalt, were collected. A traverse gravimeter reading was taken at 1A as well as several other points during EVA-1 for a total of six readings. Closure back at the LM was within 1 milligal. The Explosive Package (EP) #6 was deployed at Station 1A and EP #7 approximately half-way back to the LM.

On returning to the LM the SEP transmitter was deployed but not turned on. Back at the LM the samples collected were stowed. Approximately 15 Kg of samples, including the deep core, were collected. Four hundred and fifteen frames of black and white and color film were taken. Total duration of the EVA was 7h 12m 11s with the EVA terminating at 124:13:47 GET.

EVA-2 commenced at 140:34:48 GET, approximately 1 hour and 30 minutes late. The TGE was loaded aboard and the crew returned to the SEP transmitter where the traverse started. Traverse from the SEP to Station 2 (Nansen) went as planned. Along the way EP #4 was deployed just west of the ALSEP and three LRV samples collected. The SEP receiver was on throughout this leg of the traverse. At Station 2, temperature being high, the receiver was turned off. At Station 2, bluegray breccias and porphyritic gabbros dominated the samples. The TGE reading at this station showed a large negative anomaly, relative to the landing site, of approximately 38 milligals. Shortly after leaving Station 2 an unscheduled stop was made (Station 2A) to

The Prime crew of Apollo 17 (l to r above)
Harrison "Jack" Schmitt - Lunar Module Pilot, Ronald Evans - Command Module Pilot, Eugene Cernan - Commander
Evans, Schmitt and Cernan on the Saturn Launch gantry prior to the flight (below left).
Schmitt and Cernan train with the Lunar Rover at the Cape (below right)

The last of the manned Saturn V rockets leaves the Vehicle Assembly Building at the Cape (left) and standing on the pad the night of December 6th 1972 (above)

The launch of Apollo 17 took place just past midnight on the morning of December 7th 1972. It was the only night launch of a Saturn V. (below)

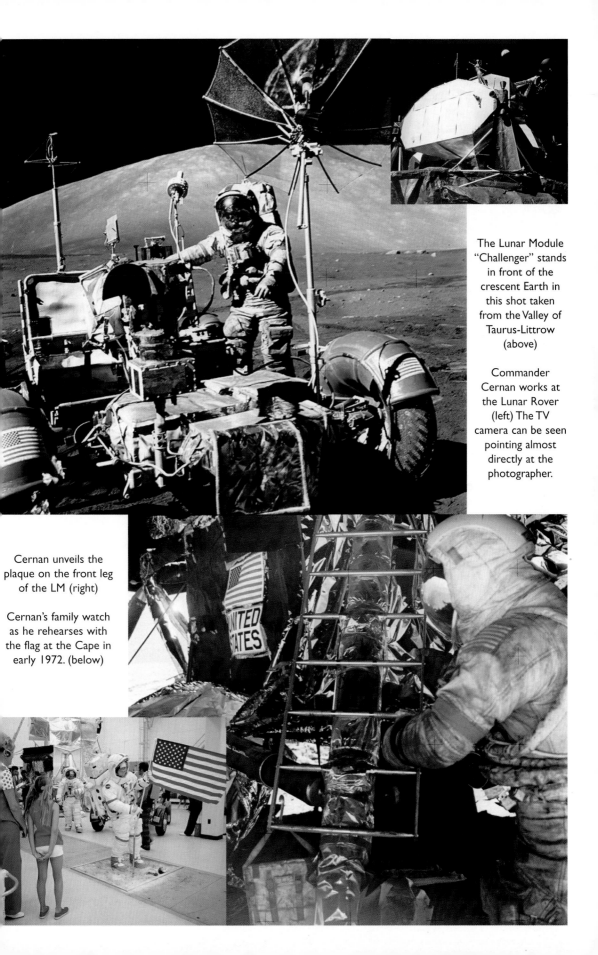

The Lunar Module "Challenger" stands in front of the crescent Earth in this shot taken from the Valley of Taurus-Littrow (above)

Commander Cernan works at the Lunar Rover (left) The TV camera can be seen pointing almost directly at the photographer.

Cernan unveils the plaque on the front leg of the LM (right)

Cernan's family watch as he rehearses with the flag at the Cape in early 1972. (below)

A misplaced hammer ripped the rear fender from the Lunar Rover which was replaced
with a jury-rigged affair of tape and a map (top left).
Schmitt the first geologist on the moon takes a sample (top right).
Orange soil was discovered on the rim of Shorty Crater during the 2nd EVA (below).

Sampling was conducted in the shadows of some of the large boulders found at Taurus-Littrow. The Lunar Rover with the gold-foil-covered TV camera is in the foreground. The crew can be seen sitting in the rover during training (inset).

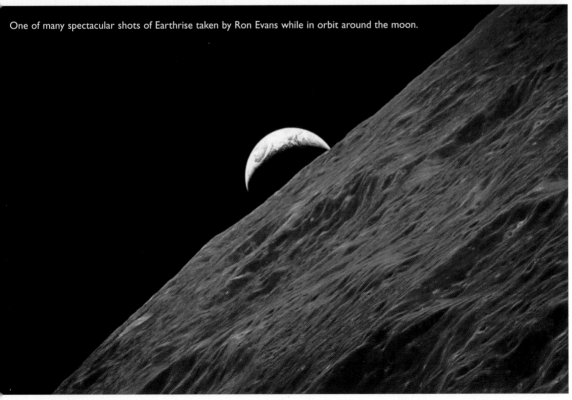

One of many spectacular shots of Earthrise taken by Ron Evans while in orbit around the moon.

Commander Cernan commented during the assembly of the flag that it was one of the proudest moments of his life. Two different angles show some of the massive mountains which ringed the valley of Taurus-Littrow. For three days Cernan and Schmitt drove the rover from one end of the valley to the other.

The Lunar Module "Challenger" sits in front of the so-called "Sculptured Hills" and the "East Massif" at Taurus-Littrow. The hill to the right is almost 20km distant. Cernan stands at the LRV with the "South Massif" behind him (inset).

Jack Schmitt works around "Tracy's Rock" at Station 6. The East Massif looms in the left background.

Command Module Pilot Ron Evans undertook a space-walk during the return to Earth. He is seen retrieving the film cartridges from the Service Module. The cameras had been used to conduct extensive mapping of the Lunar surface from orbit. (above)

An exhilarated Cernan and Schmitt share a happy moment after concluding three days on the moon. (below)

check the gravity gradient between the south massif and the valley. This reading also showed a negative anomaly (-28.8 mgal) and indicated a steep gravity gradient. Station 3, at the edge of Lee Scarp, was the next sampling stop. Here, bluegray breccia, soil and a light colored gabbroic rock were collected. A TGE reading indicated that the local gravity was very close to the value at the LM. While traversing between Station 3 and Station 4, two LRV samples were collected.

At Station 4, Shorty, spectacular orange colored soil samples were collected. The first hypotheses are that these samples represent geothermal alteration associated with vulcanism. Final conclusions will await sample return and analysis. Glasses and basalts were also collected. The TGE reading showed a slight positive gravity anomaly at this station. Further analysis is required before it can be determined if this is significant. The SEP receiver was turned on at Station 4 and was left in the receive mode until the crew returned to the LM.

A short stop was made between Station 4 and Station 5 to deploy EP #1 and take an LRV sample. A second LRV sample was taken further along the traverse. Station 5 (Camelot) fresh, angular boulders were sampled along the rim. Vesicular, gabbroic basalt was the dominant rock type. Soil samples were also collected. Returning to the LM, EP #8 was deployed slightly southwest of the ALSEP.

After what appeared to be a nominal deployment and turn-on, the Lunar Surface Gravimeter (LSG) could not be nulled. Since this problem could be caused by the experiment being off-level, the LMP was asked to verify alignment at the end of the EVA.

The LMP returned to the experiment and verified the experiment was level.

EVA-2 traverse was the longest carried out on any mission, 19.5 KM. Approximately 36.4 Kg of samples were collected during the traverse and 833 photos were taken. The EVA ended at 148:12:10 GET for a total EVA time of 7 hours 37 minutes 2 seconds, the longest EVA ever carried out. The crew commenced EVA-3 at 163:32:35 GET. The Cosmic Ray Detector was retrieved prior to start of the traverse in order to avoid exposure to low energy solar protons from a small solar flare. Once again, the traverse started at the SEP transmitter. The SEP receiver was turned on at the beginning of the traverse. While driving north to Station 6 on the north massif, two LRV sample stops were made. A large boulder was the primary sampling station at this stop. A depressed track, tracing the boulder's path as it rolled down the mountain was evident. Samples included blue-gray breccias, fine grained vesicular basalt, crushed anorthositic rocks and soils. The TGE recorded a large negative anomaly, 26.9 milligal, indicating that like the south massif, the north massif is made up of lower density rocks than the valley. On arrival at this station, the SEP receiver was turned off and not used again.

A short stop was made near Station 7. Blue-gray breccias with white inclusions and what appeared to be vesicular anorthositic gabbro were found. Between Station 7 and Station 8 one stop was made for an LRV sample. At Station 8, Sculptured Hills, glasses, basalts, and soil samples were taken. No breccias were reported although they may be present. TGE reading at Station 8 shows the sculptured hills to have the same approximate density as the north massif.

From Station 8 the traverse turned southwest to Station 10 Van Serg. Samples here were very variable. EP #5 was deployed on the southeast side of the crater. TGE reading at this station still showed a -11 milligal value. (It should be noted that all TGE readings have had a terrain correction made). Before leaving the station, the crew removed the SEP-DSEA tape recorder to prevent it from over-heating,

Between Station 9 and LM, one LRV sample was taken. Back at the LM closeout was accomplished. The LMP returned to the ALSEP site to recover the LNPE, complete photographic documentation and adjust the LSG. LSG operation was still unsatisfactory with the level beam indicating off-scale high.

LMP attempted to shock the instrument so that the beam would be free. After tapping and rocking the experiment, no noticeable change was observed. The final two EP's, EP #3 and EP #2, were deployed near the SEP transmitter. During the traverse approximately 63.5 kg. of samples were collected.

Included in this weight are four core tubes and 1 SESC. A total of 952 black and white and color photos were taken. The EVA terminated at 170:48:06 GET after a total time of 7 hours 15 minutes 31 seconds on the lunar surface.

EXPERIMENT SUMMARY

HFE - Probe temperatures continue to equilibrate. A gradient of approximately 1.2° c/meter is being measured in the lower 1 meter of the holes. All experiment functions appear normal.

LSPE - All eight EP's were detonated and the LM ascent and LM impact recorded. All functions of the experiment are nominal.

LACE - Experiments dust covers have been removed and bake-out is underway. All experiment parameters are nominal and high voltages will be turned on near lunar sunset.

LEAM - Instrument dust covers remain on. Covers over the thermal radiators will be removed within next two days. A "noise" listening mode will commence two days before sunset and continue for four days. Sensors will then be uncovered and scientific data recorded.

LSG - Experiment data still remains invalid. Studies are still underway to discover corrective actions, if possible. If status cannot be changed, the Principle Investigator (P.I.) estimates that 30% of the scientific data could be recovered. This would consist primarily of using the instrument as a seismometer.

TGE - Twenty normal and 2 bias readings were taken. Closure was excellent. A minor unknown is why there is a difference between reading the instrument on the LRV vs. on ground.

SEP - Because of temperature problems, only three traverse legs were recorded. The PI believes valid data will be on the tape from these legs.

CRDE - The CRDE was exposed for a total of 46 hours and 25 minutes, about four hours short of nominal. PI expects data to be good.

LNPE - the LNPE was deployed in the core hole for 45 hours. This was as planned. Because of the extra deployment distance from the RTG vs. what was planned, the PI expects to have excellent data. Deployment in a slight depression was also beneficial in shielding from the RTG.

INFLIGHT SCIENCE

The inflight science phase of the mission was initiated with the turn-on of the Far Ultraviolet Spectrometer (FUVS) at 86:06 GET and was terminated with the end of the FUVS operations at 302:00 GET.

Except for the Panoramic Camera stereo gimbal drive which failed 8 minutes before the end of the last photographic pass no significant instrument problems occurred and all major orbital science objectives were achieved.

Panoramic Camera
Panoramic Camera (PC) operations in lunar orbit were initiated on Rev 1 and were terminated with the last PC pass on Rev 74. During the photographic pass on Rev 15, the PC V/H sensor became erratic and the V/H manual override was used for the duration of the mission. Operations of the PC was then nominal until the last 8 minutes of operation on Rev 74 when the stereo gimbal drive failed. The last 8 minutes of panoramic photography were then acquired in the monoscopic mode.

A total of 1603 PC frames were exposed during the mission, of which 20 were acquired post-TEI. The PC film cassette was successfully retrieved by the CMP at 258:03 GET during the inflight EVA.

Mapping Camera

Mapping Camera photographic operations were initiated on Rev 1 and the lunar orbit portion ended with the last MC pass on Rev 74. An additional 30 minutes of lunar photography was acquired post-TEI. All essential MC photography was acquired. The first MC extension sequence on Rev 1 was nominal, but the second extension on Rev 13 required 3 min 19 sec versus the nominal 72 sec. The camera was left in the extended position until after operation on Rev 38 when retraction took 3 min 51 sec. Consequently, the camera was operated in the retracted position on Rev 49 with the resultant loss of stellar photography.

Fifteen minutes of the north oblique photography scheduled for Rev 65 was not acquired when the camera was not turned on by the CMP after failure to get a barber-pole at camera turn-on. This is a normal occurrence when the CSM is rolled 40°.

The camera was turned-on after the CMP consulted with the MCC and the rest of the Rev 65 oblique photography was acquired.

Throughout the mission, the shutter open pulses were missing at the lower light levels. In addition, the temperature excursions of the front lens elements were up to 7° outside nominal limits. Neither of these anomalies is expected to cause any degradation of the photography.

A total of 3554 frames of MC photography were acquired during the mission. This photography covers 8.5% of the total lunar surface area, bringing to over 17% of the lunar surface photographed by the Mapping Cameras on Apollo. The MC film cassette was successfully retrieved at 258:16 by the CMP during the inflight EVA.

Laser Altimeter

The performance of the Laser Altimeter (LA) was nominal throughout the mission. Operation was according to the flight plan except for:

1) the loss of 4 min of data on Rev 24 when the instrument was inadvertently turned off

2) 38 minutes of darkside altimetry on Rev 62 not acquired when the LA was turned off in order to allow a darkside attitude maneuver for the FUVS

3) an additional 10.3 hrs (Revs 67-72) of LA data added to that already scheduled since the LA was performing above expectations. A total of 3769 shots were fired during the mission.

A preliminary analysis of the data acquired on Revs 27-29 agree well with the Laser Altimeter data acquired on Apollo 15 and indicate that:

1) profile of nearside basins are relatively flat and are depressed with respect to the surrounding terrae

2) the farside appears to be very mountainous in comparison with the nearside (consistent with LA observations from previous missions)

3) the farside depression observed on Apollo 15 at 2605 latitude extends at least to 2005 latitude

4) the bottom of the crater Reiner is 6 km below the top of the adjacent highlands and the bottom of the crater Neper is 7 km below the adjacent highlands.

S-Band Transponder

Preliminary analysis of the doppler tracking data are consistent with the Apollo 15 S-Band Gravity Experiment observations and indicate that:

1) the Taurus-Lattrow landing site is a gravity low (i.e., mass deficiency). The value obtained from S-band observations agrees well with the value obtained by the Traverse Gravity Experiment.

2) Copernicus is a gravity low (all small craters over which doppler tracking data have been obtained are gravity lows)

3) Sinus Aestum, Mare Serenitatis, and Mare Crisium are gravity highs

4) Oceanus Procellarum is a relatively flat gravity region.

Infrared Scanning Radiometer

The Infrared Scanning Radiometer (ISR) operated nominally throughout the mission and attained all flight test objectives. In over 100 hrs of operation in lunar orbit over one-third of the lunar surface was scanned and one hundred million temperature measurements were made to an accuracy of one degree. Temperatures as low as 80°K were observed just prior to lunar sunrise and temperatures as high as 400°K were observed at lunar noon.

Preliminary analysis of the abbreviated real-time data show that several thousand nighttime thermal anomalies (both hot and cold spots) have been detected. Nighttime "hot spots" are generally associated with boulder fields or exposed bedrock near fresh impact features and "cold spots" indicate the existence of areas covered by material with exceptionally low values of density and thermal conductivity. Tentatively identified "thermal anomalies" are:

1) Kepler A, 10 days into the lunar night, is seen as a sharp spike on a broader enhancement corresponsing to its ejecta blanket

2) Kepler C, 11.6 days into the lunar night, is observed as a 132°K anomaly on a 94°K background

3) the crater Reiner is seen as a 140°K anomaly against a background of 98°K

4) a "cold spot", 10°K colder than the surrounding terrain, is coincident with a cinder cone in Mare Orientale near the crater Hohman. The ISR was operated for 4.3 hrs for instrument calibration during TEC prior to the inflight EVA. Scientific data acquisition terminated at 253:20 GET. After the inflight EVA the ISR was used to obtain data on spacecraft contamination produced by RCS thruster firings, waste water dumps, and urine dumps. These data were collected for the Skylab Program.

Lunar Sounder

Operation of the Lunar Sounder Coherent Synthetic Aperture Radar and Optical Recorder were nominal throughout the mission. However, problems were experienced with extending and retracting the high frequency (HF) antennas. These extension/retraction problems were attributed to both a faulty talkback indicator and to a low temperature of the extension/retraction mechanism. No data were lost as a result of the delays in the HF antenna extensions/ retractions.

The Lunar Sounder (LS) data were acquired according to the flight plan except for the postponement of the HF pass scheduled for Rev 55 to Rev 56 due to a low temperature of the Optical Recorder film cassette.

Ten hours 31 sec of LS data were recorded on film including:

1) two consecutive revs in the HF active mode

2) two consecutive revs in the VHF active mode

3) specific targets in both the HF and VHF modes.

Since the prime LS data is recorded on film, successful attainment of the experimental objectives cannot be determined until the film is processed, (The LS film was successfuly retrieved at 257:54 GET by the CMP during the inflight EVA). Telemetry monitoring of average reflected power has yielded some information however. In particular, there is a high correlation between the signature of the power trace and lunar features. Namely:

1) the highlands and mare show distinctively different spectral signatures

2) the highlands show a low frequency structure, the details of which are well correlated with surface topography

3) the mare show a high frequency characteristic consistent with the presence of subsurface structure. The amplitude of this structure is highest in HFI and least in VHF, also consistent with the presence of subsurface structure.

The LS was also operated in the receive only mode simultaneously with transmissions from the Surface Electrical Properties Experiment (SEP), Monitoring of average reflected power by telemetry indicate that the SEP transmitter was observed over a narrower range of angles than had been expected.

During operation of the LS in the receive only mode with the SEP transmitter off, a noise level much higher than anticipated was observed on the lunar frontside. Absence of this signal on the backside, as well as correlation of antenna orientation with noise level, indicates that the noise is of terrestrial origin. An experiment was carried out to determine the polarization state of the noise.

The LS was also operated for a 24 hr. period during TEC to further determine the levels of terrestrial noise.

Far Ultraviolet Spectrometer
The operation of the Far UV Spectrometer (FUVS) was nominal throughout the lunar oribital phase and a total of 114.5 hrs. of data were acquired. All the planned observations were accomplished. In addition, a fourth solar atmospheric observational mode was added during Rev 62. unexpectedly high background count was observed throughout the extent of the UV spectrum measured by the FUVS. This background has been tentatively interpreted as due to cosmic rays. The background noise does not degrade the accuracy with which hydrogen measurement can be made but does obscure very weak signals. For example, the background noise limits the minimum level of detection for atomic oxygen to 100 atoms/cc vs. a pre-flight estimate of 25 atoms/cc.

During FUVS operation on Rev 38, an Aerobee was launched from the White Sands Missile Range to provide a solar UV calibration concurrent with lunar orbital data acquisition. An earlier attempt to obtain a similar solar calibration was unsuccessful when the Aerobee payload viewing port failed to open after a successful rocket launch.

Preliminary analysis of farside terminator data indicate that a lunar atmosphere of atomic hydrogen does exist but that its density is considerably less than had been predicted earlier. In fact, the FUVS data indicate that if the total lunar atmospheric pressure at the surface determined by the Apollo 12 ALSEP/CCIG is correct ($P = 10^{-12}$ torr), then the hydrogen component is less than one percent of this total. Additionally, no atmospheric component detectable by the FUVS (H, O, Kr, Xe, N, C) is present in concentrations as great as one percent of this amount.

The FUVS also observed the lunar surface UV albedo to be approximately 2% with the same angular dependence of reflectivity as the visible. A surface variation in the UV albedo was also observed which may be a measure of the mineralogical variation over the surface.

During the TEC phase of the mission the FUVS was operated for 60 hrs. performing galactic scans and observing a number of galactic UV sources for extended periods. Observations were made to determine the extent of the earth's hydrogen geotail and to determine the extent of the solar atmosphere.

CM Photography
All objectives of CM photography of the lunar surface were successfully accomplished.

To supplement SIM Day photography, ten photographic strips were planned with the Hasselblad camera and color film. Five of the strips are on the nearside and 5 on the farside. No anomalies were noted and the hardware performance was nominal. Some of the film magazines had to be rescheduled because of additional crew option photography.

Under near-terminator lighting conditions, eight targets were planned using the Hasselblad camera and black-and-white film. Two of the targets were on the farside and the remaining six targets on the nearside. All targets were acquired successfully and no anomalies were noted.

In addition to the scheduled photography, the crew took photographs of lunar surface features to document visual observations. Photographs were taken over the Apollo 17 landing site using a polaroid filter, a red filter, and a blue filter.

Dimlight Photography
The dimlight photography scheduled for Apollo 17 included photography of the solar corona and of the zodiacal light. The sunrise solar corona photography, scheduled for Rev 25, consisted of seven data frames from ten seconds to one-sixtieth second duration. The corona extending eastward beyond the lunar limb was photographed as the sun moved from three-and-one-half degrees below to one-half degree below the limb. The CMP reported that this sequence was accomplished according to plan.

The sunset solar corona was not photographed because of lengthened sleep period and non-availability of the -X attitude required for its performance. The other dim light phenomenon photographed on this mission was the zodiacal light from fifty degrees eastward of the sun down to the solar corona region. This photography was carried out successfully three separate times; first, in red light on Rev 23, again in blue light on Rev 38, and finally in plane-polarized white light on Rev 49. The CMP noted that the second photography in red light, a planned sixty-second exposure, was underexposed because of inadvertent, early shutter closure.

Visual Observation From Orbit
All the objectives of visual observations from lunar orbit were successfully accomplished. Ten targets were planned for visual study and excellent comments were made by the crew. These comments will help solve geologic problems that are hard to solve by other means. Among the salient findings are the following:

1. Finding that only relatively young craters on the farside are filled with mare material. Domes in the floor of the crater Aitken are probably extrusive calcite domes.

2. Spotting of orange-colored ejecta blankets of craters in Mare Crisium, in the landing site area and on Western Mare Serenitatis.

3. Characterization of the actual colors of lunar surface units especially in the lunar maria. This will help in the extrapolation of ground truth and remotely-sensed data.

4. Verification of the extensive nature of the rings of the basin Arabia. The swirls in and west of the basin have no topographic expression associated with them.

5. Discovery of several volcanic craters under the groundtracks that were not characterized previously.

All onboard items carried in support of this task were found to be adequate, including the 10X binoculars. Only the color wheel was not used because its colors did not correspond to the actual lunar colors.

TABLE I
APOLLO 17 OBJECTIVES AND EXPERIMENTS PRIMARY OBJECTIVES

The following were the NASA OMSF Apollo 17 Primary objectives

Perform selenological inspection, survey, and sampling of materials and surface features in a preselected area of the Taurus-Littrow region.

Emplace and activate surface experiments,

Conduct in-flight experiments and photographic tasks.

APPROVED EXPERIMENTS

The following experiments were performed;

        Apollo Lunar Surface Experiments Package (ALSEP)
        S-037 Lunar Heat Flow
        S-202 Lunar Ejecta and Meteorites
        S-203 Lunar Seismic Profiling
        S-205 Lunar Atmospheric Composition
        S-207 Lunar Surface Gravimeter Lunar Surface
        S-059 Lunar Geology Investigation
        S-153 Cosmic Ray Detector
        S-199 Traverse Gravimeter
        S-204 Surface Electrical Properties
        S-229 Lunar Neutron Probe
        Long Term Lunar Surface Exposure Tasks (not classified as experiment)

In-Flight
        S-164 S-Band Transponder
        S-169 Far UV Spectrometer
        S-171 IR Scanning Radiometer
        S-209 Lunar Sounder
        — CM Photographic Tasks
        — SM Orbital Photographic Tasks
        —- Skylab Contamination Study
        — Visual Light Flash Phenomenon
        In-Flight (Continued) Other (Passive)
        S-160 Gamma-Ray Spectrometer
        S-176 Apollo Window Meteoroid
        S-200 Soil Mechanics
        M-211 Biostack IIA
        M-212 Biocore Demonstration
        — Heat Flow and Convection

DETAILED OBJECTIVES

The following detailed objectives were assigned to and accomplished on the Apollo 17 Mission:

    CM Photographic Tasks
    SM Photographic Tasks
    Visual Observations from Lunar orbit
    Spacecraft Contamination Study

Visual Light Flash Phenomenon
Impact S-IVB on Lunar Surface
Post Determination of S-IVB Impact Point
Protective Pressure Garment Evaluation
Body Metabolic Gains and Losses and Food Compatibility Assessment.

SUMMARY

Fulfillment of the primary objectives qualifies Apollo 17 as a successful mission. The experiments and detailed objectives which supported and expanded the scientific and technological return of this mission were successfully accomplished.

TABLE 2
APOLLO 17 ACHIEVEMENTS

Sixth Manned Lunar Landing
First Geologist Astronaut on Lunar Surface
Longest Lunar Surface Stay Time (74 hours 59 min. 38 seconds)
Longest single Lunar Surface EVA Time (7 hours 37 minutes 22 seconds)
Longest Total Lunar Surface EVA Time (22:05:04)
Longest Lunar Distance Traversed with LRV on One EVA (19 km (12 nm))
Longest Total Distance Traversed with LRV (35 km (22 nm))
Longest Apollo Mission (301 hours 51 minutes)
Most samples returned to Earth (115 Kg (250 lbs))
Longest time in lunar orbit (147 hours, 48 minutes)

APOLLO 17

M-933-72-17

POWERED FLIGHT SEQUENCE OF EVENTS
END OF MISSION

TABLE 3

EVENT	PRE-LAUNCH PLANNED (GET) HR;MIN;SEC	ACTUAL (GET) HR;MIN;SEC
Guidance Reference Release	-17.7	-17.6
Liftoff Signal (TB-1)	0	0
Pitch and Roll Start	11.9	11.9
Roll Complete	13.4	13.4
S-IC Center Engine Cutoff (TB-2)	2:18.8	2:18.8
Begin Tilt Arrest	2:39.6	2:39.6
S-IC Outboard Engine Cutoff (TB-3)	2:41.0	2:40.6
S-IC/S-11 Separation	2:42.7	2:42.3
S-II Ignition (Command)	2:43.4	2:43.0
S-II Second Plane Separation	3:12.7	3:12.3
S-II Center Engine Cutoff	7:41.0	7:40.6
S-II Outboard Engine Cutoff (TB-4)	9:19.5	9:19.0
S-II/S-IVB Separation COMMAND	9:20.5	9:20.0
S-IVB Ignition	9:20.6	9:21.1
S-IVB Cutoff (TB-5)	11:46.4	11:42.2
Insertion	11:56.2	11:52.0
Begin Restart Preps (TB-6)	3:11:41.3	3:02:58.1
Second S-IVB Ignition	3:21:19.3	3:12:36.0
Second S-IVB Cutoff (TB-7)	3:27:04.1	3:18:27.2
Translunar Injection	3:27:14.1	3:18:37.0

Prelaunch planned times are based on MSFC Launch Vehicle Operational Trajectory.

TABLE 4   M-933-72-17

APOLLO 17 MISSION SEQUENCE OF EVENTS
END OF MISSION

EVENT	PLANNED (GET) HR: MIN: SEC.	ACTUAL (GET) HR: MIN: SEC.
Liftoff 0033 EST December 7, 1972	00:00:00.4	00:00:00.2
Earth Parking Orbit Insertion	00:11:56	00:11:52
Second S-IVB Ignition	03:21:19	03:12:36
Translunar Injection	03:27:14	03:18:37
CSM/S-IVB Separation, SLA Panel Jettison	04:12:05	03:42:29
CSM/LM Docking	04:22:05	03:57:10
Spacecraft Ejection From S-IVB	05:07:05	04:45:20
S-IVB APS Evasive Maneuver	05:30:05	05:03:33
Midcourse Correction-1	08:45:00	Not Performed
Midcourse Correction-2	35:30:00	35:29:59
Midcourse Correction-3	66:55:38	Not Performed
Midcourse Correction-4	83:55:38	Not Performed
SIM Door Jettison	84:25:38	84:12:50
Lunar Orbit Insertion (Ignition)	88:55:38	88:54:22
S-IVB Impact	89:21:26	89:39:40
Descent Orbit Insertion 1 (Ignition)	93:13:09	93:11:37
CSM/LM Undocking	110:27:55	110:27:55
CSM Separation	110:27:55	110:27:55
CSM Circularization	111:55:23	111:57:28
LM Descent Orbit Insertion 2 (Ignition)	112:00:34	112:02:41
Powered Descent Initiate	112:49:38	112:49:52
LM Lunar Landing	113:01:38	113:01:58
Begin EVA-1 Cabin Depress	116:40:00	117:01:36
Terminate EVA-1 Cabin Repress	123:40:00	124:13:47
Begin EVA-2 Cabin Depress	139:10:00	140:34:48
Terminate EVA-2 Repress	146:10:00	148:12:10
Begin EVA-3 Cabin Depress	162:40:00	163:32:35
Terminate EVA-3 Cabin Repress	169:40:00	170:48:06
Trim Burn (CSM)	Not Planned	181:34:01
CSM LOPC	182:35:45	182:33:53
LM Liftoff	188:03:15	188:01:36
LM Tweak Burn	Not Planned	188:12:12
Terminal Phase Initiate Maneuver	188:57:32	188:55:57
LM/CSM Docking	190:00:00	190:17:03
LM Jettison	193:58:30	193:58:30
CSM Separation	194:03:30	194:03:30
Ascent Stage Deorbit	195:39:12	195:38:13
Ascent Stage Lunar Impact	195:58:25	195:57:18
Transearth Injection	236:39:51	236:42:08
Midcourse Correction-5	253:42:13	Not Performed
CMP EVA Depress	257:22:00	257:34:24
CMP EVA Repress	258:30:00	258:41:42
Midcourse Correction-6	282:13:01	Not Performed
Midcourse Correction-7	301:18:01	301:18:00
CM/SM Separation	304:03:01	304:03:50
Entry Interface (400,000 ft)	304:18:01	304:18:37
Landing	304:31:11	304:31:58

APOLLO 17 TRANSLUNAR AND MANEUVER SUMMARY

TABLE 5

END OF MISSION

MANEUVER	GROUND ELAPSED TIME (GET) AT IGNITION (HR:MIN:SEC.)			BURN TIME (SECONDS)			VELOCITY CHANGE (FEET PER SECOND - FPS)			GET OF CLOSEST APPROACH - HT (NM) CLOSEST APPROACH		
	PRE-LAUNCH PLAN	REAL-TIME PLAN	ACTUAL	PRE-LAUNCH PLAN	REAL-TIME PLAN	ACTUAL	PRE-LAUNCH PLAN	REAL-TIME PLAN	ACTUAL	PRE-LAUNCH PLAN	REAL-TIME PLAN	ACTUAL
TLI* (S-IVB)	03:21:10	03:12:35	03:18:37	346	351			10376		88:55:38 / 51	86:12:14 / 161	86:20:52
CSM SEP	04:12:05	03:43:00	03:42:29									
CSM DOCK	04:22:05		03:57:10									
LM EJT	05:07:05	04:39:00	04:45:20		3	5.5		0.3	0.6	88:55:38 / 51	86:12:18 / 150	86:19:54 / 0
S-IVB EVASIVE	05:30:05	05:03:32	05:03:33		13.9	15.1	9.8	9.5	9.4	89:21:26 / 0	86:12:00 / 0	86:12:0] / 0
MCC-1 (SPS)	N.A.			NA			NA			NA / NA		
MCC-2 (SPS)	35:30:00	35:29:59	35:29:59	0	1.6	17	0	10.6	9.9	NA / NA	86:18:09 / 53.1	86:18:14 / 52.1
MCC-3	66:55:38		NP	0			0			NA / NA		
MCC-4	83:55:38		NP	0			0			NA / NA		
SIM DOOR JETT	84:25:38		84:12:50	NA			NA			NA / NA		

NA - Not Applicable
NP - Not Performed

*S-IVB Restart

APOLLO 17 LUNAR ORBIT SUMMARY

TABLE 6   END OF MISSION

MANEUVER	GROUND ELAPSED TIME (GET) AT IGNITION (HR:MIN:SEC:)			BURN TIME (SECONDS)			VELOCITY CHANGE (FEET PER SECOND - FPS)			RESULTING APOLUNE/PERILUNE (N. MI.)		
	PRE-LAUNCH PLAN	REAL-TIME PLAN	ACTUAL	PRE-LAUNCH PLAN	REAL-TIME PLAN	ACTUAL	PRE-LAUNCH PLAN	REAL-TIME PLAN	ACTUAL	PRE-LAUNCH PLAN	REAL-TIME PLAN	ACTUAL
LOI	88:55:38	88:54:22	88:54:22	395.4	398	398	2980	2988	2988	171 / 51	170 / 52.5	170 / 52.6
S-IVB IMPACT	89:21:26		89:39:40	NA			NA			NA		
DOI-1	93:13:09	93:11:37	93:11:37	22.9	22.1	22.1	199	197	197	59 / 15	59 / 14.5	59 / 14.5
UNDOCKING	110:27:55	110:27:55	110:27:55	NA			NA					
CSM SEP	110:27:55	110:27:55	110:27:55	3.3	3.3	3.4	1.0	1.0	1.0	60 / 14	61.6 / 12	61.5 / 11.5
CSM CIRC	111:55:23	111:57:28	111:57:28	4	3.7	3.7	70	70.5	70.5	70 / 54	70 / 54	70 / 54
DOI-2	112:00:34	112:02:41	112:02:41	26.9	21.5	21.5	9	7.5	7.5	60 / 7.2	61.5 / 6.7	59.6 / 6.2
PDI	112:49:38	112.49.52	112.49.52	720	717	717	6702	6697	6698	0 / 0	0 / 0	0 / 0
LANDING	113:01:38	113.01.48	113.01.58	NA			NA			NA		
TRIM	NONE	181:34:01	181:34:01	NONE	30.	30	NONE	9.2	9.2	NONE	67.3 / 62.5	67.3 / 62.5

NA - Not Applicable

APOLLO 17 LUNAR ORBIT SUMMARY

TABLE 7   END OF MISSION

MANEUVER	GROUND ELAPSED TIME (GET) AT IGNITION (HR:MIN:SEC:)			BURN TIME (SECONDS)			VELOCITY CHANGE (FEET PER SECOND - FPS)			RESULTING APOLUNE/PERILUNE (N. MI.)		
	PRE-LAUNCH PLAN	REAL-TIME PLAN	ACTUAL	PRE-LAUNCH PLAN	REAL-TIME PLAN	ACTUAL	PRE-LAUNCH PLAN	REAL-TIME PLAN	ACTUAL	PRE-LAUNCH PLAN	REAL-TIME PLAN	ACTUAL
CSM LOPC	182:35:45	182:33:53	182:33:53	18.7	20.1	20.1	336.7	366.	366.	63 / 61.3	62.8 / 62.5	62.8 / 62.5
ASCENT	188:03:15	188:01:36	188:01:36	437.7	440.9	440.9	6062.2	6075.7	6075.7	47.9 / 9.1	48.5 / 9.1	48.5 / 9.1
TWEAK	NONE	188:12:12	188:12:12	---	10	10	---	10	10		/ 9.3	48.5 / 9.4
TPI *	188:57:32	188:55:57	188:55:57	2.7	3.2	3.2	54.8	53.8	53.8	64.4 / 46.7	64.7 / 48.5	64.7 / 48.5
DOCKING	190:00:00	190:00:00	190:17:03	N/A			N/A			N/A		
LM JETT	193:58:30	193:58:30	193:58:30	N/A			N/A	2.5	1.7	N/A	62.7 / 60.7	63.4 / 61.7
CSM SEP	194:03:30	194:03:30	194:03:30	12.6	12.	12.	2	2	2	64 / 62	62.4 / 61.1	63.9 / 61.2
ASC DEORB	195:39:12	195:38:13	195:38:13	116.4	118	118	281.8	286	286	0		
ASC IMPACT	195:58:25	195:57:25	195:57:18	N/A			N/A			N/A		

*APS only, does not include the nominal 10-sec RCS ullage (21.8 fps).

APOLLO 17 TRANSEARTH MANEUVER SUMMARY

TABLE 8

END OF MISSION

MANEUVERS	GROUND ELAPSED TIME (GET) AT IGNITION (HR:MIN:SEC:)			BURN TIME (SECONDS)			VELOCITY CHANGE (FEET PER SECOND - FPS)			GET ENTRY INTERFACE (E.I.) VELOCITY (FPS) AT EI FLIGHT PATH ANGLE AT EI		
	PRE-LAUNCH PLAN	REAL-TIME PLAN	ACTUAL	PRE-LAUNCH PLAN	REAL-TIME PLAN	ACTUAL	PRE-LAUNCH PLAN	REAL-TIME PLAN	ACTUAL	PRE-LAUNCH PLAN	REAL-TIME PLAN	ACTUAL
TEI (SPS)	236:39:51	236:42:08	236:42:08	142.2	144.8	144.9	3045.7	3046.1	3046.3	304:18:01 36090 -6.5	304:18:04 36090 -6.5	304:17:48 36090 -7.26
MCC-5	253:42:13		NP	0			0			304:18:01 36090 -6.5		
MCC-6	282:18:01		NP	0			0			304:18:01 36090 -6.5		
MCC-7	301:18:01	301:18:00	301:18:00	0	9	9	0	2.1	2.1	304:18:01 36090 -6.5	304:18:37 36090 -6.5	304:18:37 36090 -6.5
CM/SM SEP	304:03:01		304:03:50	NA			NA					
ENTRY	304:18:01	304:18:37	304:18:37	NA			NA			304:18:01 36090 -6.5	304:18:37 36090 -6.49	304:13:37 36090 -6.49
SPLASH	304:31:11	304:31:47	304:31:58	NA			NA					

NP - Not Performed

APOLLO 17 CONSUMABLES SUMMARY

TABLE 9

END OF MISSION

CONSUMABLE		LAUNCH LOAD	FLIGHT PLANNED REMAINING	ACTUAL REMAINING
CM RCS PROP (POUNDS)	T	233.2	178.5	No Data
SM RCS PROP (POUNDS)	T	1341.7	680.5	717.0
SPS PROP (POUNDS)	TK	40539.8	1731.8	1630.0
SM HYDROGEN (POUNDS)	U	78.9	18.3	18.0
SM OXYGEN (POUNDS)	U	955.8	377.7	349.0
LM RCS PROP (POUNDS)	T	631.2	158.0	127.0
LM DPS PROP (POUNDS)	TK	19487.9	708.1	1237.5
LM APS PROP (POUNDS)	TK	5247.5	246.6	228.2
LM A/S OXYGEN (POUNDS)	T	4.6	3.6	4.6
LM D/S OXYGEN (POUNDS)	T	93.4	45.2	48.5
LM A/S WATER (POUNDS)	T	83.3	53.0	43.2
LM D/S WATER (POUNDS)	T	407.5	67.0	32.4
LM A.S BATTERIES (AMP-HOURS)	T	592.0	255.0*	260.0*
LM D/S BATTERIES (AMP HOURS)	T	2075.0	380.0**	490.0**
LRV BATTERIES (AMP-HOURS)	T	121/121	71/71***	80/78***

TK TANK QUANTITY	* After LM Jettison
T TOTAL QUANTITY	** After LM Liftoff
U USABLE QUANTITY	*** After EVA-3

LSP EXPLOSIVE PKG EVENTS

TABLE 10

E/P No.	Chg Wt	TB Time *	Deploy Time	Distance to ALSEP Geophones	Day	DETONATION TIME (CST)				
						Minimum	Maximum	Probable	Nominal	Actual
6	1 lb	90:45	Mon (12/11) 22:69 (122:05:48)	1.10 km	Fri (12/15)	17:17 (212:23:48)	18:11 (213:17:48)	17:49 (212:55:48)	17:44 (212:50:48)	17:49 (212:55:35)
7	½ lb	92:45	Mon (12/11) 23:36 (122:42:40)	0.65 km	Fri (12/15)	19:54 (215:00:40)	20:48 (215:54:40)	20:27 (215:33:40)	20:21 (215:27:40)	20:19 (215:25:01)
4	⅛ lb	90:45	Tue (12/12) 18:29 (141:36:04)	0.20 km	Sat (12/16)	12:47 (231:54:04)	13:41 (232:48:04)	13:01 (232:10:04)	13:14 (232:21:04)	13:06 (232:15:45)
1	6 lb	91:45	Tue (12/12) 23:00 (146:07:00)	2.95 km	Sat (12/16)	18:18 (237:25:00)	19:12 (238:19:00)	18:49 (237:56:00)	18:45 (237:52:01)	18:42 (237:49:52)
8	¼ lb	93:45	Tue (12/12) 23:57 (147:03:39)	0.15 km	Sat (12/16)	21:15 (240:21:39)	22:09 (241:15:39)	21:45 (240:51:39)	21:42 (240:48:39)	21:46 (240:52:50)
5	3 lb	91:45	Wed (12/13) 21:32 (168:39:28)	2.20km	Sun (12/17)	16:50 (259:57:28)	17:44 (250:51:28)	17:20 (260:27:28)	17:17 (260:24:28)	17:17 (260:23:56)
2	¼ lb	92:45	Wed (12/13) 22:05 (169:12:12)	0.40 km	Sun (12/17)	18:23 (261:30:12)	19:17 (252:24:12)	18:45 (261:52:12)	18:50 (261:57:12)	18:45 (261:52:02)
3	⅓ lb	93:45	Wed (12/13) 23:18 (170:24:46)	0.35 km	Sun (12/17)	20:36 (263:42:46)	21:30 (264:36:46)	21:14 (264:20:46)	21:03 (264:09:46)	21:08 (264:14:29)

LM ASCENT   Thursday   12/14/72   188:03 GET   (4:56 PM CST)
ASCENT STAGE IMPACT   Friday   12/15/72   195:58 GET   (12:51 AM CST)

* TB Time - Nominal Time to EP Detonation after Timer Activation.

CSM LOCATION AT DETONATION
#6 - Backside of moon
#7 - Overhead +1 minute
#4 - Backside of moon
#1 - 1 hour after TEI
#8,5,2,3 - After TEI

TABLE 11
SA-512 LAUNCH VEHICLE DISCREPANCY SUMMARY

Failure of Vehicle Automatic Test Sequencer (VATS) to initiate S-IVB prelaunch LOX pressurization.	Open
S-IVB forward #2 battery voltage drop.	Open
S-IC remote digital sub-multiplexer replacement.	Open
S-II helium injection bottle pressure decay.	Open
S-IVB helium bottle pressure deviation.	Open

TABLE 12
COMMAND/SERVICE MODULE 114 DISCREPANCY SUMMARY

Spurious master alarms.	Open
Mission timer behind other timers,	Open
Spacecraft fragments observed before and after CSM/SLA separation.	Open
H2 tank upper pressure limit shift.	Open
Fuel cell current oscillations.	Open
Inoperative tone booster.	Open
Fuel interface pressure measurement fluctuations.	Open
Glycol temperature control valve failure to maintain glycol temperature.	Open
High suit pressure; 5.26 psi.	Open
Excessive mapping camera extend time; 3 min, 21 sec.	Open
HF lunar sounder boom 1 limit switch failure.	Open
Velocity/altitude sensor improper data to pan camera.	Open
HF antenna 2 boom deployment slow at 194:18 GET.	Open
Instrumentation dropout of several parameters for two minutes at 1944:22.	Open
No CMP EVA warning tone at 256:22 to 257:22.	Open
Primary Radiation flow control switched to secondary at 277:08 GET.	Open

TABLE 13
LUNAR MODULE 12 DISCREPANCY SUMMARY

Gimbal drive actuator drive timer in error at 109:04.	Open
Three guidance and navigation restarts prior to powered descent initiations.	Closed
Battery 4 read 0.5 volt lower than battery 3.	Open
Cabin pressure increase above regulator A lockup pressure at 163:31 GET.	Open

TABLE 14
LUNAR ROVING VEHICLE 3 DISCREPANCY SUMMARY

Left rear fender extension lost on EVA 1.	Open

TABLE 15
APOLLO 17 CREW/EQUIPMENT DISCREPANCY SUMMARY

UV spectrometer photomultiplier tube dark current excessive.	Open
Mapping camera film motion exposure discrete drop outs,	Open
Excessive surface electrical properties receiver temperature at start of EVA-2.	Open
Lunar surface gravimeter null failures.	Open
Out-of-spec OPS regulated pressures at 171:70 GET.	Open
ALSEP 5 signal fluctuations in downlink.	Open
No CMP EVA warning tone at 256:22 to 257:22.	Open
Panoramic camera motion compensation ceased operation at 233:52 GET.	Open

MSC-07631

NATIONAL AERONAUTICS AND SPACE ADMINISTRATION

APOLLO 17
TECHNICAL
CREW DEBRIEFING
(U)

JANUARY 4, 1973

PREPARED BY
TRAINING OFFICE
CREW TRAINING AND SIMULATION DIVISION

This document will automatically become declassified
90 days from the published date.

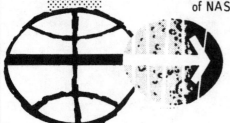

MANNED SPACECRAFT CENTER
HOUSTON, TEXAS

1.0 SUITING AND INGRESS

CERNAN — Except for one small item, the entire suiting and ingress and all equipment supporting it was nominal. There were no complications or problems. Suit circuit checks and cabin closeout were rapid and complete and, to the best of our knowledge from inside the cabin, went very well.

EVANS — The anomaly Gene mentioned was my insuit drink bag. Unfortunately, I didn't try it out prior to putting on the helmet. I wish I would have now, because the waterbag itself had gotten twisted sideways underneath the neckring instead of hanging straight down from the neckring. The tube was crimped and I was unable to get any water whatsoever out of it.

72-H-1503

2.0 STATUS CHECKS AND COUNTDOWN

CERNAN — Ground communications with the spacecraft and all the launch preps for a nominal on-time launch went well. There were no spacecraft anomalies or problems during the launch prep. All systems checked out well. Controls and displays went well through T minus 30 seconds, when there was an automatic sequencer hold due to a potential problem that the ground support equipment saw on S-IVB pressurization. However, to the best of my knowledge, the S-IVB was GO on the cockpit displays. The S-IVB pressures were nominal, but, nevertheless, we had an automatic hold in the sequencer at T minus 30 seconds. From then on, for 2 hours and 40 minutes, we had a series of 20-minute recycles. I don't know exactly how many now. Did we ever get down to 8 minutes one time in the count?

72-H-1502

EVANS — No. Once we got started below 20 minutes, we went all the way.

CERNAN — The problem turned out to be apparently in the software of the ground support equipment. The workaround was caught up, checked out through the Cape and Marshall, and once the count picked up, we had two azimuth updates.

72H1501

EVANS — We had two azimuth updates, because the first recycle was more than 20 minutes, wasn't it? It was more than 20 minutes and we recycled to that point and then they found out that they weren't going to be able to pick it up again in 20 minutes. And we stopped at 20 minutes and made the second azimuths.

CERNAN — The point here being, both azimuth updates in the spacecraft went well. The CMP put them in the computer. The computer took it. I watched the IMU torque. After each one of those, they had to reset the GDC, which worked fine. So we launched with a good GDC following the platform. The only difference was a small roll angle, and it was reversed, because we had gone through 90° on the azimuth change. But that didn't really bother anything because the roll came in on time in a reverse direction. It was a small roll that culminated in just a few seconds. The count and lift-off, through the yaw and the roll program, were nominal once we got through T-0. Distinction of sounds in launch vehicle sequence countdown to lift-off - I think the only thing that really comes across in there is that at some point you get a good vibration. At some point in the countdown, you get a good vibration as you're sitting up there. It's not part of the CSM's operation, so you're not sure what's going on. And this happened in the CDDT and, of course, all we did was check and find out we were doing something with the booster.

EVANS — When they ran through some gimbaling programs.

72-H-253

CERNAN — The major portion of the launch count has to do with checking out the systems, so the commander stays very busy and many times on separate loops. The entire EDS system checked out very well. We only checked it out once in the initial count and during most of the recycle we stayed in EDS AUTO and then we de-armed EDS AUTO but still maintained a manual EDS capability to abort during that recycle time. We picked EDS AUTO as part of the T minus 20 recycle for final lift-off.

3.0 POWERED FLIGHT

CERNAN — The S-IC ignition - The lights started going out at 7 seconds, and somewhere around 3 seconds they were completely out. You could feel the ignition. You could feel the engines come up to speed. Just prior to lift-off and during the first few seconds of lift-off when we were near the pad, both the CMP and I could see the reflection of the engine ignition out the left-hand window and the hatch window in the BPC. We could not see the fire but could see a red glow through the windows reflecting apparently off the surface. Ignition was like a big old freight train sort of starting to rumble and shake and rattle and as she lifted off. We got a good tower clear. As you go through max-q, as in the past, it gets very rough and much noisier, but I don't think we ever had any trouble hearing each other in the spacecraft. I had my intercom very high and all my S -bands and tweaked everything up prior to lift-off. We went through max-q and the only unusual thing going through max-q, considering wind components that we had was that I saw 25 percent on the ALPHA going through max-q. The

72-H-298

72-H-305

yaw needle was right on, but the pitch needle had dropped to a degree and a half at the most. I guess I didn't really expect it because of the predicted wind components. After we got through max-q, you could still certainly tell the bird was burning as we pressed on toward staging, but it got much quieter and it was very evident that you were through max-q when that time came. We had center engine shutdown on time. We had staging on time. I don't think it's ever been recorded on a daylight launch before, but as soon as the S-IC shut down during the time involved in recycling and getting the staging sequence going and the S-II lit off, apparently the trailing flame of the S-IC overtook the spacecraft when we immediately went into that zero-g condition. And, for just a second, as the S-II lit off, we went through the flame. It was very obvious. We could see it out of both windows. I particularly could see it out of the left-hand rendezvous window of the BPC. It was not a smoke; it was not an orange fireball; it was just a bright yellow fire of the trailing flame of the S-IC; and it happened for just a split second. Then we got off on the S-II and things got very quiet and very smooth and was a very long, quiet, smooth ride.

72-H-1514

EVANS — I really wasn't watching the lights because I guess I didn't expect the thing to shake quite as much as it did. To me, I felt like I was really vibrating. I wanted to find out what was making me vibrate. I wasn't expecting that much vibration when the S-IC lit off. At lift-off, again, once it got vibrating, I didn't feel the yaw. I was watching the needle on the thing but didn't feel the yaw, though. The shaking increased a little bit up to max-q and then there was a different type of shaking. It was more of a vibration, I think, going through max-q. And there was more noise associated with going through max-q. Of course, with the shutdown of the S-IC, I think that was about 4-1/2g.

CERNAN — We pushed 4g.

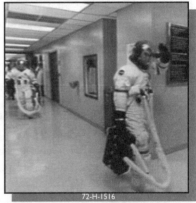

72-H-1516

SCHMITT — Just pushing 4g on the thing and it quits just like that. I was prepared for it because Gene had said, "Hey, brace yourselves because it is going to happen," and it happened all right. It just flat quit when we went from 4g to 0.

CERNAN — The great train wreck.

SCHMITT — I think in all those booster cutoffs, it's hard to see how rapidly the g-level decreases. I guess the only other comment I have is that I think that it is good to do a lot of simulation about malfunctions during launch, but up through max-q it is a little bit unrealistic to think that you are going to analyze a malfunction in the spacecraft.

72-H-1466

CERNAN — To sum up the S-IC, I personally didn't think it was any different than my previous ride on the S-IC and up through this point being a night launch really didn't make any difference at all. The only thing I did different that I hadn't really thought a lot about until I sat on the pad and began to think about staging was, just prior to staging, I took my hand off the abort handle and held the support arm rather than the translation control handle until after staging. I did this just a couple of seconds prior to staging. I had talked about it with John Young a little bit prior to the flight and it turns out that's what he did, also. Probably a good thing. The S-II ignition was very smooth. We got skirt sep right on time. I could feel skirt sep going. We had tower jett, which was really sort of spectacular at night. I think the LMP is going to add something to it, but from the left-hand rendezvous window, I could not only see the flame, but the inside of the BPC seemed to be lit up. Of course, it doesn't stay there very long; it's gone in just a split second. But it was a very spectacular sight at night to see that tower go against the blackness of space out there. We could see guidance come in very definitely. It was not as big a pulsation as I've seen on the simulator but I did see the needle and the spacecraft did change its attitude slightly. You could see the mixture ratio shift. It was just a long, smooth, quiet ride. Inboard cutoff was right on time. You could feel it, a definite physiological feeling. Of course, the g-meter saw it also. The S-II cutoff, as Jack said, is again very sharp, almost instantaneous, from almost 4g to 0. But on the S-II, although it's sharp and a very hard hit, you don't unload the entire stack like you do when you're on the S-IC. The staging was very smooth. It did not seem like an exceptionally long time before we separated and the S-IVB lit off.

S7248730

SCHMITT — On the tower Jett, I wouldn't say a split second. As a matter of fact, I was surprised it lasted as long as it did. It was a few seconds.

EVANS — I couldn't see the rocket go. All I could see was an orange glow out the center window.

CERNAN — While we were on the S-II, we would see no indication of light from the engines. We were just thrusting out in the darkness of space. I tried to see stars for potential mode IV and, of course, at that time, mode II abort and turned the lights down on the left side once or twice. But even with the lights down (we had the LEB lights relatively low), in my estimation, it would have required all the lights in the spacecraft to have been off and certainly more than a few seconds to become night adapted to be able to see through the windows and pick up stars that would have been able to help in an abort situation had you lost the computer and the SCS. We had looked, potentially planned to use those stars in an abort condition if we had to. We had excellent constellations to look at. They obviously were there, but I could not see through the low glow reflection on the window even with

S7254813

S7255482

our lights, floodlights, turned almost all the way down. I even went to the extent of trying to shield my eyes on the S-II and looked out the window and I still could not pick up anything that I could have recognized for an abort. I also could not pick up any night horizons during that point in time which I thought I might be able to base on seeing where the stars cut off and where they do not.

SCHMITT — We had another indication of that during entry when we were looking for a night horizon and finally saw it, but it was extremely hard to find.

CERNAN — We got lit off on the S-IVB, and, unlike the flame we flew through on the S-II, we did not do that on the S-IVB. I don't know where the reflection came from, but I could see the reflection from somewhere out the forward window. Either it was the S-II trailing flame trying to overtake the vehicle but didn't quite make it, or it was S-IVB ignition reflecting off the S-II because there's no atmosphere up there at that point. But I did not see a flame, but a residual back light out that window just for a short period of time, either right at staging or just at S-IVB ignition. As I think back, my best guess would be that the same thing happened on the S-II, that the trailing flame, when you go from 4g to 0 instantaneously, tends to overtake the vehicle. But in the case of the S-II, it's not nearly as big a pattern and just didn't quite make it up the stack. I just saw some of the glow of it. That's my best guess. After the S-IVB ignited, we never saw anything except the APS firing throughout that burn. You could see the mixture ratio shift.

SCHMITT — But PU shift, both vehicles, was surprisingly noticeable.

CERNAN — Communications throughout the booster phase were excellent. I never had any problem hearing either Stony or CAPCOM. Controls and displays performed super. Crew comfort through powered flight - I felt very comfortable throughout the entire flight in orbit. As far as I'm concerned, there was no pogo on the burn.

EVANS — No, none.

CERNAN — Summing up the birds. If you want to put them in more layman terms, I think the S-IC acted and performed like some big, old, rugged, shaky, big monster. It has to be noisy, has lots of vibration, and smoothed out somewhat after max-q, but still was a rumbling bird. The S-II was a Cadillac: quiet, less than 1g flight most of the time until we built up our g-load prior to staging. It was

72HI513

quiet, smooth, had very little noise, or feeling of rumbling or anything else. The S-IVB: a light little chugger is probably the best way I can describe it, which is not different than I remember it in the past. It just sort of rumbled on, not anywhere near the extent of the S-IC, but just sort of continued to rumble on through the burn. After a while, especially during TLI, it got to be a very pleasant, warm feeling that she was burning like she should burn.

EVANS — Chugging, I think, has two different connotations. I felt the S-IVB was more of a very light rumble in the background, something that is kind of rumbling as opposed to chugging. A chug to me is a bang-bang type thing, and to me it was more of a rumble.

SCHMITT — I agree, it may be a sense of rumbling but the ride was smooth. I could sense some activity behind it, but I wouldn't have said that it was chugging.

CERNAN — I'll modify chugging to say it was a hummocky chug, just a rolling type. Nothing different, and, as I say, the best recollection, similar to the S-IVB I had the opportunity to ride on before, but probably even more steady and continuous flow of light rumbling.

4.0 EARTH
ORBIT AND
SYSTEMS
CHECKOUT

CERNAN — Evaluation of insertion parameters - We got a good onboard orbit. Ground gave us a GO for orbit. The postinsertion systems configuration systems checkout and the complete spacecraft and booster preparation for TLI went extremely smooth and extremely rapid. By the time we came back over the States on the first pass, we were ready and the spacecraft was ready, and we were configured and could have gone on a TLI-0 without any hurrying and scurrying whatsoever. From that point on, when we got our GO on the booster and a GO for TLI-1, it was an Earth-orbit, an extra Earth-orbit ride to sit back and just monitor our systems in the spacecraft and see what we could see from Earth orbit in terms of viewing. It was an extra 90 minutes of the flight that, if you really had to do without, you could have. And it was not hurried. It was very comfortable, even progressing toward the TLI-0.

SCHMITT — Let me add just a couple of things. One thing that we had because of the later launch was a number of LOS/AOS updates to plot which did not interfere with our getting through the checklist. The checklist, I had a feeling went more slowly than it ever had. But, as Gene says, still with plenty of time to meet the zero up time and to have essentially a whole daylight pass to just relax and look at the Earth. We had one note here. I didn't even remember until I read it here that in the ECS checks the hydrogen pressure indicators, or part of the indicators, were reading about 10 percent lower than we expected. But, as I recall, it may have been expected.

EVANS — The optics cover jettison worked as advertised. We jettisoned the optics cover in the daylight and you could see the two covers flipping off straight down the optics path.

CERNAN — I think everyone reacted normally to weightlessness. There was no feeling of disorientation or vertigo or any other disturbances at that point. The CMP is the only one who left the couch prior to TLI and that was for his P52.

EVANS — I didn't get that fullness in the head at that point at all. That wasn't until we'd been up there for 5 or 6 hours.

CERNAN — Launching at night, we just had a somewhat different view of the Earth than most other flights have had. The first real view we got of being in orbit at that point was pretty spectacular because it happened to be Earth sunrise and that's a very intriguing and interesting way to get your first indoctrination to Earth orbit.

SCHMITT — The transcript contains some descriptions, by all three of us, of sequences of that sunrise which, in the color banding, may be of some significance for other people.

5.0 TLI
THROUGH S-
IVB CLOSEOUT

CERNAN — The TLI burn from the ground targeting point of view and targeting went just as written. We went down the checklist and cue card without any problems or any anomalies, without any changes except to the manual. We had a change to all our manual angles to monitor the S-IVB burn because of the late lift-off. We wrote those down on our cue cards and were going to use those in case we had to take over during the burn. We had to change to the nominal and we rewrote both of those on our cue cards. That's the only basic change I think we had.

SCHMITT — The communications all through Earth orbit were excellent, as I recall. There was no difficulty getting the pads up. They came up expeditiously and well read. We actually gained a little time because we didn't have television. But we didn't need it. We could have configured it for use. If there's ever any attempt to do weather observing from Earth orbit, in the low orbit like that, you're going to have to have a very clear plan of where you're looking at what time you're looking in order to make reference as to where you are because you're moving so fast. You can't really keep track of where you are and specifically in terms of weather observation. Later on, once you get the whole globe in view, it's a relatively simple thing to pin down to within a few degrees of latitude and longitude where you are looking on the Earth.

CERNAN — On all these lunar missions, we've never really done much in Earth orbit except get prepared for the TLI burn. Future Earth-orbit flights need this continual map update, you're right. You have to do that. As I think back to 3 days in Earth orbit, unless you continually follow a map and a map update as to your rev as you progress around the world, what part of that world you're looking at is very difficult to follow except the precise piece of real estate you're flying over.

AS17-14822688

SCHMITT — The lunar orbital operation is somewhat different because you stay in the same groundtrack much longer I think.

CERNAN — The S-IVB performance was outstanding. She lit off on time and burned for 5 minutes and some odd seconds as I recall. And we had shutdown on time. The residuals and the EMS on the spacecraft are written down somewhere, but they were all very nominal, very excellent. We stayed in IU. As the S-IVB maneuvered, we flew through a sunrise during TLI, which in itself was also very interesting, very spectacular. We had nominal S-IVB performance after shutdown; and maneuvering to the sep attitude, we went through checkout load NOUN 17 and NOUN 22. There was again no noticeable pogo. The S-IVB sounded and performed just like it did on the insertion phase burn and I'll let the CMP pick up the separation and the transposition and docking.

SCHMITT — We all were very aware of PU shift.

CERNAN — I guess I could have called that or I was looking forward to seeing it. It is on my checklist. It's on my cue card and I've looked for it and I've seen it in the simulator.

SCHMITT — It just didn't register in the simulators, I guess. And the other thing flying through that sunrise, it did to a small degree interfere with visibility in the cockpit.

CERNAN — It didn't bother me from the standpoint of monitoring on my side at all.

AS17-14822714

EVANS — As far as the separation from the SLA, it was nominal. There's a louder bang than I expected from pyros. This is the first time that I really noticed that in the plus-X translations, or in any translations as far as that goes, you get about 0.4° per second rates within the dead band. As opposed to the simulator, it has about 0.1° per second on any of the translations maintaining attitude. Formation flight was great. The S-IVB by itself was as steady as a rock out there. No problems. I couldn't tell it was dead banding or moving at all. I came in relatively slow, about 0.1 ft/sec, somewhere in that area. Docking was nominal. As soon as he got capture on the thing, there were no rates. Everything was steady. I didn't have to handle the translation controls or null rates at all. We went directly to hard dock. There's more spacecraft movement during that period because I feel that the COAS and the docking target were off a little bit. And I don't say misalined, but it's a little bit off. But, of course, it was in limits and was no problem.

CERNAN — When we went to retract, we got our big ripple fire - bang on the latches, so we had a relatively good hard dock. We only got one gray indication on the talkbacks. The other one was barber pole.

SCHMITT — There's a lot of descriptive material, I think, in the transcript on that. As I recall, we got two pulses in the ripple fire. It seemed like there was one or two latches and then the ripple fire.

CERNAN — I just recall a woomph!

SCHMITT — I think, if you look at the transcript, we said that there were two pulses to it.

CERNAN — Subsequent inspection of the latches showed that there were three latches which were not made entirely. One of them, as I recall, had to be recocked. Anyway, it turned out that once we got those three latches (which at that time looked like they were operating properly) reset, we got two barber poles on the talkbacks. Ultimately, latch 4 was found to be unseated on the ring, although, at that time, it looked nominal. The attitudes given us were excellent; we were able to watch the S-IVB maneuver. We were able to see the S-IVB vent and it all went well and nominal.

SCHMITT — It was very clean as far as any debris or anything coming out during the docking phase, and I could see a few little things that were bouncing around inside around the LM, particles of some kind. It was nothing like previous flights where they had a lot of debris. It was very clean.

CERNAN — As I recall, we undocked and separated just a little earlier than had tentatively been planned, but that was no problem because we were ready to do it.

6.0 TRANSLUNAR COAST

EVANS — The IMU realinement and optics calibration - We've mentioned the visibility of the stars in talking about the systems in the section on systems. Systems anomalies - We already hit that one. Heat flow demonstration - it worked great. There were no real problems on it. It was a real time operation with ground. Everything is recorded on the down-link.

SCHMITT — There was some problem with the orientation of the experiment. As I recall, you reoriented it between the two experiments. I never quite understood why there was that problem. It was a checklist problem or something.

EVANS — The problem was something about the orientation of the radial experiment with respect to the X-axis. I pointed the radial experiment along the X-axis. It was supposed to be perpendicular to the X-axis, but it shouldn't have made any difference in the results anyhow. PTC - We got it started and had no problem. Cislunar navigation or navigational sightings - It's already mentioned in the systems part.

SCHMITT — You mentioned apparently you had a very good Delta-H determination - horizon determination.

EVANS — The P23s worked out great. The vehicle is heavy enough that you can control it quite easily with minimum impulse. I used the EMP on P23 so that once you had the star in the field of view and all lined up you could recycle through the program without getting all the maneuver data on the thing. While it was recycling, I could just watch the spacecraft and not let it drift too far out of field of view. When it came back in, I would maintain the star in the middle of the crosshairs of the sextant and maneuver the spacecraft so I could get the substellar point and maintain the substellar point. As it turns

out, I guess the resulting Delta-H is within the limits that are recorded in the E-memory. Midcourse correction - I think that's all recorded on the down-link. There should be nothing anomalous about that. Photography - Jack, I guess you've taken most of the pictures on the translunar coast.

AS17-14822727

SCHMITT — Most of the photography came to GET within a few minutes. It was almost a continuous effort at the beginning of the day and maybe in the middle and at the end with some irregularities - getting a continuous record of a very nice view of the Earth and the weather patterns. We had about three-quarters to two-thirds Earth through most of the translunar coast period. And that should be in the photographic logs on the ground. High gain antenna performance - through the whole mission, not just translunar coast when I was using it, it was perfectly nominal. The ground did most of the calling on it. Between omnis and high gain when they didn't call, it was easy enough to get the high gain to peak up. Usually in MANUAL and WIDE and either AUTO or REACQ depending on the occasion, it seemed to work very well. I wasn't aware of any high gain anomalies.

EVANS — Daylight IMU realine and star check - Again you can't see the stars through the telescope. Most of the time you can't see the stars through the telescope. However, if you have a good alinement and it shows up in the sextant, there's no problem. ALFMED

experiment - I think that's all recorded on the downlink. The one thing that I might add to that is that prior to this time I hadn't seen a light flash. So I put it on anyhow and sure enough the light flashes are there. And that's all recorded.

AS17-14822742

SCHMITT — In the experiments notebook, where the LMP was taking notes on the ALFMED experiment comments, it was necessary in this translunar coast period because we were on omnis and PTC. It is very difficult with two guys observing to take notes if they both start seeing marks at the same time. Interestingly enough maybe even for the experiment the marks seem to come in batches. They'd be periods of quiescence, then both of us would start seeing marks. So the notes are relatively incomplete and, hopefully, the DSE plus the down-link will fill in all the gaps. It's feasible to take notes but they will be incomplete compared to the verbal description.

EVANS — CM/LM Delta-P - Nominal. Orbital science photos - We really didn't have any on translunar coast.

SCHMITT — Nothing was called out. We used about a half a mag on the Earth, maybe more.

EVANS — More than that. We used a full mag before we got to the Moon. and tunnel pressure was okay, no problems. Removal of the probe and drogue - Went as advertised. Worked great. Odors - Every time I got up in the tunnel after docking or anytime, there was always a musty burned odor or something. It's hard to describe.

SCHMITT — Like a powder burn.

EVANS — Kind of like a powder burn, I guess. This was there both in lunar orbit docking and transearth docking. This was the second day we were out when we finally went up in the tunnel. Every time I opened up the tunnel, that's what it smelled like. We didn't mention the SIM door jett. I guess I never did see the door. You guys wouldn't let me up to the window.

SCHMITT — Yes, we saw the door. I didn't get it right away. I was supposed to be taking pictures out of the window.

CERNAN — I saw it right away out of the hatch window. You should have been taking pictures out of the hatch because it wasn't immediately obvious out the window. It came off just as clean as a whistle, with almost no tumbling until it got 20 or 30 feet away from the spacecraft. Then you could see that there was just a little roll and a little pitch as it drifted on away, but very very little. Not a great deal of debris and garbage as I recall came off with it either. You could probably sum up all the pyro operations by saying there are absolutely no questions. They're just good, solid, hard thuds, including SIM door jett. Just a big solid bang, really not that much different than some of the other big bangs when you separate the spacecraft. They're just all big, hard, solid clunks.

SCHMITT — I don't remember what Apollo 16 said about it. Apollo 15 was suited, and they commented they didn't even know it went.

CERNAN — I'm surprised at that, even suited. It was a very definite jolt to the spacecraft when the door was jettisoned.

7.0 LOI, DOI, LUNAR MODULE CHECKOUT

CERNAN — The only thing leading up to LOI that had to be changed in the Flight Plan or in the Cue Cards, since we did a 2-hour and 40-minute clock update which by the way went perfectly, was the fact that I had to replot all the LOI abort parameters on the card. But the words came up very smoothly, and we just replotted the curve and changed the numbers. We had all our LOI abort constants and numbers for the new LOI configuration. SPS burn - I thought the SPS burn was very smooth. We had an on time burn. The burn report came back to you, and the residuals and everything were just as nominal as could be. It was just a short little "g-thud," if you want to call it that, at ignition; throughout the burn, it went smoothly. Jack, you got anything about either one of the LOI or DOI SPS burns?

SCHMITT — They were all auto shutdown. We covered the problems before. Gravitational Effects on the Spacecraft Attitude - That was on rev 1, wasn't it, where we had the pan camera going and we had it all figured that we had one jet firing and the gravitational effects were supposed to keep the spacecraft within that dead band. Sure enough, it did. At least, we didn't get out of the dead band at all and didn't have to change the DAP at all on the first rev. This is to keep the jets from firing into the mapping camera.

EVANS — Communications - We never had any problems with communications at all throughout the flight.

CERNAN — PGA Donning - Our PGA donning practice was a worthwhile exercise. It takes a lot of work to get the suits unstowed and stowed, because putting the suit on in zero-g is just a little bit different. Unlike the previous flight or two, none of the three crewmen had any problem in donning or doffing their suits. I'd say donning is easier in zero-g than doffing. The CDR and the LMP helped each other with the zipping on every donning and doffing, as we've done in training. We had no problems at all. I'm glad that we were aware of the problems that Apollo 16 had. I think we were more conscious of the potential problem that existed when zipping the restraint zipper. We were conscious of it and had sort of trained in a direction to cover all bets on being able to zip up. I would say that in zero-g the zipper was a little bit more difficult to zip, but certainly I can't really say it was a problem for either one of us.

SCHMITT — The only problem was that little blue donning aid always got in the way.

CERNAN — Every time we zipped it, we hooked the zipper coming around.

SCHMITT — One time, I can't remember whether it was on the surface or in orbit, I got something in the lower portion of your outer zipper, and we lost maybe 5 minutes while I worked that over. Other times, I learned that you just have to move through that smoothly, and it's no problem.

CERNAN — Tunnel mechanics and pretransfer operations - They all went as advertised.

SCHMITT — We might mention that we did take that extra film magazine over there. That was purely because that was preplanned in our minds, an extra 16-millimeter magazine, mag EE, because we felt that we just didn't have enough film to get the orbital CSM/LM activities in addition to the planned activities for descent/ascent and lunar surface.

CERNAN — If you're going to use that film during that period of time, it's better to have it in the LM than in the command module, and if it weren't used, you could always bring it back and use it in the command module. That worked out fine. I think we used it all.

SCHMITT — We used it but there was something wrong with the mag though.

CERNAN — We had a gear strip in the mag, apparently.

SCHMITT — It showed a half a magazine of film usage, so we did something with it.

CERNAN — The condition of the CSM thermal coating was excellent.

SCHMITT — Comm checks - We did have an S-band comm problem initially. I talked to some guys a little bit last night about it. As near as I can tell, it was primarily the combination of two things: (1) Up-link data dropouts which were causing the problem on the lockup, plus (2) some phasing, when I would switch antennas when they would just about have lockup. I think we're going to have to wait until we get with the communications people in the systems debriefings to really work that out exactly what was happening. It was a combination mainly of ground problems of getting lockup plus the unfortunate switching on my part. The transfer and restowage of equipment were nominal. I can't think of anything right now that was a problem.

8.0 ACTIVATION
THROUGH
SEPARATION

CERNAN — That's probably one of the most nominal parts of the mission. It reall
went smooth. We oscillated on the timeline. We'd get a little ahead, and then we'd get
little behind; we'd pick up a few minutes, then we'd lose a few minutes. Basically w
worked around the nominal timeline. I certainly wouldn't want to shorten it any, but w
came to the milestones on time and met the ground at the right places.

EVANS — Prior to LOI, we manually pressurized the SPS, which was no problem. It wa
because of the oxidizer helium. In all the sims we never got suited in the spacecraft. W
never have all three guys in there trying to get suited and going through the sims. So th
sims for the CMP were fat, dumb, and happy. There is all kinds of time in the sim; yo
could go out and get a cup of coffee and come back and still pick up everything. It's no
that way in the real world. You get into the real world out there and you work your ta
off trying to keep up and get things going and get suited. When I'm scheduled to do th
P52, the CDR and the LMP are down in the LEB getting suited. There was no way I coul
do the P52 at that time. By the time I had a chance to do the P52 at the sep attitude
the optics were looking down at the Moon, so I'd have to manually roll and do som
pitching to get the optics back up in the air, in the daylight, until I could get picapar t
work. And when I finally got the P52 on, I had a little bit of a problem getting my suit o
that day. There was evidently an "S" or something right in the back part of the crotch.
had a heck of a time getting the zipper across that little S-band thing by myself, whic

was back there where I couldn't pull it through
with the lanyard. I finally backed it off the other
way to make sure everything was all clean and
cleared out. A little squishing sideways and a
contortion here and there, and I finally got the
zipper all the way around. I think the rest of it was
nominal. I was down in the LEB when you guys
lowered your gear, but I could still feel the clunk
in the CSM.

72H1612

CERNAN — You could feel it in the CSM?

EVANS — Yes.

CERNAN — We could feel it in the LM, and we
could also see the forward gear and the ladder.

EVANS — Once I went back up to the couch, I could see the gear sticking out, too.

CERNAN — Which one?

EVANS — Whichever one is over there.

CERNAN — Did it have the ladder on? Probably not. I think the ladder is on one c
the Y-struts.

EVANS — I think so.

CERNAN — Well, anyway, that's interesting. I didn't know you could feel it over there

EVANS — Yes, I could feel it when you dropped.

CERNAN — In the rendezvous radar or the landing radar self tests (the transcript wi
have it) there were some residual numbers in the registers that I had not seen befor

during these tests, when I brought up VERB 63. They didn't affect the test. The tests came out very well, and there's only one other slight anomaly in the rendezvous radar and that was during the rendezvous radar test. It was either on this rendezvous radar test or the rendezvous radar test prior to lift-off - I think it was shaft. I did not get the cyclic oscillation in the DSKY on shaft. But the interface was good, and I'm not sure what the particulars of that problem were. At undocking we had P47 running in the LM, and I got zero in all three registers, zero residual velocity as a result of the CSM soft and open total undocking. Systems operation throughout that time was normal. Vehicle performance was as expected, in terms of attitude control. Lunar landmark recognition - We were able to be in attitude and recognize and look at the landing site on that first pass when we went over. The MSFN relay worked. Generally throughout the flight, I think MSFN relay is more of a pain when you've got good VHF with the other vehicle than it is anything else, because you end up getting a repeat on the voice. I recommend against MSFN relay when you can use direct VHF voice.

SCHMITT — Yes, I agree. On the systems, I was surprised that the component lights in the test positions were very dim. But when they are activated by the caution and warning system, they are bright. I guess I never realized that before. Purely academic interest at this stage.

CERNAN — The secondary glycol pump start up was, I recall, a somewhat ragged start up, as if the pump was slightly cavitating for about 15 seconds. Then it was smooth. There was no subsequent indication of the problem with the secondary loop because we didn't use it subsequently.

SCHMITT — Referring to that radar test, it was the PGNCS turn on self test. I had a 400 in R-2 initially, and I had never seen that before in PGNCS turn on. That's what I was referring to about something different in the registers. It was on the initial PGNCS turn on, and self test.

CERNAN — All the alinements went well. One thing that we discovered - the gimbals apparently were mistrimmed on the descent engine prior to lift-off. Someone is going to have to resolve whether that's true or not. The pitch and the yaw gimbal trim in the DAP were reversed as to our checklist. When I inquired about it, I found out that our checklist was correct, which gave me an impression that the gimbals were both mistrimmed. But they were so close to each other that the ground indicated we should press on and we should see no reaction to that mistrim, and to start up. We did and we did not see any indication of the mistrim. If pitch and yaw had been separated quite a bit, I'm sure we probably would have had to go through a retrim of the gimbals during the DPS throttle check.

SCHMITT — One clarification comment with respect to the AGS. I mentioned yesterday that I thought it was a Z-gyro that indicated greater than spec calibration. It was the Z-gyro, just slightly greater, about 0.4.

9.0 SEPARATION THROUGH LM TOUCHDOWN

EVANS — In optics tracking, I tracked RP-3 which is about 5 minutes prior to the subsolar point. And that's too close to the subsolar point to be doing any optics tracking. As soon as I got to the TCA on that thing, I completely lost the visibility of the landmark. So the only good marks on that are going to be prior to TCA.

9.1 COMMAND MODULE

CERNAN — What was that landmark?

EVANS — That was the landmark for updating the mapping camera film. Actually, it is a recalibration of those particular points that had been tracked on previous launches. The circularization burn was a good burn. The only anomalous type thing on that is that the residuals prior to trimming were plus 1.70 and minus 0.6. The minimum impulse is a 4-second burn. It underburned in the minimum-impulse case by 1.7 ft/sec. It turned out to be no problem. The ORDEAL worked as advertised throughout the flight.

9.2 LUNAR MODULE

CERNAN — Prep for PDI - We just went out of the Activation Checklist into the Timeline Book. There are no notes concerning any anomalies. We stayed on the time line and as I said, we met the milestones with the ground. We came around the horn for PDI and established comm, and the ground had a load waiting for us. We had no NOUN 69 prior to P63.

SCHMITT — We did have the communications problem prior to PDI. The thing that started it off was the ground started up-linking on the omni, which they had never done before in the sims that I remember, unless it was a situation where we hadn't gotten the steerable. They started on the omni. I was not watching that, and I switched out of the omni to the steerable in the middle of the up-link. That started the problem, which apparently was compounded, as I found out last night. Anyway, the Goldstone antenna went belly up somehow, and the men who talked to me last night still do not know how. Somebody may, and I'm sure it'll be worked out. The up-link did get in and all you had to do was proceed on the VERB 33. We did have a good up-link, and that whole thing was in there but nobody's quite sure how it actually got in there. The ground surprised us by coming up almost immediately with that up-link, which we've never seen before. In fact I expected it would be quite late.

CERNAN — We'd seen them come up fast, but they always waited for the steerable.

SCHMITT — Yes. That's what caught me by surprise. At any rate, we got it in there and there was no subsequent problem. The comm thing did delay us, and we were running a little bit behind the time line. Let me mention one thing on the DPS start which I didn't mention yesterday. And that is monitoring 471 in the AGS showed essentially no Delta-V accumulation in Y. That was a good idea, although it was unnecessary.

CERNAN — As far as the start was concerned, the LMP confirmed ullage. I had my physiological cue, and I knew we had ullage. I was prepared to back up the ullage and back up the start, but we got an automatic performance in both.

SCHMITT — It was very clear that the SHe tank had opened up within a few seconds. We got our first jump in pressure a lot sooner than I expected.

CERNAN — All the pyro functions prior to PDI in the LM we could verify with a physiological cue. We could feel, and/or hear all of these functions.

SCHMITT — And this was suited.

CERNAN — In some cases, it was suited; not all. I covered the performance of the engine. The PGNCS performed admirably. I called up the NOUNs I needed: 68s and 92s loaded NOUN 69s, and she just spit them out just like she always has.

SCHMITT — The SHe pressures during descent held low. About 30 psi, as I recall beneath the predicted number.

CERNAN — NOUN 69 was plus 3400 feet, and that sounded very familiar, as I recall. Didn't we almost always in the sims have a plus? Even the nominal ones have a plus. I that the problem they had? As it turns out, as soon as I pitched over, I took it right back out to get to our landing area.

SCHMITT — Is that right?

CERNAN — It was almost exactly the same number, which means that their targeting was essentially perfect, because the planned landing area was about at least a crater diameter short of Camelot.

CERNAN — That's where we pitched over and that's where we would have landed, which was the planned, targeted landing area. We did not say anything about DOI-2. DOI-2 was slightly smaller than we'd seen in the past, because of the orbit degradation we were in. I think it went down to something like 11 miles, but the DOI-2 just went super. We got the residuals down to 00 and 0.1, something to that effect. We saw a 7.0 perilune out of the PGNCS and a 6.7 out of the AGS, which is exactly the type of thing we expected. We went around to PDI in good shape. We got excellent radar and VHF ranging correlation during that radar checkout.
VHF Ranging and Radar Tracking - Everything was nominal during PDI right through pitchover. We got throttledown on time. We watched the computer and followed NOUN 92. The computer was happy, the GDC was happy, and everything was just perfect. At 13 000 feet, I could look over the edge of the window and see the South Massif. At 13 000 feet, I knew we were coming down in the valley because I could see the South Massif, and I could tell that we were in the valley or coming into it. At 13 000 feet, I had the impression we were level with the top of those mountains. (Laughter) I really did. We pitched over, the needles dropped, pitchover occurred, 64, everything was nominal. Our target point was about a crater diameter short of Camelot. I used LPD frequently. I don't know how many times I used LPD, several clicks back, a couple left, a couple right. I just flew it where I wanted to fly it. I brought it back to an area in the vicinity and to the right of Poppy. As soon as I did that, I just sort of tumbled in on that area and did some more LPDs to finally what I'd call a suitable landing site. That suitable landing site became more evident the closer you got. Initial LPD changes to bring the landing site back east were just gross to change the area. Once I had my area, I started tweaking it up to find what I considered a blockless and level area. I ended up taking over in P66 just a little below 300 feet. The reason I took over is that I wanted to slow our forward velocity down. I did not want to go any farther west, because there were more blocks and more hummocky terrain. As a result of all of our aft LPDing, we ended up (1) with a great deal more fuel than we might have anticipated, between 7 and 9 percent, I believe, and (2) the rate of descent, H-dot, was a little bit higher than normal, because of our steeper descent in the latter phases of the braking and landing. But as far as the CDR was concerned, they were very comfortable rates of descent. The LMP passed them on and said they were a little higher. I knew where we were. I think the most significant part of the final phases from 500 feet down, as far as the CDR was concerned, was that it was extremely comfortable flying the bird, either LPDing in P64, and/or flying manually in P66. I contribute that primarily to the LLTV flying operations. That's why the rates of descent and what have you were just very comfortable. I kept a good rate of descent down through 200 feet, slowed it down at a little bit over 100 feet to 1 or 2 feet per second, and then started it on down again. We started to get dust somewhere around 100 feet.

SCHMITT — In my window, I didn't see dust until about 60 or 70 feet.

CERNAN — The dust layer was so very thin that I could definitely see through it all the way down. It didn't hamper our operations at all. When I was satisfied that that was my landing site, I made sure we had between 1 and 3 feet per second on the crosspointer forward velocity, and to the best of my ability, zero left and right. We continued on down with about 3 feet per second to landing. I saw the shadow come right on up to me, and this is very well done in the simulator. When it passed on under me, I was expecting a blue light. It seemed like it didn't quite come, when the shadow passed on under me for just a split second or two. We got the touchdown light. I had planned to say, "I potato, 2" and then push the stop button. But I didn't. As soon as we got the touchdown light, I, like most everybody else, hit the stop button. Then things just went "plunk." We plunked down with a relatively good thud, I'd say. Visibility through the final phase was excellent. The tendency, once you redefine your landing area, is to become a little bit less concerned with your peripheral landmarks out there, because you know now about where you're going to go. You get more tunnel vision, and you are concerned with finding these specific touchdown points within that landing area. That's effectively what I did. I had no Sun angle problems. At that point in time, estimation of distances didn't mean much, because I was concerned more with what was right down below me and in front of me. I can't say enough for what I consider the accuracy of the guidance. Manual control of the spacecraft was hard and firm, different certainly than the command module operation but exactly what I expected the LM to be. The simulator, I think, does an excellent job of controlling the firm good solid ATTITUDE HOLD, RATE COMMAND capabilities of the LM. I'd say that I touched down with about 1 to 3 ft/sec forward, and 0 left and right, and about 3 ft/sec down. We'll just have to find out what those numbers were. I don't know. The fuel remaining was between 7 and 9 percent. From the CDR's side, the systems were excellent.

AS17-13420384

10.0 LUNAR SURFACE

CERNAN — Postlanding powerdown - We got the verb for STAY at T1, T2, but we got a GO for at least a T3, and we started right through the checklist and the power on configuration. Based upon the review of the Surface Checklist, there were no anomalies in powering down the spacecraft. We just followed right on through. PGNCS and AGS worked fine. Z, once again, had a higher than spec gyro count. It was nothing serious though. Eat and rest period - We had an eat period on the surface. As we were beginning our EVA-1 prep, we took some pictures out the window. We just followed the checklist, and, all told, we ended up getting out some 30 minutes late. I'm not sure why.

AS17-13420425

AS17-13420435

SCHMITT — Part of it was that P57.

CERNAN — Oh, we had to do a P57 over because we reversed the marks on a spiral cursor, which was just an onboard problem on our part. So, we did the P57 over, and we lost several minutes. We sort of never lost any thereafter, but we never made them up either. Suit doff and don - This will cover all the EV prep and post activities. We both found, LMP and CDR, that donning and doffing the suit in 1/6g was relatively easy. Once again, we had no problems zipping up the suits. In the course of doffing, and prior to getting the suit fully off, we mutually lubricated each other's open zippers and all the connectors. When we doffed the suit, we went into a drying mode as the checklist suggests prior to the sleep period. I'm really glad we did because our suits stayed relatively fresh and clean on the inside. We doffed our LCGs every day and slept in CWGs rather than the LCG. And I'm glad we did that because it was much more comfortable. We made it a buddy system in the entire donning and prep when it came to the suit operations, except for putting on the gloves. We found it easier to put them on in parallel and get them locked and verified locked.

We actually, each individually in almost all cases, put our own glove dust covers and ring dust covers on. Maybe we had to help each other once in a while. And contrary to some of our initial desires, we decided to go ahead and put those dust covers on for every EVA. After the first EVA, we found out what the dust problem really was.

AS17-13420473

SCHMITT — One of the tabs on the LMP's dust covers did break off on the first prep.

CERNAN — But besides that, we never used that donning lanyard that we had available. We never needed it. I can't really say anything else except that the doff and don went pretty much as we both expected it to. We obviously took extreme care of our suits - the best we could - because we had to use them several times. I think that care paid off because even at the integrity check of the CM/EVA, the suits were tighter than a drum. I think the wrist connectors, even with the dust covers, were tending to get a little bit stiff.

AS17-13420481

SCHMITT — Yes, mine were very stiff.

CERNAN — But nothing ever really froze up on us. LM vehicle systems operations - There weren't many systems operating during the lunar surface activities other than the EPS and the glycol system. We set it up per the Flight Plan. We updated the PGNCS periodically. It was all nominal operation.

10.1 FIRST EVA, MASSIF

CERNAN — First EVA prep activities - And all I can say about the PLSS donning and checkout verification, cabin depress, communications checkout, and power transfers, is that it just

AS17-13420530

followed the checklist and went nominally. The only thing that we might consider as a deviation is the fact that the CDR left his O2 hoses off during most of the donnings because I felt I didn't need them with the water cooling from the spacecraft. It was easier to get them put out of the way early, and there was certainly adequate airflow. We left the flow on through the hoses to keep circulation in the cabin during that time. I felt very comfortable and less contained by having those two hoses out of the way. All I had dragging from me was the water hose and the comm hose.

SCHMITT — LMP wore the hoses most of the time to partly have a convenient place to put them. Also, I like the airflow.

CERNAN — And they're more out of the way of the LMP because they're on your side. EVA 1 - We just commenced the egress very slowly to get familiar, but basically there

were no real problems with the egress. I felt you had to get down a little bit lower to the floor than I'd seen in the airplane, but once you understood where you had to get, getting out was no problem at all. Everybody knows that the LM cabin is very small, and you're restricted. You cannot move very fast or get out of each other's way very easily. So when you did have to turn your back to change valves or switches or circuit breakers, you had to move one at a time to get out of each other's way. Once we found out what those requirements were, we were able to work together very well and stay out of each others way most of the time.

AS17-13720910

SCHMITT — Yes. Let me comment about the LMP's egress and ingress and general activities a little bit that I've done in the 1/6g airplane in the mockup. They seem to be more difficult and more constrained in the LM than they were in the airplane. I don't know exactly why. Part of it may have been in the pockets. I kept finding I was hanging my leg pockets up on those things. I don't remember whether I had those on in the airplane or not.

AS17-13720979

CERNAN — The key to ingress was to get all the way in and then bend my legs up. As soon as I bent my legs up, all of a sudden everything broke free. I think it was that the pockets were hanging out on the sill, and as soon as I bent my knees, it took the pockets off the sill, and I just slipped right in. I didn't learn that until the second egress. Work on the platform and on the porch was fine. We got the MESA deployed. The LMP egressed. We got the LRV deployed.

SCHMITT — Cosmic ray was deployed nominally. LM description and plan - There wasn't much to say. I had the impression maybe the strut was stroked, but that was discussed and photographed.

AS17-13720990

CERNAN — The whole EVA, as we call it, "closein," went so close to our EVA closeins and eventually closeouts at the Cape that even I was amazed. It turned out that I got to the flag just about the time I always got to the flag, and you were ready. It just couldn't have been a better reproduction of the training activities at the Cape. I think the transcript and the television better describes and debriefs that portion of the EVAs than we could by just sitting here and saying everything went nominal by the checklist because that's essentially what happened.

SCHMITT — You've heard all about the ALSEPs and the LTG problem in real time. It's on the transcript. It was something in the dome removal strip. We pried it off with a hammer. The ALSEP traverse surprised me in that the package seemed heavier than I had expected.

CERNAN — You lost a block.

SCHMITT — I lost a block. It just came off the Velcro. I may have hit it with my leg. Really the dust was so deep and soft that the blocks were relatively ineffective, and I ended up putting a rock underneath one corner. ALSEP deploy - In the LMP's point of view, it was slower than I expected it to be. But, everything got deployed. And the geophones were faster as we expected.

AS17-14522157

CERNAN — The heat flow went very well. It just went bang, bang, bang. Really the only difficult thing in 1/6g is that fact that you cannot bend over very easily to pick things up. I used the drill for a brace almost every time I had to get the wrench off, as you saw and heard in the transcript and pictures. Every time I found out that I reduced a work output and reduced the frustration when I set the drill in the right place, leaned on it, and took the wrench off. The only little thing I had some problems with was with the core and the bore; you have trouble in 1/6g with the gloves on to aline the threads and make sure they get all the way seated on the following bore or core prior to starting to drill. I had a couple of problems with that, but eventually I got them all. I never rammed a thread down with the drill. I always had it all the way flush, which preserved the bores, of course. The whole operation just went well. You saw it; you heard it. We followed the procedures. The TGE could have been taken on and off very easily on the Rover. The only thing that we didn't anticipate about taking readings when it was off the Rover was again the same problem. You have to lean down to get to anything, and the TGE is very low. It's very difficult to get down there and make anything but a swipe at the buttons when it's on the surface. I'm very glad we did not have to

AS17-14522165

AS17-14522183

take it off the surface for all the readings because it made it much more convenient. It was not a problem of taking it on and off. It was a problem of pushing the button once it was on the surface.

SCHMITT — The ALSEP photos were not taken in the normal way. I think that by the time we had finished our second and last traditional revisit of the ALSEP, a fairly good collection of photos had been obtained, both on specific request from the MOCR and also random photos I took while they were thinking.

CERNAN — The whole EVA-1, all the way through the station 1 activities and the SEP deploy, although there were modifications in it, followed the checklist. The best debriefing is the transcript and the TV. I don't think there's anything we can really add to that or any of the other particular stations that hasn't already been said in real time.

SCHMITT — Let me mention again, for the record, that the geophone module package did not constrain the geophone's lines very well. But the net result was a good triangular deployment of geophones, even though they are not anchored at the base of that triangle.

AS17-14522224

CERNAN — We go into ingress and the EVA closeout was again pretty much as planned without anything worth talking about other than what was heard and seen.

SCHMITT — I don't know whether you've been told yet or not, but both the SRCs have excellent backings.

CERNAN — Good.

SCHMITT — Number 2 has the best they ever had.

CERNAN — I took pains to make sure that that thing was sealed. They did have excellent backings? That's good. EVA post-activities - Again, the refurbishing of the PLSSs went as was written in the checklist, both with, oxygen and water. Apparently, we got them completely refurbished for every EVA because the total time we were able to accumulate on them in the second and third EVAs. I never had any problems throwing the CDR's PLSS back in the recharge station.

AS17-14622367

SCHMITT — Let me go back to the EVA closeout. The transfer of the gear up the ladder, by hand was not difficult, but it was more difficult than I had expected. Getting the EVA pallet in ahead of me looked like it might be a problem, but I found that by pushing the hatch full open and putting the pallet off to the right, I still had plenty of room to move around. I put it to the

AS17-14722523

right, next to your stowage area, and it was out of the way. I got in and then reache over and undid it. Taking the gear off the pallet took longer than it did in training. It wa a more difficult job.

CERNAN — That whole transfer seem to go very well, the transfer into the cabin an transfer back out of the cabin.

SCHMITT — Tool management reminded me of that for some reason the lefthand pocket down low on the left leg was essentially not used. I couldn't get to it easily. I was able to get to the right pocket and I did stow odds and ends of samples in there occasionally, and once or twice, the hammer. In general, it was only the right-hand pocket that was useful to me. Tool management was as we had trained, with the exception that as the EVA's progressed, the spring-loaded latch that locks the scoop into a given position in the detent ceased to function very well.

CERNAN — EVA post activities - You got anything to add?

AS17-14722526

SCHMITT — We did that in parallel with other activities.

CERNAN — We approached that relatively casually but with the idea of getting to bed on time, and for the most part, I think we had a little fat in there. Where we didn't we still preserved the 8-hour sleep period because the next day was not necessarily critical, except the day of launch, on which we wanted to get up on time. Performance comments, equipment - I cannot say enough for the PLSS operation. Cooling capability was there tight as a drum; communications were excellent; and the suit performed well.

SCHMITT — The only problem we both had was in the gloves. Just general fatigue and also continual pressure against the nail there bruised under the nails.

72H1542

CERNAN — That pressure against the nail areas was not a pressure caused by short gloves for me. It was just because of use. You required so much dexterity during the ALSEP deploy that it was apparently a pressure that got you across the top of the hands or the top of the fingers, but it was not a fore and aft pressure for me.

SCHMITT — But you still got some bruises under your nail? I don't see any other way to get that but by pushing against the nail. There was no way to avoid it either.

72H1572

10.2 SECOND EVA, SOUTH MASSIF

CERNAN — Here again, you can talk about the prep activities. We were obviously smarter. Some of the things you do in EVA-1 do not have to be done during EVA-2 because they're only done once in terms of stowage and what have you. We had some OJT on EVA-1, and EVA-2 just went right down the line. We got the cabin depressed, got out, and went to work. I cannot say anything about EVA-2 egress or equipment transfer or anything else.

SCHMITT — Yes. I don't want to waste time on the traverses because I plan to do that with the tapes.

CERNAN — I've talked about Rover mobility and capability and the requirements of the driver for continuous attention and that became very evident on EVA-2.

SCHMITT — Although I made reference to most of the little memory jogs we had in the Cuff Checklist, it turned out they were not specifically necessary to have them in the checklist since our continuous observation and discussion of the surface covered those things as a matter of course, if they were there. I think the most important thing that they did was to force us to review cuff checklists prelaunch to learn, train, and think about the kind of problems they were referenced to. In the actual operation, most of those discussions took place relatively automatically.

72H1576

CERNAN — The CDR's navigation page used in traversing to each station was probably one of the most useful things I carried on my cuff checklist. It kept me very much aware of the general heading I had to go and general large features we were looking for. I just think it was extremely useful. Because of the terrain and the inability to travel on a straight line for very long periods of time, I primarily did not navigate on heading. I primarily navigated to a point. And so the particular points that were shown for jogs in the traverse, or for Rover samples, or charge deploys, or for stations were most valuable to me, because I navigated to a range and a bearing and didn't worry particularly about the exact heading. That seemed to work out very well. And that's why we never, on any of the three EVAs, followed our tracks back to anywhere. We crossed our tracks a couple of times but we never covered the same piece of real estate twice. Performance of all equipment after EVA-2 was excellent. Going into the EVA-3, the prep, again, was familiar.

72H1579

10.3 THIRD EVA, NORTH MASSIF

SCHMITT — Station 3 - We both did most of that station separately. Gene was working the double core as planned and I was doing sampling. I got a little inefficient at the start because I didn't have a bag to put samples in. Once I got a bag, it was a little hard to handle because I was on a side slope. But in the time that we spent there, I think it turned out that Gene got an excellent double core canned and I got on the order of 10 or 11 documented samples, both surface and trench samples at the edge of that crater. I would still have

72H1596

a hard time evaluating now whether we could have operated more efficiently together or separate in that particular case.

CERNAN — EVA-3 closeout was nominal. It was modified because the LMP had to go back to the ALSEP again. As far as I'm concerned, the recovery of the neutron flux, parking the Rover, turning off the SEP and going through all that worked very well. Here again, any modifications to those closeouts are really not bad at all because we used the checklist as a reference and not as a cookbook. We understood what had to be done and what had to be closed out so that we could accept modifications and also pick up each other's task. And we did that quite frequently on the closeouts. We could see what the other guy was doing, and picking up the other guy's task occasionally, when you had a free moment or an easier reach, was a very simple thing to do. That comes from having done this together many, many times. Probably the most difficult job of all the closeouts was trying to dust the suits. It's a difficult and awkward position. It's hard to make fast sweeping movements in a stiff suit. We did our best, and I think probably the time spent was well spent. But I think also it was a bit more time than we had anticipated. The real-time transcripts will show just how much time and effort was spent in dusting. Both of us found that our lower limbs and boots could probably be better dusted by jumping up and down on a ladder or clapping your feet together on a ladder, which, incidentally, the CDR had to do in every case because he was the last one in. His feet were always in the dust prior to getting on the ladder. But I think that worked out pretty well. Third EVA was pretty much operationally like the second. We worked on slopes on both EVAs. On the third we did have the Rover on the slope. That didn't seriously perturbate the operations. I intended to rake larger areas for samples than I had planned to, but that was mainly because we weren't getting very many samples per rake swipe in most places. I think the only place we got a large number of samples was at station 1. After that we were dealing with no more than 10 in raking over a very large area in any of the other rake samples. But that's clearly documented in the samples. I don't know how many LRV samples we actually took, but it wasn't a problem. And the sampler was used whenever I worked around the LM or went out to the ALSEP or anything. As a result, I picked up maybe a half a dozen more samples just because it gave me something to carry a sample in.

72H1600

72H1601

72H1604

CERNAN — The only piece of hardware I remember that broke was the bag fastener on your camera.

SCHMITT — Somehow or another I strained that and I taped it on in the cabin between EVA-2 and EVA-3. That taping job, using the food-pack tape, worked very well. We had no further problems. The EVA-3 post comments are the same. Equipment jettison went smoothly with no problems. You had the feeling that if you had an infinite amount of oxygen and water, you could have used those PLSSs indefinitely. Good systems.

CERNAN — In closing, as obvious and as always true in the past, the efforts put forth on the surface of the Moon, or any place else, are based upon a great deal of work by a lot of other people. In general, the most significant group of people that supported us in excellent fashion, and probably the best I've ever been associated with, is our team led by Dave Ballard. Those guys continually went out of their way to make sure that things were done right. I just can't say too much for the effort that they expended. They performed in a super professional manner. Without that team and the training, the debriefing that we've just gone over here for the last 2 days might be a lot different. The

success of Apollo 17 is due to a lot of people. In particular, the LM activities went so smooth. The LM stowage, in which there were a few changes right at the end, the interior cockpit stowage and the exterior descent stage stowage, was really in outstanding shape, and it was due in no small part to the efforts of Terry Neal. Terry's had a great deal of experience in the past on previous flights, and that experience really showed itself. He was a tireless worker. He supported every activity without being asked to at the pad, and came back and told us what he had to support. He kept us informed. He made sure that people who were in charge and responsible for all the training gear had all the knowledge to keep it up to speed, based upon flight configuration of gear. He was concerned about the type of details and things that the crew is either too busy to handle or certainly would have let slip by. He's the guy that got the job done for us so that when we got up there, to unstow the gear and to put it to work, it was not only like we had planned it to be, but it was all there and it was properly and professionally done.

72H1606

72H1607

SCHMITT — Your statements are certainly echoed in my mind with respect to the entire team. Every time something needed to be done there was somebody there who had already done it, generally. It wasn't a question of asking. It was a question of doing, or of utilizing the results of the team's effort. Terry Neal certainly made the lunar surface stowage and equipment operation, both in flight and in training, outstanding. There is no other word for it. We had no difficulty at all in learning where the equipment was and how to use it in its storage locations. I'd also like to congratulate the EVA operations group for their work in putting together three, very complex Cuff Checklists, and in keeping a general trend of

72H1614

training going that was just about at the right level. We reviewed the various EVAs in a reasonable sequence. And by the time we launched, I think we had enough of a feeling for what was in the cuff checklists that we really, as you said earlier, only used them in the review and that can't be to anybody's credit but the people who organized the training program.

CERNAN — And the entire support team - it wasn't a case of them keeping up with us getting ready for the flight, but a case of us keeping up with them. Because they were going to be ready for the flight and they made it a point of making sure that we were going to be ready also.

SCHMITT — I think it's also worth mentioning that we have nothing to give but praise for the ability of the suit technicians not only to keep our gear in working order and up to date with the changes that might be coming along, but also in training us on how to use the gear. That is perhaps not in their job description. No small part of our ability to get in and out of the suits, and understand what you can do and can't do with the suits, in terms of doffing and donning, goes to the four guys who were our suit technicians.

72H1647

11.0 CSM CIRCUMLUNAR OPERATIONS

EVANS — Operation of the spacecraft - The CSM solo operations are essentially nominal. One time on the back side of the Moon, after I'd done the zodiacal light, where you had to switch to CMC free during the pass to prevent any jett fires and then you switch back to auto, I missed the switch back to auto and proceeded on into the waste-water dump and urine dump. Unfortunately, I locked the spacecraft control switch and CMC free. The waste-water dump evidently puts in quite a torquing force or perturbates the spacecraft such that I was getting a master alarm with the gimbal lock light. As soon as I had the caution and warning, I checked back and found that it was getting close to gimbal lock. I switched to SCS, and it backed away from gimbal lock. Then I pushed back to auto and got back to P20 attitude. Navigation, normal state vector updates - When the down range error got to about 30 000 feet, I let go and shifted up a state vector. The RQ model being used over in mission control to project the orbital decay didn't work quite right, so I ended up with the orbit not decaying down to the circular orbit prior to the plane-change burn. I ended up making the high adjust maneuver or trim burn to bring the orbit down to 63 by 63. The trim burn was performed about an hour before the plane-change burn. Trim burn was a 9-foot-per-second RCS burn.

AS17-14522254

AS17-14522273

LM acquisition - Nominal in all respects. The thing that is somewhat of a surprise to me, and I should know this, you get molded into a false sense of security by doing rendezvous in the CMS. You look through the telescope, and there's a big blob of light. The telescope is indicating where the LM is. In the real world you look in the telescope, and you can't see. It's very hard to see 150 miles away. As a matter of fact, the LM was at about 80 miles before I actually saw the flashing light in the telescope. As I went into darkness, I could see the flashing lights in the sextant. I did not get LM acquisition prior to going into darkness, and I did not have it in the first part of the rendezvous. I did not have the Sun in the sextant. There was no Sun in the telescope, and it was about 3 minutes prior to spacecraft sunset before I had the Sun in the sextant and in the telescope. I could not pick up the LM in either the sextant or the telescope. Once I had picked the LM up in the sextant, I had no problem from then on.

Update pad and alinements - No problem. We kept the P30 pad in R-11 where it was always available in case I needed it. I always realined to different REFSMMAT. In translunar

coast or transearth coast, I always switched to SCS minimum dead band and gyro torqued. As I picked stars on the dark side of the Moon, I would coarse aline to the new REFSMMAT. It might be interesting to note that on PDI day prior to LM separation, there's a P52 scheduled about the same time the LM crew is getting suited up. I delayed the P52 until they were in the Challenger. By this time, the SEP attitude pointed the optics right down to the Moon. The PICAPAR didn't work, so I just started the spacecraft roll and kept recycling the 404 alarm until I finally was able to get it to work. After I got to P52, I maneuvered back to the LM sep attitude.

AS17-14522285

Lunar sounder boom deployed - We had a little test to extend and retract the booms. Extend worked okay. Retract and HF 1 never did get the gray. The antenna retracted to the extent that there was no problem for RCS or SPS burns. In trying to retract prior to plane change, we looked out the window and could see it start back in. The extensions on HF 2 - Number 1 always extended all right, but number 2 would go out for a little way and stall. We retracted it for 5 or 10 seconds and then switched it back to extend until it deployed fully. At any rate, with a little bit of work, we got the booms in and out.

14722465

Monitoring lunar activity - I did not attempt to monitor it but I could put on VHF and talk to them. I was usually operating during VOX during the solar periods, so I just left the VHF off. Prior to lift-off, we had MSFN relays activated that worked real fine. Lunar sounder pad experiment - No problems. Everything worked fine.

SIM bay daily operations - On the mapping camera, the first extension took longer than anticipated, so it was elected to leave the camera extended throughout that day. It took about 4

15223274

minutes to retract when we retracted it. On one of the mapping camera oblique passes where we were starting at the spacecraft sunrise terminator, I went to operate and got the barber pole. The Malfunction Procedure is to go to standby, which we did. We left it in standby until we just about came up to AOS. At that time, the barber pole disappeared. Evidently it was caused by the mapping camera area being too cool. And as soon as I got the gray indication, I went to operate and had no problems the rest of the time. Laser altimeter - It seemed to work fine. There were no anomalies.

Pan camera - There were no anomalies that I know of with the pan camera. There was some concern at one time if it was getting a little warm in there and also some concern as to whether the lens had really stowed.

UV spectrometer - As far as I know, we got outstanding data. The information that was passed up to me indicated that there isn't as much hydrogen in the atmosphere around the Moon as was originally thought.

IR scanning radiometer - It worked real fine. We're still getting good information, and we were getting good information on the way back. It was on most of the time.

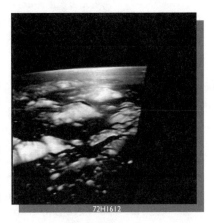

72H1612

The SIM bay photos - Let's see, that means photographs by the LM. It seemed to work all right. The Challenger was easily maneuvered around to the right viewing attitude. They got some good pictures. Sunlight was okay.

Dim light photography - The dim light photography was the zodiacal light and the solar corona. It was okay. Terminator photos - Hopefully, those are going to come out. I used a lot of Nikon film for terminator photos. We should have a lot of 35 mm stuff that was not planned or not scheduled in the Flight Plan. We used the Nikon with a red filter and a blue filter and took three shots with the red filter and three different shots with the blue filter of the landing site area. We also used two different polarizing filters in one direction and then in the other direction. That information should be in the Flight Plan. In each case, the zodiacal light with the filters worked out real fine. The timing and the settings worked correctly. I've got it noted in the experiment checklist that I had the wrong setting for half a second. I ended up on the 1-second mode. I think that was in the polarizing part. In any event those pictures should be good. In sketching the zodiacal

light as you come up to the spacecraft sunrise, I think we probably didn't get the longest streamers that are just half a second or quarter of a second prior to the Sun popping over the horizon. In each case of the zodiacal light passes, the sequence ended 7 to 10 seconds prior to spacecraft sunrise. I think we probably missed the longest streamers. I didn't really observe this phenomena until the last day of lunar orbit and didn't have the opportunity to take a hand-held target of that particular phenomena.

S7253472

Solar corona - The sequence worked real fine, no problems.

Earthshine photography - We worked it differently than it is indicated in the experiment checklist. I used Aristoteles and Copernicus starting out with a 1 second and taking two 1-second exposures. As we rotated around about every 30 seconds, it ended up a little closer than I thought. We were passing up the target too fast because we'd never get everything. The timing sequence may not be correct. It may not be exactly 30 seconds between each one. We would cycle down the exposure setting to 1 second, 1/2 second,

1/4 second, 1/8 second, and 1/16 second on Aristoteles and Copernicus. We'd leave it on one-sixteenth of a second following Copernicus and switch over to window three and pick up Reiner Gamma and do the same type of sequence. Then we stopped on 1/8-second exposure and carried it out until the end of the film mag.

Orbital science photography - It worked according to the Flight Plan. We would have the initial setting, and on the orbital monitor charts, we would have the inpoints and then pick out specific craters and have these noted on the chart as to change settings. I did notice that it is very easy to bump and change the camera settings as you bounce around in the spacecraft - trying to keep track of the camera pointing as you try to maintain your own equilibrium. A couple of times at the end of a particular sequence, I noted that it had changed from what I had started with. The orbital science photography was accomplished with no particular problems other than trying to maintain a constant camera setting. We had two magazines of what we call CM option or option-photography colored film. Those two magazines were completely filled up with just targets of opportunity.

Plane change 1 - I previously mentioned the trim burn part of plane change 1. Plane change 1 was a little larger than anticipated because of nondecay of orbit. Plane change 1 is where I had 0.7 ft/sec and it seems to me like an X. I did not trim it because we were only trimming Y. There was also a plane change where I ended up with a different roll because the pan camera was looking right into the Sun. No real problem. If I were going to trim anything, I would trim Y and Z just to make sure I didn't perturbate the apogee and perigee orbit. To keep the pan camera out of the Sun, I went into P40 trim and utilized that roll angle. Communications were outstanding. Maneuvers done to support the lift-off presented no problem.

Rest and eat periods - I never got to sleep on time. It just took a great amount of time for one man to go through that.

Presleep Checklist - to go down and chlorinate the water, take the panel off, pull the return valve and clean the hoses - it just takes a lot of time to get it all done. But, there's no real problem.

TPI backup - My TPI solution agreed quite well with the Challenger, no problem.

Midcourse backups - I ran into a bit of a problem. I ended up with 5 ft/sec as a Z-value, and the LM ended up with 1 ft/sec. I don't understand why there's that much difference between the two midcourse solutions. Of course, the Challenger made all the burns during rendezvous and braking, so I didn't have any problems there.

Prep for docking - There is no time to get all the cameras and things squared away prior to going into rendezvous, so I strapped the TV monitor to the XX strut by the CDR's couch and utilized it during the rendezvous and braking phase or final phase of the burn. I used a P79 to point the X-axis out the LM. And once it got close, I essentially pointed the spacecraft such that the LM was always in the center of the TV field of view while coming in for docking.

12.0 LIFT-OFF, RENDEZVOUS, AND DOCKING

CERNAN — LM powerup and launch preparation went well. We did not do the P22. Everything else just went as advertised on the LM. She powered up beautifully. The lift-off was normal. Obviously, we got all our pyros, and we lost no changeover, Parker valves, or anything. Very soon after lift-off, we had apparent loss of comm, a lot of noise in the S-band. It turned out that we were down-linking, but there was something wrong with the up-link. So the CDR watched most of the guidance and would call out, in the blind, altitudes and GOs and what have you as we pitched over and pressed on up. For about the first 2 or 3 minutes, the lunar module pilot had to concern himself with trying to get comm back.

SCHMITT — Apparently, Goldstone dropped the up-link. When they were getting it back, I was switching omnis, and for a while there, it was just completely out of phase. They had a continuous down-link on us.

CERNAN — It was a very inopportune time, I might say, because it happened just right after ignition. I think that's something, though, that the INCOs are going to be able to clarify. We certainly can't give you the details. It's just that there was essentially no comm on all the antennas.

72H1544

CERNAN — We flew into a trajectory that appeared to be nominal. AGS showed us slightly out of plane. As a result, our tweak at 9 ft/sec was minus 4, minus 9, and plus 1. We burned out X, Z, Y, in that order.

SCHMITT — It was about 7 ft/sec, a little over 7 ft/sec.

CERNAN — It looked like we might have had a g-sensitive drift in our Y-accelerometer in the PGNCS. The tweak was excellent because our rendezvous was just as nominal a rendezvous and as nominal a trajectory profile as I've ever been involved with. The drift in accelerometer did not bother us anywhere else in the tracking or in the rendezvous at all. Rendezvous navigation followed the checklist; we got right off the form very well. We got all the updates into the AGS. The residuals in the TPI burn were greater than what I had expected. We did not record them because I wanted to get them nulled out just as soon as possible. I don't know the tenths, but they were minus 7 in X, and they were 4 and 4, and I'm not sure whether they were plus or minus in Y and Z. They were large, larger than I'd expected. They were minus 7 and 4-point something and a 4-point something. We reduced those to less than 0.2 ft/sec. From then on, we continued to plot right through the midcourses right up the pike on a nominal trajectory.

72H1543

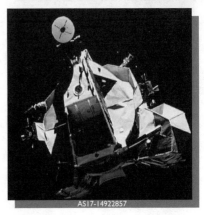
AS17-14922857

SCHMITT — The comm was good. I have a couple of comments about the AGS. Early after insertion, I always checked the accelerometer. They looked real good. About 5 or 10 minutes later (I can't remember exactly), I looked and I'd accumulated maybe a foot and a half per second in X. I did a gyro cal, and after that, there was no significant accumulation in X. It went very well. I did that without talking to the ground, but I felt I had an understanding with them on that. On the TPI solution, the AGS was essentially within 2 or 3 ft/sec, a good TPI solution after six marks. The insertion solution was not very good. It was off by a number of feet per second in X and even more in Z. This was the first one of 17 marks. The PGNCS recycle and PGNCS final were very close, within a couple of feet per second.

CERNAN — Midcourse Solutions - The first midcourse solution agreed effectively all systems, except AGS out of plane was a little bit high. The decision was made to burn the onboard PGNCS solution out of the LM, which was minus 1.2, plus 0.4, and plus 0.3. We continued to track right up the pike. Midcourse 2 came up, and we again prepared all the solutions. The AGS out of plane was still a little bit high and actually in the opposite direction from the PGNCS. We had a slight variation in the CSM solution in Z. I don't know why. It came up with plus 5.4 ft/sec in Z. So we really didn't get a very good correlation between the CSM and the LM on the second midcourse. But the PGNCS was still performing, the radar was still performing, and based upon our trajectory plot and based upon our following a nominal inertial line of sight rate, we decided to burn the onboard PGNCS solution in the LM. It was minus 0.4, minus 0.7, and minus 1.6. From there on out, we just continued to follow the inertial line of sight angles. There was very little tweaking in either Y or Z. We just sort of floated right through the braking gates. At 1 mile, I think we took about 6 or 7 ft/sec off to hit 30. We met all the gates as prescribed and just came moving very slowly into the final stationkeeping. We went into a formation flight around the CSM. We got a good inspection of the spacecraft and the SIM bay, the report of which is in the transcript. Everything looked good to us. The command module maneuvered to the docking attitude. The LM just took its docking attitude, gave stationkeeping control to the command module, did pitch and yaw maneuvers, and stood by for docking.

EVANS — One of the noticeable differences between this docking and the docking with the S-IVB is the fact that the ascent stage did dance a lot more than the S-IVB did. The S-IVB is steady as a rock. The LM dead band would change attitude, and you'd try to follow it. On the first attempt, I must have had less than 0.1 ft/sec, just barely closing. I was just taking it nice and easy. We made contact and did not get capture. As soon as we didn't get capture, it was obvious we were closing too slowly. We backed off a couple or 3 feet, renulled the rates, initiated the closing rates, and got capture. As soon as we got capture, both vehicles went to CMC FREE. I looked out, and I had some rates in the CSM and I'm sure that the LM had rates also. He must have had.

CERNAN — We went FREE. Upon capture, the LM went FREE. The CSM trying to null the rates ended up perturbating the LM and giving us rates.

EVANS — We finally gave up on that mode and had the LM go to ATTITUDE HOLD. Once you get ATTITUDE HOLD, the CSM could null the rates. We got it lined up and attempted the hard docking. There was no problem. The probe retract came back. This time, it didn't sound like it was as much of a ripple fire. It was more of a "phhtt." It was a quicker hard dock than it was the previous time.

CERNAN — I want to say something about the visual sighting during rendezvous. From the LM, I was able to see the command module when it was sunlit at somewhere around 100 miles. I definitely defined that that was the command module. After the command module went into darkness, I could not pick up his tracking lights until we were well within about 40 miles. I could not pick up the docking light, the rendezvous light, of the command module until we were well within 40 miles. It was initially a very dim, faint flash. I was able to verify on board that the LM tracking light was working. I finally figured out how; it was reflecting off the underside of the EVA handrail on the left forward side of the LM. I could see the LM tracking light flashing. There were some particles we took with us that stayed with the spacecraft, and you could see the sequential flash off the particles as the result of our LM tracking light.

SCHMITT — Regarding the television and photography from the LM, we'll just have to wait and see how it turned out. I took a lot of footage. We put it on not only the ascent mag, but we put it on the other mag. That includes the SIM bay. Right or wrong, we did have a Hasselblad on board, so we have a lot of Hasselblad photography.

13.0 LUNAR MODULE JETTISON THROUGH TEI

CERNAN — Postdocking Check and Pressurization - The general comment I want to make about the postdocking operations is that both pilots in the LM took their helmets off to keep the dust off, primarily. The commander took off his gloves almost immediately after insertion, and flew the entire rendezvous that way. Jack took his off some time later.

SCHMITT — I kept mine on for some time. I can't remember exactly when I took them off. I did most of my preinsertion work with the gloves on, because I didn't want to take the time. I wanted to get that initial AGS solution. I could get that fairly rapidly with the gloves. I didn't take the gloves off until maybe 10 or 15 minutes after insertion. I kept the helmet on all the way through most of the transfer, just to avoid breathing the dust. I had the sinus irritation on the surface.

CERNAN — The commander kept his helmet on throughout the rendezvous and docking. I took my gloves off after insertion and left them off. As soon as we were hard docked, the commander took off his helmet. As I look back at that, because of the dust debris in the LM spacecraft, I'm sorry I did. I could have left the helmet on, and I would have had a lot less eye and mouth type of irritation. You knew you were in a very heavily infiltrated atmosphere in the LM because of the lunar dust. I don't know how much lunar dust previous flights had, but I think we saved a great deal of grief by sweeping all the dust we could find on the floor into the holes and putting our tape covers over those holes. I think that had to help a great deal. There was an awful lot of dust on the floor that we didn't see. The commander had his helmet and gloves off all throughout the entire transfer. We handled the transfer the way we'd planned. The LM pilot did most of the preparation of the gear in the LM, and the commander stayed in the tunnel and passed things on. The inventory was going on in the command module side and on the LM side, both. We vacuumed each other's suits the best we could and everything else that got supposedly transferred, unbagged, or uncovered.

SCHMITT — In spite of the CMP's comments to the contrary, I think we got things remarkably clean. There wasn't an awful lot of dirt in the command module coming back.

EVANS — That's true.

SCHMITT — In contrast, he may have thought it was dirty, but I was surprised we were able to keep the level of contamination in the command module down.

CERNAN — After I took my helmet off, I could go halfway through the tunnel and stick my head up in the command module, and it was a totally refreshed, unpolluted atmosphere up there. It never did get polluted.

SCHMITT — I think having that vacuum cleaner running in the LM had a lot to do with keeping the flow in the other direction, filtering out the air.

EVANS — We never did vacuum in the command module because it just wasn't necessary.

SCHMITT — The suits were noticeably cleaned by the vacuum cleaner. You could tell you were pulling stuff off them, although they were still dirty. Every subsequent time we handled them, we got our hands dirty. I think most of the free dust was taken care of.

CERNAN — We effectively stayed on the transfer list. I say effectively, throughout the transfer. However, some things got transferred out of order and temporarily stowed in the command module. We effectively used the transfer list not as a cookbook recipe type of thing, but as an inventory list. We inventoried it several times from both ends and were satisfied we had everything transferred. We then pressed on with the LM closeout. The LM closeout went nominal. We got back into the command module, and the LMP closed out the LM. For convenience, the commander went back and closed out the LM hatch and put in the command module hatch. Because of the slow tunnel vent, or the long duration of tunnel vent, the commander stayed in the tunnel, the LMP in his seat, and the CMP in the left seat. We suited up and prepared for our integrity check. As soon as the LM tunnel vent was complete and we were satisfied with the integrity of the hatch, we went into the suit integrity check.

EVANS — I bet it must take at least three or four times longer than the simulator did for the tunnel vent.

CERNAN — I think that's going to be applicable to Skylab. They're going to have to vent before they undock, I think. The tunnel closeout was easy. We had no drogue and probe which were stored in the LM for LM jett. We just followed the checklist, and it all seemed to happen just as advertised.

EVANS — We got a little bit intrigued with the LM jettison. It was great. It just sailed out there nice and pretty, and we got a lot of good pictures of it. We should have been maneuvering. We ended up getting into P41 after jettison for sep burn, a little bit late. That was no problem either, because we just trimmed the residuals for P41 and got a good sep burn.

CERNAN — Cleaning control in the command module was excellent, considering all the dust and dirt that just seemed to adhere to everything in the LM. When we got back in the command module, with the exception of the suits, and LMP and CDR, everything was clean. Everything was clean because everything was bagged before we brought it over - bagged and zipped. We never did open anything once we got it zipped up. So the command module stayed exceptionally clean throughout the remainder of the flight.

SCHMITT — In the bagging of the decontamination bags, I made a special effort, after requests prelaunch, to pull those zippers as tight as I could. They should be pretty tight.

EVANS — High gain always worked good; omnis and S-band were good. Photography went as advertised. We had lots of targets of opportunity. SIM bay operations have been mentioned before. TEI updates, normal. Sextant star checks were good for TEI.

CERNAN — Every one all through the flight was good, which made me feel real good. I made sure I got it on those last few. I wasn't going to change any mode of operation. I made sure I got it on TEI. Just to make you guys feel at home. I figured you'd think I didn't do it right, if I didn't get the master alarm.

SCHMITT — The TEI, at 1/2g, or whatever we were pulling there, seemed like more than that.

EVANS — It sure did; it seemed like it was really pushing you back in the seat.

SCHMITT — Ron and I both started out holding our heads up and eventually relaxed them back on the couch.

CERNAN — I guess we must have had the spacecraft pretty well stowed, or tied down. I briefed the CMP and LMP, and, as I recall, those kind of burns back on Apollo 10, lots of things start moving through the spacecraft and find their way to the aft end of the spacecraft because of the g-load. Much to my surprise, all we had was an initial thud as we moved away from the station, and we didn't have any gear flying through the spacecraft.

SCHMITT — I found a white tag, wetwipe.

CERNAN — Other than maybe one or two of those things, in looking back, I would have expected more gear to come from somewhere, but we prepared for those burns pretty well.

EVANS — That reminds me of all this water condensing on the ECU unit, the pipes, and what have you. When we put our suits on for the EVA the next day, your suits were noticeably wet. When I pulled the PGA bag up, it was damp down underneath the PGA bag. As a normal procedure, we should have, either after the burn, probably before the burn, made sure we wiped up the water in the LEB.

CERNAN — Our suits were damp when we put them on, but I could not find any real water down there.

EVANS — There's always water down there in the ECS. I just assumed that's where all of it came from. There's not a puddle of water. Like I said, it's just damp.

CERNAN — It's almost as if it was colder down in the LEB, and water was condensing all over the suits. It wasn't as if they were in a puddle.

EVANS — The simulator is set up such that in roll dead band, it goes over to one side of roll dead band and just kind of stays there. During the TEI burn, it was bouncing back and forth from one side of the dead band clear over to the other side of the dead band. When it's bouncing back and forth, the roll rate is up around, oh, 0.4° per second, arcing back and forth across the roll dead band. I'd like to mention chlorination at this point.

CERNAN — Without fail, almost every chlorination leaked. Sometimes large quantities of water, other times just small quantities of water.

EVANS — Water or chlorine?

CERNAN — A combination. Where it leaked appeared to be around the bag. It was the cylindrical chlorine dispenser that was continually wet. It was not where the dispenser fit into the needle or where the needle adaptor fit into the spacecraft. It was within the chlorine dispenser itself. Chlorination was a case of always cleaning your hands with chlorine because you always had it available down there within that dispenser. In some cases, you had a larger quantity of water that had to be wiped up with a tissue. That plagued us throughout the whole mission. It turned out not to be a serious problem because we learned how to handle it. That was one system anomaly that hadn't really been brought up.

EVANS — In two cases, I'm almost positive, it did not puncture the ampule. The reason I believe that's correct is that, when you started to crank the outside of the cassette down to push the chlorine into the water system - it was very hard to turn. If you tried to force it, you could force it on down there, and I'm sure that's a good way to break an ampule on the thing. In two cases, we took the bayonet fitting loose again and put it back on there, and in both cases, then you'd start to squeeze the chlorine out of the ampule into the system, and it would turn easier.

CERNAN — We got the chlorination done. We didn't miss any injections of chlorine, and we didn't miss any of the buffer samples. I guess we got the job done; it was just a little bit messy. The chlorine was evident because the CDR eventually peeled all the outer skin off his right hand. I'm convinced it was due to the chlorine, and had nothing to do with the EVA.

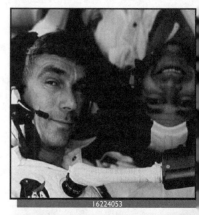

16224053

14.0 TRANSEARTH COAST

CERNAN — Passive thermal control was what I would call unusual attitude because of the UV/IR requirements. These unusual attitudes did two things. They required us to remaneuver the spacecraft several times and enter and exit PTC several, several times, which, in itself was not a problem, just additional coordination. Coincidentally, most of these particular PTC attitudes were within 30°, certainly 45°, of gimbal lock most of the time. We were looking at the red apple a good portion of the trip home. Some of those attitudes where you actually were in attitude or PTC in these relatively unusual positions, change the equilibrium heat load on the spacecraft. RCS quad temperatures were all right, but you could see it in helium package temperatures and, most noticeably, you could see it on the change in condensation from the tall hatch to the forward hatch. The tall hatch eventually, for most of the way home, ended up to be very dry. The second day out on the way home, the center hatch got soaking wet to the point that we even took a dry rag and wiped off some of the latch components and some of the gearbox components, externally. Not that it did much good, but there was just that much water on there. I think this is all due to the PTC attitudes required for the SIM bay experiments on the way home.

AS17-15223311

SCHMITT — It was cold in the spacecraft, too.

CERNAN — Oh, yes, it was cold in the spacecraft.

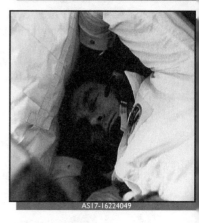

AS17-16224049

SCHMITT — Not as cold as the commander thought it was.

CERNAN — Cold enough to warm it up, on the commander's orders.

SCHMITT — We mentioned we warmed it up on the ground's suggestion of an extra invertor and going to MANUAL on the temp gain. I think we discussed that.

CERNAN — Ron, all your REFSMMAT changes, your platform torquings, all those went very well, I thought.

EVANS — They were great, went really well.

CERNAN — All the way back home, it was just changing attitudes, changing attitudes, changing attitudes, with the exception of the EVA day, which we'll cover here shortly.

EVANS — CSM EVA - On EVA prep, we really didn't have any problem. We didn't know of any at that point. The EVA prep went right down the line, essentially. It was well laid out within the experiments checklist. We checked things off as we went, and stayed pretty much on the timeline. We started about a half hour early, and finished a half hour early.

AS17-15223374

CERNAN — We were a half hour early throughout the whole thing, and we lost that half hour in opening the hatch. We turned out to be exactly on time. Where we lost that half hour was on a comm carrier change. Post-EVA - One thing that helped us immensely on what ended up to be, I think, a very fine entry stowage was that we backed off after EVA and took a good long look at the long-range stowage as well as the post-EVA stowage. We really started housecleaning, cleaning up the cabin, and effectively stowing some of the articles that were not going to be used any further in the mission for entry at that time. Our entry stowage really started with the EVA timeframe period, and I think that really helped us out in the long run. The only change to the prechecklist and postchecklist was the order in which we doffed and donned suits. It was very evident there were certain convenient ways, because of the way the suits were stowed and the way that people fit into the checklist, that when we donned the suits. The commander was first, then the LMP and the CMP donned last. It worked out very fine. The CMP had less work to do in his suit, which also aided him in the long run. In doffing, the LMP was first, then the CDR and then, the CMP. That wasn't exactly the way it was called

AS17-15223391

AS17-15223393

for, but that's the way it worked out. We stowed our suits in the L-shaped bag prior to putting the center couch back in. This was another good decision, I believe, in helping us get the suits stowed back in that L-shaped bag.

EVANS — Cabin depress - No problems. Normal depress. Hatch opening. Even though the cabin was completely depressed, we were reading zero pressure. As soon as I opened the hatch, there was enough residual pressure, or something, inside the spacecraft that it actually tended to pull the hatch out of my hand.

CERNAN — Because your suit is bleeding into the cabin all the time, so you never truly get zero.

EVANS — That's right, you never truly get zero. The dump valve was still open, and if I had not been hanging onto the hatch, it would have blown it all the way open.

CERNAN — That's not unexpected because it's exactly what we had on the lunar surface. We completely dumped the LM. I'd still have to break that hatch loose and hold it open about 6 or 8 inches until things just vented. Then, I could let go of the hatch and open it all the way. If I didn't it would slam back, closed. It was basically the same thing. You have to open that door and really let things get down to zero.

EVANS — When I opened the hatch, all of the little ice crystals started flowing out. A pen went floating by, and something else went floating by - wasn't quite sure what it was. There's all kinds of little particles and pieces that start coming out through the hatch.

CERNAN — I looked specifically for the scissors. I didn't see any scissors go out that hatch. I hate to say it. Ron, I'd like to say they went out the hatch, but I sure didn't see them go.

SCHMITT — Sure you didn't see them go?

EVANS — I caught the one thing that started to go by me, and I put it in your pocket. Once all particles and junk were out of the way we pushed the hatch open. We disconnected the counterbalance with the tool E. So, that we locked the hatch in the open position, so I just shoved it open, it went beyond the center position and locked in the open position with no problem.

Egress - I had a tendency to float up against the MDC. I had to cautiously duck to get my face as close as I could to the bottom of the hatch in order to get the OPS past the MDC and get on out. TV and DAC installation worked fine. I could hang on with the right hand on the hatch, the great big D-handle on the hatch, with the TV pole in my left hand. Worked out real fine. Just stick it in there and line it up; make sure it was locked in, then climbed on up the pole to turn the TV on. I turned the back on. You couldn't see the light on the thing, but you could feel the camera running once you turned it on. You could touch it and you could feel it vibrate a little bit. The lunar sounder cassette retrieval should be on the air-to-ground tapes. Most of it was no problem. The pan camera cassettes were next. No problem on the pan camera cassettes. It's obviously a bigger mass, and it's quite apparent when you try to move that big mass around. It is heavier and weighs more than the other things. It's easy to move, but it is it takes a little effort to get it started. You know that if you ever get it started in one direction and it's going to keep on going and you have to stop it. I just tried to keep it under control. Mapping camera cassette had the same problem I had in the SIM bay c^2 f^2. That was getting the thermal cover off. It stuck underneath the mapping camera laser altimeter door. I gave it a big jerk and it came off. SIM bay inspection - That's all covered in the air-to-ground tapes. TV/DAC removal again was real simple. You just had to squeeze the lever and TV came out. It was easy to hang on to with one hand and maneuver the TV around and point it toward the Moon. I didn't have to worry about shining it into the Sun. I tried to again hang on with one hand and point the TV around toward the Earth. The Earth was maybe 15° away from the Sun. I tried to be a little more accurate. When I did that I really lost control of my body position. I was trying to maneuver the camera. You need both hands to maintain your body control. Comm during EVA was loud and clear for me throughout the EVA. There was a lot of background noise; I'm sure it was coming over the VOX circuit.

CERNAN — It didn't appear to me that anyone on the ground had trouble reading you.

SCHMITT — One thing we did because it was bothering us I turned the VOX sensitivity down about two notches. That really improved the comm performance.

EVANS — I don't know if it made any difference or not, but I got the impression that it did help.

CERNAN — Comm into the cabin was excellent. I never had any trouble understanding with that hissing in the background.

EVANS — Ingress - It seemed to me it was easier than egress. For some reason, hatch closing was harder than I'd anticipated. Maybe this is the same reason in that I must have been exhausting into the cabin all the time. That hatch would come closed to within about an inch of closing on the outer edge. Then it took an effort to pull the hatch closed so you could activate the latching handle so that you could get the latches over center. Of course, once you got the latches over center, it was real easy, a couple more cranks on the hatch for closing. Repress was normal.

SCHMITT — All I did was work in the hatch area. I want to emphasize what everybody's always said that you do your best work when everything's going easy. Move yourself in small increments to where you want to go. You can turn and dip and raise yourself out. I think it's also useful for any hatch or port operation to have somebody available to push you out on your tether towards where you want to go. It just eases the operation. With the struts and everything available there, there was never any feeling there that I could not have a way to control my body position. Sometimes it took a few seconds to get it where I wanted. The one thing, invariably, everytime I went back inside I had the 90° disorientation for a few seconds until I got the perspective of the cabin again. I'd say okay, that's right. Then I would go back outside and come back in, and once again it seemed that cabin had rotated 90° to my perspective. It's just something that's no problem, it's just a change of perspective. For some reason, I experienced it several times. I guess the biggest problem working in that angle for me, attitude, was I had the Sun full face.

EVANS — You had the Sun in your eyes most of the time.

SCHMITT — It made it hard to look in detail to see what you're doing. You were clear image; you were there. I could see every major operation, but I could not see specific details.

EVANS — I had no awareness whatsoever that I had an umbilical on my back. I never got the feeling that the umbilical was restricting my movements. I didn't even know that it was there. Did you observe at any time, did the umbilical ever get tangled around.

SCHMITT — No, the umbilical was easy to tend. There may have been one. I had a vague impression that I asked you to hold up, or maybe I did not say anything, I just moved you away from a handhold or something. The umbilical didn't seem to slink around. You seemed to have everything you needed on it.

EVANS — I did not even know it was there. Being tied to the umbilical does not restrict your movement or give you a feeling that it is restricting your movement at all.

Transearth - I did not see a light flash.

SCHMITT — That evening I did see them again falling asleep.

EVANS — I did, too.

SCHMITT — So then, it was just that period during the actual experiment for some reason they were not visible.

EVANS — We never really utilized the waste stowage vent to get rid of any odors out of that waste stowage compartment. It was always a crime if you were in that area, if you got real close to it.

SCHMITT — The cabin generally turned over the atmosphere in pretty good style. I got saturated sometimes with gas and it took a few minutes to clear. The cabin did good job.

CERNAN — Flight Plan updates were super. The Flight Plan was excellent. Changes were held to a minimum, and we really did not change any part of the entire flight except a few dates, times, and attitudes.

CERNAN — Entry preparations began after EVA and continued all through the next day. We had very little final stowage to do on the final entry morning, just those things we had to leave unstowed until we got out of our sleep restraints. Basically, we just had to tie the big bags down. Final entry preparations went by the checklist. If anything, we stayed about 5 minutes ahead throughout the entire checklist, including separation and activation of the command module RCS and .05g, which came on time. Communications I thought were very good through this time. I understand the ground heard everything we said right through blackout. As soon as we came out, they still had ARIA, and they could still read us. We could have read them, but they never transmitted anything.

15.0 ENTRY

EVANS — Prior to midcourse 7, we did a null bias check and also an EMS Delta-V test. The Delta-V test had been going at about minus 22.2 or 22.1 at the end of the 1 seconds. Then prior to midcourse 7, we ended up with a minus 27. We'll have to check the air-to-ground tapes, but it still was within limits. We'll check the air-to-ground on the actual values of this, but it failed the null bias check by a considerable amount. Since did that, I went through an extra EMS entry check. It passed that EMS entry check. I can say for sure whether the .05g light was on during test 1 or not. It was on during the second EMS check. As a result of that, it was determined by the ground that the accelerometer in the EMS was probably putting out a couple of extra pulses. It was decided to change the entry checklist so that we would not put the EMS to normal until .05g time. This is what we did and the EMS functioned correctly throughout the entry. Entry parameters are on the air-to-ground tapes and also on the frames. .05g light came on, The RCS sounds were a little bit louder than we'd been practicing with in the simulator, I thought. I've also mentioned the drifting and the cross coupling and minimum-impulse SPS.

CERNAN — Is it louder or more of a bang?

SCHMITT — It was less than the LM and more than the service module. That's a good way to put it.

CERNAN — Banging on a solid can.

EVANS — Communications blackout - You'd never know it from inside the spacecraft. Ionization - Ionization is bright. It was very bright, very bright.

SCHMITT — There seemed to be an early glow. Now, whether that was ionization or the initiation of the fireball, I don't know.

CERNAN — They're one and the same.

EVANS — They're one and the same, I think.

SCHMITT — Yes, but with the true fireball, it would seem to me that that would be something that you really couldn't look at. I couldn't look at it; it was too bright. I couldn't stand it.

EVANS — You couldn't look out the rendezvous windows at the fireball because it was too bright. I felt like I should have put on my sunglasses in order to be able to see. That intensity only lasted for about 10 seconds, maybe a little longer.

CERNAN — It was longer than that.

EVANS — It's hard to remember for sure. Peak-g - The one thing I can recall about peak-g is that I definitely could not see the peak-g value on the EMS because I couldn't see where the pointer was on the EMS. I determined peak-g by looking at the g-meter. I could read the g-meter, and it was something just less than 7. You're pretty well pinned to the back of the seat at peak-g. You definitely have wrist action with no problem, but trying to raise your arm took a lot of effort. I don't think you'd ever get your arm up if you didn't already have it up at 6g. Guidance Termination - No comment.

CERNAN — Let me talk about guidance for a minute. The CMP was in the left seat, monitoring the EMS, g's, and what have you. I was in the center seat, monitoring CMC and passing bank angle information so we could come to a logical conclusion about giving the spacecraft over to the CMC for guidance. I had the impression after peak-g that the two of us were very close to convincing ourselves that the CMC was not going to roll the spacecraft.

EVANS — That's right.

CERNAN — It seemed like it took a long time for CMC guidance to roll the spacecraft back out of peak-g. I had the impression that you were just waiting for me to say, "Let's take it back," and we would have taken it. It seemed that it was longer than the simulator. It was a long time before the CMC made its first initial roll command, almost too long. In another couple of seconds, I think we might have taken it over.

EVANS — We might have taken it over. I think the reason that we felt that way was because most of the runs that we ran in the - simulator were nominal runs where you get about 6.1g. If you get 6.1 or 6.2g, you do reverse the bank angle to a one-eighth roll quicker than you do if you have a higher peak-g. We were pretty close to the 7. I don't know if you ever saw 7 on the DSKY or not.

CERNAN — No, I never saw 7g, to my knowledge.

EVANS — I never did see 7.

CERNAN — I saw 6.64 or 6.65, something like that, but that's about as much as I ever saw. It just occurred to us that the CMC was never going to get around to rolling 180°. Once it did roll 180° to the best of my recollection, it never rolled except from left to right. It never rolled across the top again. It went from 90 or 100 one way, and 70, 80, 90, or 100 the other way but never made the complete turn again. It just rolled left and rolled right, rolled left and rolled right.

EVANS — Visual sightings and oscillations - The one thing I forgot to look for was that in the simulator from about 90 000 on down to 50 000, it starts pitching. I don't remember if we ever got that pitch rate going or not. I think most of my comments should be on the air-to-ground tapes throughout the entry. Those would be more appropriate than something I might recall at this point in time.

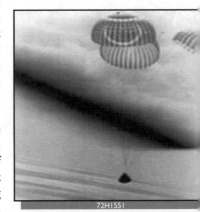

72H1551

SCHMITT — I think we all had about the same impressions. My standard comments for launch and entry are that there are certain periods of time that lasted for several minutes where I don't think you would be too extensively doing malfunction analysis and problem solving, particularly during peak-g. I think you're mainly concentrating on the g-load, and it would be hard to move your arm anyway to take care of any problem with the switches or otherwise. I'm not saying you shouldn't simulate it. You learn a lot of systems and that sort of thing, but I don't think you can anticipate doing work during that period of time.

S7255834

CERNAN — I thought the drogue deployment was violent. I thought the spacecraft oscillations were quite violent. I'm not saying that there was that much difference than I personally witnessed in the past. I just forgot to expect the violence of the oscillations on the drogue.

SCHMITT — I was watching the drogues, and they were moving just as hard as the spacecraft. I think that the drogue movement was being transferred to the spacecraft.

CERNAN — We had all drogue deploy, all main deploy, and once we had the mains, apparently we had two good parachutes.

72H1550

SCHMITT — I watched the full main deployment, and I could see all three reefed parachutes after deployment. They stayed reefed probably about the amount of time you'd expect them to. Then you could see the reefing lines start to go, and the two parachutes that were on my right filled fairly quickly and seemed to push the left parachute away and out of the main slipstream. It filled much more slowly. It was clear to me, and it should be in the photographs, that the reefing lines were free. The parachute was not filling. Then gradually, it filled completely. I would say it was 15 or 2 seconds before that other parachute filled completely. It was sluggish. It just got pushed out of the way and couldn't get the full flow of the air to fill it.

CERNAN — And I think that some of the people on the recovery team said that they saw the two parachutes plus the streamer.

SCHMITT — I wouldn't call it a streamer. It was just unreefed. It was just still reefed.

EVANS — I'd like to make a comment about the dynamic visual cues of rotation. Throughout the entry, I didn't really feel that I was rolling. I didn't get a feeling of dynamic roll other than the fact that I was watching the needles. There were no centrifugal forces involved in that operation until we were on the main parachutes. When we were on the main parachutes, I felt like I was lying on my back on a revolving table.

SCHMITT — I suspect that might be because of the higher g-loads when all these other things were happening. I don't know how much you were looking at the horizon as you rolled, but that's all I had to look at.

EVANS — I could see the roll. I had the visual sensations of it, but I didn't have the dynamic feeling of roll until we were on the main parachutes. While we were on the main parachutes, the roll was not continuous in one direction. It was rolling in one direction at 15° to 20° per second, and for some reason, it would reverse and go back the other way. The rolling sensation on the parachutes was kind of a wind and unwind type of a roll.

SCHMITT — The DAC operation was normal. We took a little bit of extra footage early of the horizon which I hope turns out. I don't understand why you turn the DAC off after you're on the mains. There's no reason to. We ended up with some unexposed footage.

CERNAN — Oh, you did? I thought you let it run out.

SCHMITT — No, the checklist said, "Stop DAC," and I stopped DAC and it was a little while later that I wondered why I stopped the DAC.

CERNAN — I'm sorry. I thought you let it run out.

SCHMITT — I don't know if we would have gained anything by it except some more pictures, but there was no reason to turn it off.

EVANS — Communications - From 90 000 feet until about main parachute deployment, I had a time trying to hear Jack. There was a lot of background noise.

SCHMITT — That's right; I remember that.

EVANS — It just gets noisy in the spacecraft from about 90 000 feet on down. Once you get the altimeter off the peg, I had a time hearing you call out.

SCHMITT — I was shouting, too. I realized you were having trouble hearing. There was noise. It must have been air noise coming through the hull.

EVANS — It was something.

CERNAN — You were on VOX that whole time. You could have been keying, and that-noise could have been coming through your VOX. I'm not sure. That takes care of entry, which was a good one.

SCHMITT — Let's mention ECS. I never was uncomfortably warm in the cabin at all even through hatch opening.

CERNAN — We cooled the spacecraft effectively. Just normal powerup of the ECS systems cooled the spacecraft down prior to entry, and it was comfortable. Even after we landed when it normally does warm up because of humidity et cetera, it was still very comfortable. I never thought it got hot or extremely humid throughout the whole recovery operation. The altimeter read about 100 feet when we hit. We'd been warned that we might hit with 17 feet on the altimeter. We made callouts all the way down on crew condition, altitudes, and the DSKY read-out in terms of position. They had a visual on us all the way down. We were right next to the ship, apparently right at the zero aim point.

72H1552

16.0 LANDING AND RECOVERY

CERNAN — We hit with a pretty good thud. As soon as we recovered from the thud, the LMP went for the main parachute release breakers and I hit the switch. The parachutes, apparently from the lack of a great deal of wind, just rose petaled in an almost 120° position around the spacecraft. We had no tendency of ever going stable II, partly because of the seas and the wind and also because we released the parachutes in a hurry. We proceeded to go through the postlanding checklist. In addition to what we said about the temperature and humidity, I think the postlanding vents certainly did help. We had that running. We had communications with Recovery all the way down on the parachutes. We monitored the recovery all the way through by communicating with the recovery chopper. Spacecraft status was excellent. We followed through the checklist; and, although it was not needed, the checklist calls out to inflate the bags after you've been on the surface 10 minutes. We inflated the bags for 7 minutes after we'd been on the surface for 10. It is a good idea in spite of the fact that they were not needed because it does give you that added protection of staying in stable I in case you might end up going over. There was no seasickness.

72H1553

72H1559

SCHMITT — We did not put the postlanding vent ducts up although they were available. There was plenty of air moving in the cabin from the normal ventilation. You could feel it. You could feel it move. I don't know about you fellows, but I had plenty of air.

EVANS — I had plenty of air coming across on the left couch, too.

CERNAN — The CDR climbed out of the center couch, went down to the LEB and got the cosmic ray prepared and available. We stood by for hatch opening. When the hatch opened, we received the bag with the lifevest, the cosmic ray protector box, and the temperature gage. We put the temperature gage on, the cosmic ray was stowed in the waterproof package, and we put on our vests. When we were ready to open the hatch for the final time, we powered down the spacecraft via the checklist and panel 250.

SCHMITT — I would call the touchdown a very sharp crack rather than a thud. It is an obvious sensation. It's not one that seriously jarred you or hurt you in any way. It was a sharp and abrupt stop. I think it might have taken me 2 or 3 seconds to start making a motion towards the breakers. There was enough jar to say, "Okay, I better recover from it," and then I reached for the breakers. The windows fogged up inside almost immediately and there was also material on them on the outside. It looked like something other than moisture on the outside. It was sort of a brownish yellow. I had no motion sickness at all, but I didn't really care whether I got up out of the couch or not for a while. I didn't have the desire and that's about all I can say as far as any change of feeling from zero-g to one-g. I noticed that my neck muscles seemed to be really working to hold my head up. It was much more than normal and this persisted for about 24 hours. It gradually went away until about 24 hours later I felt perfectly normal raising my head.

72H1560

72H1568

EVANS — I guess I didn't even notice the transition from zero-g to one-g. I didn't pay that much attention to it.

CERNAN — I didn't really notice any difference either. I particularly got up on the LEB to see if I would but I didn't.

SCHMITT — On the egress, my lifevest did not inflate automatically. That might be worth looking into, because apparently they were a new set sent out specifically because the first set sent out had not gone through inspection. These presumably had and they still did not inflate. Only one out of the three inflated all right.

72H1570

EVANS — Another point I want to make is that if we're going to put that temperature gage in I would recommend that you send in a roll of tape, a bungee, or something so you can strap it to the strut. We just happened to have a piece of tape in the LEB so we could tape it to the strut.

SCHMITT — Crew pickup for the LMP was exactly as I'd been told it would be. We practiced on Apollo 15.

CERNAN — I don't think there is any other comment on crew pickup other than to say that it was done in an outstanding manner.

EVANS — I concur. It was good.

17.0 TRAINING

EVANS — CMS - The crew station was always in good shape. Some of the interior storage was boxes, but the items of storage equipment that needed to be used were always there. Jerry Stoner and his crew kept it in excellent orbital storage most of the time. If we wanted it restowed for a SIM, for lift-off, or anything, all we had to do was let him know and they were in there all hours of the day and night to get everything squared away.

72H1422

Fidelity of the CMS - I've mentioned the differences in the actual vehicle and the CMS in the various other sections of the report. They are minor.

Availability - The CMS was always available any time I wanted it - more than I could use it in some cases. The people involved in the CMS training - knowing full well that Apollo 17 was their last shot - were outstanding in their desire to continue training and to put out their best efforts in insuring that I was trained and ready to go.

72H1507

Visual systems - I didn't seem to have any problems with that. The biggest problem was in the star ball. Every once in a while it would get fouled up. That and the sextant drive were a little bit jerky but it worked great.

Software - F computer was going all the time. If it ever conked out, they fixed it fast.

SCHMITT — I echo all of Ron's good words about the quality of the training and the dedication of the troops down there. From the systems point of view, I think the fidelity of the systems was all that was required and was generally very high. Only those differences that were spacecraft peculiar were the ones that were not simulated. Where there were other comments to be made, they have been made in conjunction with systems work. Availability was fine as far as I'm concerned. The visual systems were good. The only ones that really concerned me were the entry visuals and they were certainly adequate, although they do not give full representation in drogue and main deployment.

72H309

LMS - we never really stowed the LMS. The gear necessary for general training was perfectly adequate. All of our crewstation-type training was done in the mockup of the LMS. We've also mentioned the fidelity of the training and wherever there was differences, the L&A and the AOT were excellent representations, based on a little bit of comparison that I did. AGS software in flight was just like the AGS software in the simulator with one well-known exception. You get your displays faster in the simulator than you do in the flight but this was never a problem. My work with the PNGS is limited and confirms what the CDR has already said, that the PGNS in the simulator and PGNS in flight are essentially the same.

SCHMITT — The LRV navigation simulator - The main usefulness I received from this simulator was working over the traverses and understanding and knowing what we were supposed to do. The comparison of driving on the lunar surface with that of the simulator was very poor. I think the problem is that the simulator has to give you a much higher point of view. The simulator is 20, 30, maybe 40 feet higher above the surface. When you're down at 4 or 5 feet, as you are in the LRV, it's a different world. It makes a big difference in what you recognize. The other side of that coin is that once we started moving on EVA-2 and EVA-3, there was never any difficulty on the lunar surface of recognizing the larger features that we had seen on the LRV simulator. So it worked out very well. CMS/LMS simulations - In general, the integrated work we did always went very well.

EVANS — I think it did. We lost 1/2 day on an integrated sim.

SCHMITT — I think the few places where we ran a little bit behind in flight were those portions we never really simulated, such as suiting operations and the tunnel operations. They went smoothly, too. I think not simulating them in detail was a good decision, and I don't think it affected our operations.

EVANS — I don't think so, either. SNS - I don't know how you can ever overcome this, because in the SNS you are training the crew and you're also trying to train the MOCR. There isn't that much for the CMP to do to keep busy all the time from a training standpoint. I don't know if that's necessarily bad or not.

SCHMITT — We had a few excellent SNS, from the standpoint of fairly continuous activity. In general, Ron's comments are valid for Apollo 17. If my memory serves me correctly, Apollo 15 SNS were much more active. And I don't know why there might have been that difference.

EVANS — On Apollo 14 backup crew, I was more in a learning stage at that point than I was on Apollo 17. In Apollo 17, it was more of a review stage for training than anything else.

SCHMITT — I think that's a natural point.

EVANS — DCPS - We tried to get it once every 2 weeks, which we did in the first part of the training cycle. The last 3 months we were lucky to get into the DCPS once a month. I feel it's a necessary part of the training and should definitely be continued. The CMPS was shut down after Apollo 16, so all of my rendezvous rescue procedures and training was accomplished prior to Apollo 16, with a final review of the rescue book about a month prior to the Apollo 17 launch.

SCHMITT — I'd like to make a general remark about CMPS and LMPS type simulators for future programs. If you ever have a program where you're bringing in a new group of people to fly your spacecraft, this type of facility is extremely valuable. It gives a new man a chance to train without the constraints of simulator ties. He doesn't have the pressure of other crewmen looking over his shoulder and evaluating his performance. He can figure out how to do things, what a simulator really is, and what many of the more standard procedures are. I think it's a very valuable type of simulation. When you're dealing with a large pool of experienced crewmen, then that type of simulator is not necessary. This type of simulator develops habit patterns which are necessary in order to move on to the total mission simulators. Let me go back to SNSs. I had a feeling - and again I'm comparing with 15 - the total readiness of the combined MOCR and crew team came up more slowly than it did on 15, sometimes more slowly than I expected it to. But, at the end, I had the feeling that we were every bit as ready as a team as we were on 15. There was a lot of Skylab work for a lot of people there and I think that may have affected the rate in which we came up.

EVANS — Command module egress training - The mockups over in building 5 were utilized from an EVA standpoint for the CMP. The probe and drogue mockup was utilized several times. The last was a review and a final check. This is an absolute necessity for the drogue operations and also for the command module EVA operations. You need to utilize the mission simulator once or twice to tie in the systems procedures with the mockup procedures.

SCHMITT — The lunar module pilot's egress training was largely accomplished on Apollo 15. We did procedures reviews or mockup reviews and I did not get into the water tank for Apollo 17. The launch pad final walkdown came at a good time. The normal training we did in the hypergolic building was standard and excellent. I think it was good familiarization.

EVANS — It was good familiarization and also a must.

SCHMITT — It's a confidence builder and I think you ought to do it. Altitude chamber work tends to give you a little bit of egress training just because you have to deal with a real vacuum. I think that also is something you just pick up but that adds to your total readiness as far as egress is concerned.

EVANS — The water tank is where I received most of my EVA training. The water tank is a pretty good representation of zero-g. It was a lot better for me than the zero-g airplane. I became sick in the zero-g airplane every time except one. I never became sick in flight and never felt like I was going to be sick in flight. Every time I got on the zero-g airplane I always wanted to get as much done as possible before I started throwing up.

I don't have too much confidence in the zero-g airplane even though I flew in it four or five times.

CERNAN — I'll make some general comments about the CMS training. The CMS, from a hardware point of view, supported our mission in an excellent fashion. I think the crew at the Cape made themselves particularly available and were a vital part of the training. They did an outstanding job. The CMS is always limited in a visual and a dynamic system because it is a fixed base simulator. I think within the capabilities that it has to reproduce the visual, we received a good preview of what this flight was going to be about. It was mentioned earlier that for launch and reentries there are certain periods of time that you can not do in the real world that you can in a simulator because of the dynamic g-forces. It was also mentioned that this method of training in solving systems problems during those phases is still an excellent way in which to train as long as you realize that there are certain phases in the dynamic portions of the mission in which you will not be able to exercise the freedom that you can in the simulator.

The LMS from a hardware as well as an individual instructor support point of view supported our mission, in outstanding fashion. The entire system was excellent. The L&A, from a gross recognition point of view, was a duplicate of Taurus-Littrow. When we pitched over, it was almost like being in the L&A, except you very obviously got the realistic three-dimensional feeling. All of the software practices we used in the simulator were used in the spacecraft. The duplication of the spacecraft's software on the ground in the simulators was outstanding because I never had any problems or overloads. Everything performed just as advertised. I want to mention something about the LRV navigation simulator. It's a very good area familiarization simulator. I anticipated it would be a real great navigation driving simulator, but it's really just an area familiarization simulator in terms of driving from station to station and completing your EVA traverses. I think its major shortcoming is you never get the feeling of size or distance on this simulator, because on the Moon you have to at least double or quadruple your estimate of size and distance. You do not get that on the lunar Rover navigation simulator. You do not get involved in what it takes to drive the lunar Rover on the simulator and I don't just mean the 1/2-g effect. I mean the effect that in the real world when you drive the Rover you are continually avoiding rocks, holes, and craters. Some you can see and some you can't quite see. It's a continuous requirement to watch where you're going. The duty cycle of the controller is almost 100 percent. You do not have this requirement on the lunar Rover navigation simulator, and it's a little unrealistic from that point of view. We didn't spend that much time on the LRNS and I'm thankful we didn't. I thought it would be more valuable. The simulations we had with Houston and the integrated sims at the Cape went very well. We had very few hardware problems. The LMS in the last couple weeks had hardware problems now and then, but the people were able to recover and we only lost 1/2 day, and I think we made that up.

WARD — The backup crew essentially lost a day.

CERNAN — The DCPS in Houston was used extensively until 3 or 4 months before the flight. I'm very glad I did that because it was not just abort training, but it was abort and booster familiarization work. I felt very comfortable in flying the aborts as well as the manual takeovers on the booster. The rest of that training was done at the Cape in the CMS and I never felt anything but at home and quite knowledgeable about that part of the training. CM/LM egress - The altitude chambers speak for themselves. We did the launch pad work. We did water tank and not Gulf egress work and I heartily recommend that. Systems briefings went hand in hand with our simulator briefings from the simulator people. We did a lot of those very early and then just kept up the speed as we felt we needed them throughout the last 4 or 5 months with the other training.

SCHMITT — I spent a lot of time, the first 6 or 8 months, with the flight control division people going over the various systems that I was concerned with. I found this very valuable, not only in learning the systems but in learning how they were thinking about these systems. Once we were at the Cape, most of that kind of training was done directly with the simulator people, who did an excellent job. I was in fairly continuous phone contact with the Flight Control Division people to whom I talked earlier. This combination kept my system's knowledge pretty well up to date. I think it was an excellent way to approach the problem.

CERNAN — Simulator training plans - Eight to ten months ago, I sat down with the training coordinator and the senior simulator people at the Cape and asked the people there to go into the back simulator training history of the entire crew, because each crewman had a little different background. We found out where we all stood in terms of our simulator background looking forward to our future total simulator requirements. We tried to emphasize scheduling to fill in our weak spots. We reviewed this periodically - about every month - just to see how our training was going. This type of review with the simulator people did two things. I made them work out a particular schedule which we did our best to live up to, and it gave them a schedule that they could work on, plan on, and get ready to brief on. It made sure that we covered all areas which we could have skipped if we just randomly went out and told them what we wanted to do. In addition, I asked them to make sure the backup crew did not go off in one direction while we went off in another. The backup crew ran within the same time frame, the same type of training that we ran. In addition, they verified all our flight procedures and checklists. They could uncover possible errors or shortcomings in the procedures, due to their experience, that we might not. I think that all paid off. In the end we had all the important squares filled. The initial simulator requirement time was now an academic number because we knew exactly what we had done and where our strong points were. We reevaluated our entire simulator background and found out we were in pretty good shape. We probably spent more time in science training, both in the mechanics of ALSEP development and SIM bay operation as well as from orbital geology to geogeology work than any other crew in the past. And at the time it seemed like we were expending almost too much time in this area. But, in retrospect, I've got to say, it was time very well spent although it was time that had to come at the expense of something else. But I think those things were reasonable in terms of our previous training and background and not compromise the entire training and readiness for the flight.

SCHMITT — Let me add a comment to science training. We made a very special effort and many people in the Science and Applications Directorate went out of their way, particularly people associated with contractor support, to see that we had extensive exposure to the lunar sample. I think that in itself also paid off handsomely in recognition of rock types on the lunar surface. Those people are to be complimented doing what in a time of tight budgets - is a difficult thing to do and let us see the lunar rocks. They also supported very frequently with 2- or 3-hour discussions on various lunar problems which also was above and beyond the call of duty.

EVANS — Orbital geology - From my standpoint, three people were indispensable in this respect. Dick Laidley, Jeff Warner, and Farouq El Baz, each in their own little areas. Dick Laidley was indispensable in that he had been the pilot and the CMP's geologist so to speak for Apollo 16 and for the field trips involved - getting ready, knowing where to go, how to follow flight plans, what to expect, what photos to take, and this type of thing - he's indispensable from that standpoint. Jeff Warner took over and organized the rest of the scientific briefings for the CMP, got these squared away, and participated in the field trips from the low-altitude standpoint and also with the site specialists. Farouk came into his own along towards the end of the training cycle when we were involved primarily in the crew familiarization and training of the lunar geology itself. I think in each case we had the right amount of field trips. We just about exhausted all of the field trips that were available, since we got an early start on them. Even though you like to get a refresher field trip along toward the end of the training cycle, there just doesn't seem to be time to get it in. The lunar geology should also begin close to the end of the training cycle and continue on up to launch, which it did. You really don't need to make a recommendation any more, but El Baz should have been on the primary contact list.

SCHMITT — Well, I think that would go for any activity, in Earth orbit or anywhere.

S7250271

EVANS — It's hard to work through a window. It can be accomplished, but it's a lot easier to be side by side. One more for identification training. We had good landmark maps. We only had four or five of them and so not a whole lot of time was involved in that because they were pictured quite well. SIM bay training - In the early part of the training cycle, I was essentially following the manufacturing and the design really of the lunar sounder, so I was somewhat involved in the initial part of that. And then you get down to the final stages of it and work through the ASPO people and the FOD people as well as the people on the CMS who keep you up to date on the nitty gritty and the systems diagrams and that part of the training. And it was sufficient and adequate.

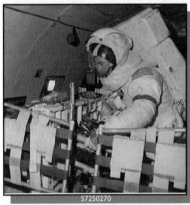

S7250270

SCHMITT — The 1/6g aircraft - I always felt that there's an important but limited area for the 1/6g procedures. I think if nothing else it paid off in evaluating the LRV sampler and convinced us that it was a feasible way to sample. I think for general familiarity with part tasks that could be accomplished in 20 seconds or so that the 1/6g aircraft was extremely valuable and those people did an outstanding job of supporting us. I mean Jack Slight and the Air Force and NASA in general. I also think the K-bird is a good vestibular trainer in spite of the fact it's very uncomfortable. I think it's probably worth doing a couple of times - for myself, anyway, to keep my vestibular system in some kind of condition. One-g walkthroughs - Well, mainly that's familiarity, and that's exactly what it did and it was extremely useful. Field trips - If we checked the number of trips, we probably had a little less than 15 or 16 did, but nevertheless they were well organized. They were mostly to

S7250268

brand new areas, so the people should be complimented on coming up with trips that had never been done before. That was mainly in order to keep the LMP from seeing a lot of familiar terrain. The support we received, the cooperation of the U.S. Geological Survey and the Science Mapping Directorate, was outstanding. I think all problems that existed several years ago and continued to exist to a limited extent were just about gone if not completely gone. The groups are working together extremely well and I

understand continue to work together through the mission in various capacities i
supporting our operations on the surface. We appreciated it very much. The LRV traine
- The LMP did not have too much to do with that. He went through the norma
familiarization to drive it and knew the systems remarkably well. I think the one unit w
got the most out of was the Grover, which was a U.S.G.S.-built machine that we use
on field trips. It had a lot to do with getting us used to the problems and advantages o
using a four-wheel drive vehicle for geological explorations. In particular, I know the CDI
would comment and he may yet that the Grover was good for emphasizing the amoun
of time you have to spend in driving versus what you would normally expect to see o
the LRV simulator. The CSD chamber work was extremely valuable. The two runs on th
PLSS and going through the EVA prep and post operations in the chamber and using ou
flight PLSSs, and backup PLSSs was some of the most valuable EVA training we receive
in my opinion.

EVANS — I think I can just pretty much second that although my training was strictl
on the umbilical, the O2 umbilical and OPS. In both cases, the first one was strictly
familiarization and confidence-type builder, knowing that you can survive and move in
vacuum with all of this equipment. The second run was more a refamiliarization with th
equipment and also, as far as I'm concerned, a necessity.

SCHMITT — The two CSD runs were probably all you needed. We did one early an
one late, and I think that was excellent scheduling. The people in the chamber should b
complimented for the quality of the training. The schedule is to be complimented fo
having it in there because it really topped off the EVA prep and post training that w
received.

CERNAN — And the late one really meant something in terms of us remembering hov
much pressure there was to put water connectors on. Also, it gave us a closed loo
matrix on handling all our EMU gear. As it turned out, we changed out the commander'
PLSS. The Grover, I think, was very useful for extending our geology training and puttin
us in the right environment in terms of distance to cover and getting on and off and wha
have you. The dynamics of the vehicle were nowhere what the real vehicle is, but it wa
certainly an advantageous device to have for field training and geology without question

SCHMITT — I mentioned on the Grover, and see if you agree so it's clear on th
record, that you commented several times that the driving tasks as termed to workloa
was comparable in certain kinds of trips.

CERNAN — That's a good point. Well worth mentioning again is the fact that even o
Earth terrain the guy in the left seat is not going to do much geology. He's going t
navigate and he's going to pay attention to the driving task. The Grover brought tha
home very clearly. I convinced myself that that was going to be the job in a real worl
The one-g trainer that we had down at the Cape I think certainly did more than a
adequate job. I felt very much at home in a hardsuit in the Rover in 1/6g because of th
work we had done with the one-g Rover trainer down at the Cape. The reach capabilit
the control capability with the hand controller, studying the low-gain antennas, th
surprising reach on LRV sampling, and taking the sample out of the container bag an
reaching over and putting it in the LMP's bag was almost exactly like the one-g trainer.
personally felt that simulating zero-g contingency EVA training was not worth the tim
doing. You'll never know, but I still feel that way.

EVANS — I concur with you, Gene.

CERNAN — Now, you've had some EVA training. Tell us.

EVANS — Walking the handrails was a piece of cake. I felt confident in everything that I was doing out there.

CERNAN — And it's a case of exercising the procedures in 1g. There was no question in my mind but that we could have transferred if we had had to with the training background that we had.

EVANS — The next item there is the EVA prep and post training. Jim Ellis had the procedures essentially all squared away from Apollo 16. He made a few modifications to account for the differences in the stowage so all I had to do was to come in, follow through the procedures, and get trained and do it. I think we had the right amount of EVA preps.

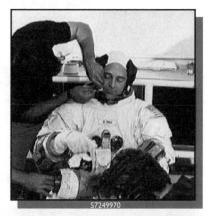

CERNAN — All of this EVA, both command module and LM prep and post-EVA training - we really walked in the footsteps of the guys who had prepared and exercised the procedures in training and in the real world. And we really altered them very little. We just based our training efforts upon their experience and it paid off. They were good procedures. They worked well in flight, and we did not make many small personal changes to these at all.

SCHMITT — I did a couple of extra mockup and stowage training exercises and I'm glad I did them. It made me more generally familar with where things were in the command module and I can't say that I really needed it but I felt a lot better once we were up there that I know where A-1 and A-2 were. I didn't have to keep asking Ron quite as often where things were. So if you have the time, it's still useful to the whole crew if you have that familarity.

EVANS — I'm also a firm believer in having the CMS fully stowed in whatever orbital operation you're doing, because this spacecraft is a lot different when it's fully stowed than when it isn't. Photography and camera training - This flight was essentially review for me. I was familar with the cameras, the photography, and this type of thing.

CERNAN — I think that's the case for all three. But we did have a session or two anyway in that area. Let me say something about the lunar surface experiment training. This goes to the SIM bay, too. When we first got introduced to the new experiment packages, we started out by having briefings by the PIs. It gave us a chance to meet personally and know each other's basic objectives. I think it gave the PIs a feeling that we were interested personally and professionally in carrying out to the greatest extent possible every objective of their experiment. I think, when we launched, every PI was satisfied that everything humanly

possible had been written into the Flight Plan to meet the objectives of their
experiment. It was a very, very good relationship and I'm very glad we did it. The LLTV
besides being a very enjoyable machine to fly from the pilot's point of view is just one
of those things I feel just makes a landing on the Moon that much easier. Puts you in a
familar situation. The dynamics of actually being out there on the front of that LLTV are
slightly different possibly than the real lunar module, but the roll and pitch and rates of
descent in our actual lunar LM landing were not new and different because of the LLTV
experience. I consider it a very valuable piece of
time spent in preparation for our lunar landing.

SCHMITT — The LMP doesn't have too much
activity with respect to lunar landing. We might
use manual throttle from the LMP side, and the
simulator, with Gene more or less in a GCA
mode, showed that that was a perfectly adequate
way to land the vehicle. I'd also like to
compliment the use of the helicopters. The LMP
continued to fly those because, in general, that
kind of two-handed control has a direct feedback
into handling two hands, ACA and PTCA in the
LM. Whether you're landing or not, it gives you a
two-handed coordination proficiency required to
perform those tasks to a fine degree. Particularly,
I found that in LOI aborts, where we were doing
manual attitude control, that the more I flew the
helicopter the more finely I could control it for
those particular maneuvers.

S7248891

EVANS —   In the planning of the training
program I thought it was outstanding. I didn't at
any time in the training program have to worry
about, "What do I have to do next week." It was
all taken care of. All I had to do was look at the
schedule and say, "Hey, this is it." And press on. I
don't believe I ever felt I was doing something
unnecessarily. Nor did I feel like I completely
missed anything that I should have been doing.
You can never get enough training. However, I felt
I was confident and ready to go at the right time.
That same logic can be tied into the fact that the
trainer requirements were organized many
months ago and we took a good look at them.
Once they were established and down on paper,
you didn't really have to worry about whether
you were getting extraneous training or not
enough training at that point in time. You knew
you were eventually going to get what you
needed and it did work out that way. The last week

S7248888

or two you begin to vary from that a little bit, based upon what you want to emphasize
maybe eliminate something you think that you're very familiar with. I flew a lot of manual
descents and manual launches and manual TLIs the last week or so just to make sure
that that would not be the thing that would keep us from going or doing the job. But
beyond that, we followed the original planned training program.

SCHMITT — The LMP concurs with all of that.

8.0 COMMAND MODULE SYSTEMS OPERATIONS

EVANS — The first one is subsystem modes. All we did was utilize the nominal modes of the inertial subsystem. We had no problems, with the exception of one. And I think I should let the CDR talk about that one. The ISS worked real fine. The drift on the platform was phenomenally low. Most of the torquing angles were, within a 12-hour period - .0 blank, blank, or something most of the time. With the LM aboard, I seldom saw any stars through the telescope, so I had to rely on PICAPAR to put a star in the sextant. Once the star was in the sextant, you could assume that it was a correct star because we always had a good platform. If we had needed to do a realinement or a P51, I think the only way we could have done that was to get an initial alinement on a Moon/Earth type of system. The only degrading factor about the optical subsystem was the focus of the sextant, and you just couldn't quite bring the reticle of the sextant into focus. If you cranked the reticle brightness all the way up, you could see the center of the sextant. However, if you cranked it down a little bit, it was hard to see. And even with full bright, it was a little difficult to see, so you never really had it in focus. If you pointed the sextant close to a bright object, for instance, close to the Moon or close to the Earth or close to the Sun, the reticle brightness would be completely blanked out and you'd end up with a black line and two sets of reticles. One would be a heavy black line, and then there was kind of a ghost reticle behind that which wasn't superimposed on the heavy black reticle. I don't know what that was, so I always used the heavy black reticle.

SCHMITT — It's not of major importance, but it's interesting that you were continually saying that it was hard to pick groups of stars and to identify groups of stars in the telescope when you could look out the window, as long as the Sun was on the other side of the spacecraft, and identify constellations with no problem.

EVANS — This was particularly true on translunar coast. Even if the Sun was behind you, the reflection off the LM's radar box or RCS quad would interfere with the telescope's field of view. Now on transearth coast, if the optics were looking down-Sun, you could pick up constellations; however not as bright as looking out the window. Around the Moon, even in earthshine, which was very bright, you could pick out constellations. However, they were considerably dimmer than they were in the double umbra on the back side of the Moon. On the back side of the Moon with the double umbra, you could look out and almost see constellations as well as you could by looking out the window, but not quite as well.

SCHMITT — I looked once or twice through the optics at the Earth. It appeared to be an excellent Earth-viewing system.

EVANS — Yes, it is. A couple of times, I observed the Moon through the sextant. However, the field of view through the sextant was so small that you had to look through the telescope first to see where you were and then look through the sextant because you couldn't recognize the general features at all.

SCHMITT — Trim displays and SPS displays - Were they all what you expected?

EVANS — Evidently, from where I was sitting in the cockpit, plus 2 and minus 2 on the gimbal drive check always ended up a plus 2.2 and a minus 1.8 from my left seat viewing angle. The trim values were always just a bit higher than what I thought they should be, which didn't bother me much either. I finally got so I would set the SCS gimbal trim position just a bit higher than what I thought they ought to be.

SCHMITT — During those checks, each movement of the gimbal was indicated by an increase of amp loads on the buses.

EVANS — The other thing that's more noticeable in the vehicle than in the simulator was the feeling of the dynamic motion every time the gimbals move. You also got an indication on the rate display. If you were in the 51 setting, rate display would go up to maybe a tenth of a degree before it would null itself out again. You could definitely see the spacecraft banging back and forth within the dead band. The CMC SPS TVC - I think the greatest difference in that field and the simulator was in the roll rates involved in the SPS TVC. In all of the burns, the roll rates were almost always up around 0.4° per second within the dead band. In the simulator, it always stayed on the same side of the dead band. But, in the actual vehicle, it would hit one side of the dead band, bounce back, and go back across again at about 0.4° per second and hit the dead band on the other side and then come back. So it was oscillating back and forth across the dead band, whereas we had the simulator pretty much set on a 0.1° per second. I think it always hits one side of the dead band. The pitch and yaw rates during the TVC/SPS burns seemed to be almost steady, very little change. When you had the yaw change during TEI, it was a nice gradual change. Rates were not noticeable at all; just steady as a rock.

EVANS — The only part utilized in the SCS system was attitude control during the TVC checks, and I guess the other time was when NOUN 20 got its glitch. We switched to SCS control, which took care of the rates right off the bat, caged the BMAGs, and maintained attitude quite adequately. I guess entry was the only other time I used SCS minimum impulse on the command module only, after command/service module separation. And, in that case, you always had residual rates, which wanted to yaw the vehicle to the left. You had to continually yaw it to the right, and in pitch when I was trying to pitch down, it would continually decrease the pitch-down rate. It was essentially evolving its own body pitch up. But, minimum impulse, control of the command module only, is quite adequate. It's a little bit different than the simulator in that in the simulator, roll control is the one you can't quite get with one minimum impulse blip back into zero roll. In the vehicle's case, it's yaw. You'd either give one blip, and it would go to the other side. You'd give it a blip back the other way, and it'd never end up with a zero rate in the yaw axis. However, I never did try it with the single ring authority. The only maneuvering I did again is the minimum impulse on the command module in SCS. Thrust vector - We never did any SCS/SPS burns.

EVANS — Power up/power down. We never did power down. It was powered up all the time. We never did power it down. Attitude hold worked quite well. Whenever you switched to SCS control and you had all 16 auto RCS selects on, then you had a continual bang, bang, bang, back and forth within the roll dead band. If you put limit cycle on, that kind of knocked it down a little bit. But, of course, the best way to control SCS in roll is to use two-quad authority and not four-quad.

The Delta thrust switch - I would always wait until average g on the computer before bringing the Delta-V thrust A switch. All burns were started on Delta thrust A first. Or, if it was a single-bank burn, it was Delta-V thrust A. The Delta-V remaining counter and rocker - the EMS Delta-V worked real fine. The difference between the actual spacecraft and the CMS is that in the CMS, you can see them count up and down, but in the command module when you held the button down to maximum increase or decrease, the last three digits remain solid. It really counts up, so you wouldn't have time between each of the counters to see the numbers change. It stayed on whatever number it was on. Actually, it just sat there as an eight, a constant eight all the time because it was whipping through there so fast. The Delta-V test worked all the time. There was always a minus 22.2 or 22.1 on all of the EMS Delta-V tests except the one prior to entry, which ended up as minus 27 or something. That's when the accelerometer was picking up some extraneous counts and counting a little bit more than it should have.

SPS thrust direct ON switch - I never did use it.

Direct ullage button - I never did push the direct ullage button. I don't even know if it worked.

Thrust ON button - I never did push the thrust ON button, so I don't know if it worked, either. Engine thrust vector alinement - I don't recall any attitude deviations or maximum rate changes because of thrust vector misalinement at the initiation of any of the burns.

SPS chamber pressure indicator - During the LOI burn, the bank A indicator came up to about 87 percent, which was a little lower than anticipated. I was expecting somewhere around 95 or 97, somewhere in there. That's about 10 percent lower than expected. When I turned on bank B, I got the nominal 5 percent increase, and then throughout the LOI burn, the chamber pressure just gradually increased a little bit and finally got up to about 97 percent at the end of the burn. The other anomaly on the chamber pressure indicator was that after the LOI burn, we noticed that we were down around 5 percent, and then later on, sometime in lunar orbit, it ended up back at zero again, and I'm not sure when it went back to zero. On all the rest of the SPS burns, the chamber pressure on bank A would always come up to about 86 to 87 percent and then a 5 psi increase when you put on bank Bravo.

SCHMITT — PUGS - The PUGS was essentially nominal in general. Apparently, there were some sensors out, so it was erratic in its sensing of the LOI burn. It tended to hang in the decreased area. I went to decrease and left it in decrease for the LOI burn, and it seemed to try to keep going low. After DOI, we didn't see any real change to it. Then I guess it acted pretty much the same for you on circ. When I came back in, it looked like it had decreased more than when I looked at it last.

EVANS — Well, on the plane change burn, the ground called up saying to start it in decrease. So I went to decrease, and I think it didn't have time to stabilize at all because we ended up with a gage with data of 400 decrease.

SCHMITT — When we picked it up after that for TEI, we started TEI full decrease and left it there. It was, in fact, low and gradually worked itself up until it was almost balanced about 30 low at the end of the burn.

EVANS — That's with the switch in full decrease.

SCHMITT — And that corresponded with what we read on the gages. So, I guess you'd say it ended up nominal, but it was a little erratic during the burn, and that may have been the result of those sensors.

EVANS — I think I'd better back up there a little bit on the chamber pressure indicator. Evidently, the chamber pressure indicator had a bias on it on the low side because we were definitely getting full thrust.

SCHMITT — Yes, in checking the V go versus time chart, you were ahead of that on V thrust. You were getting more than the nominal thrust and that corresponded with our cutoff time.

EVANS — Service module RCS - We had no anomalies with any of the quads. The audible cues are not like the simulator, but you can hear some back there. You hear something that's more than the clicking of the solenoids on and off. I don't know what it is, but it's more than the solenoids.

SCHMITT — As a matter of fact, it was like somebody in another part of an old house turning on a water faucet and when it's turned off, you hear the water pound against the faucet. I think it probably was the plumbing more than the solenoid.

EVANS — Whenever you started a maneuver or were in a maneuver, you could always tell because the vehicle would vibrate a little. The vehicle would move around and when it got to its position, it would sit there and shake a little bit, and then when it started moving, it would also shake a little bit, more than I had anticipated that it would do. The command module/RCS - We had no anomalies there.

SCHMITT — I guess I was impressed by both the service module and the command module firings at the amount of unburned fuel and/or oxidizer that was propelled out of it. My impression was that the command module gave more afterburn material than the service module, but that may be because I was closer to the command module and I watched it. Also, in regard to the service module evasive maneuver after separation, it was very clear what was taking place. You could see those particles streaking out.

EVANS — I guess the other thing is, whenever an engine fired at night or on the dark side of the spacecraft, you always got a white flash.

SCHMITT — Fuel cells were perfectly nominal as far as I could tell on board, and the ground didn't call anything. The one switching anomaly we had was that, in the process of some activity, one of the fuel cell pumps got turned off, and I don't have any idea how that happened. It obviously happened on my watch, and they caught it within 5 minutes.

EVANS — Well, I turned the laser altimeter off one time when trying to maneuver to a position to see out the window. My feet were flailing all over the place, and I kicked them off with my toe. I must have. So, I think that's probably the same thing that happened with the fuel cell.

SCHMITT — In high power loads, we did see some caution warnings on the O2 purges which I didn't expect to see, but it was just barely triggering the high flow. The ground called saying not to worry about it. The batteries were nominal as near as I could tell. never did quite figure out whether the ground was concerned about the vent pressure after charging because it hung at 0.6 for a long time and then gradually creeped up. They didn't seem bothered enough about it to discuss it with us, so, I ignored it too. It was always within limits. The only thing that I was a little bit surprised at is that they left the batteries uncharged longer than I had expected. I expected more calls. We never really got our entry batteries charged up until the day before entry. We had time to charge them before then. It seemed to me that after TEI we waited a long time to get the batteries back to charge, but that's a minor point. We launched with our batteries down more than normal, so that's probably what started us off in a real-time call in the battery charging because we left battery A charging - I think it was A first - for a long time practically 24 hours, I think. I'd have to go back and look, but it was a long time recharging. And when we put it on, it carried over 2 amps on the battery charger, which is impressive because in the simulator you never see more than 1. The battery charger was, as far as I could tell, perfectly normal. Caution warning - Very soon after insertion we got something like 7, if I remember correctly, before we had comm with the ground again, spurious master alarms. It gradually became evident to us that it was associated with switching panel 2 switches. Ron hit it with a helmet once. It was with the neckring and he got an alarm. During the pressurization for the first LM entry, we got a couple more. I thought it might be associated with a higher pressure cabinet. That's the only other correlation with that anomaly. I guess after that, we never saw it again.

EVANS — Well, one thing we want to mention is that we never got a caution warning light.

SCHMITT — The gages - There were no anomalies or power levels that I jotted down at various times to keep track of any possible shorts. Things seemed to be perfectly normal, and anytime it jumped up, you'd always be pretty confident that you could look up and see that the O2 meter should have come on. There was one time when I thought I saw a major blink on the lights when we weren't expecting any power surge. The ground couldn't find anything on their records, and I suspect it was my imagination. AC, nothing - I was surprised as I always am, and I've seen it in the chamber that the AC-1 voltages were as low as they were. They are right down at the lower limits, but they're not below the limits. We've asked that question in chambers, and nobody ever worries about it, so I'm sure that's standard. AC inverters, perfectly normal - We did run inverter 3 for a while as a heat source when the cabin got cold during some of the weird SIM bay/transearth coast attitudes in conjunction with manual control of the mixing valve. Main bus tie switches, no problem - One surprising thing was the first time I put them on at launch, during the first try at launch, the fuel cells apparently were performing much higher than I was used to seeing on the GSE in the chamber. Very little current was drawn from the batteries. My normal mode of monitoring the bus ties to see if they got it or not was to watch the battery amps, and in those particular cases, I got no indication (for a few seconds anyway) of any amps off the battery. Gradually, a little came up and you could see there were 2 or 3 amps. Then when we checked the gimbal motors. (The batteries were not a good place to check the gimbal motors.) I went to fuel cell amps to check gimbal.

EVANS — I always used the fuel cell amps or the O2 flow.

SCHMITT — On the simulator, I always used the battery flow. That's another thing I might mention is that the H2 and O2 flow in the fuel cells aren't any good for that because they're too sluggish, much slower than the simulator. The simulator reacts instantaneously to changing loads, whereas the real fuel cells are quite sluggish in their reaction. The sensor bus switch, we turned off once for EVA. The cryogenic system - The ground was playing games with the H2 fan because of a thermistor shift, scale change or something like that. We did a lot of manual switching on their call, but that was no inconvenience whatsoever because they had a sleep configuration that they could go to and it didn't bother us at all. One thing, it seemed that we, at least as far as the tank pressure was concerned, carried our H2, 1 and 2 tanks, with us after the service module jettisoned. I don't know why.

EVANS — This is a good point to mention, the surge tank was biased a little bit low, too.

SCHMITT — That's right, but we were alerted to that. It performed just as the alert specified. Cabin lighting controls - One thing we didn't use initially but used later on transearth coast was the fixed position being brighter than the maximum on the restat.

EVANS — I used that a couple of times trying to get enough light in there to use a camera.

SCHMITT — Split bus operations - They worked fine.

EVANS — Oxygen masks - We never took them out of the bag. With the gassy situation in there, I was tempted to, but we never did. We very seldom utilized the cold water dispenser.

CERNAN — Most of the cold water came out of the gun.

EVANS — It was handier; that was the reason. You wanted to keep the hot water hot so you keep it going. The water-gas separator stayed on the hot water tap, and we always ended up with gas bubbles generally of about 1/2 to 3/4 inch in diameter in the hot drink or hot beverage, things like that. In the cold water, drink gun, seemed like there were a lot of very small bubbles - just little bitty ones maybe a centimeter in diameter that would end up in the drink gun.

SCHMITT — The hot water bubbles were bigger.

EVANS — Suit circuit - No problems. We mentioned the difference in the bias on the surge tank already.

CERNAN — The waste management system was all right as far as the CSM was concerned. I still think it's a poor system from a standpoint of hygiene in waste management control. I made that statement in different sections and I'll specifically say any time you use a condom-type system you want to make the valve end of the condom of a larger diameter so that whenever you reroll it for the next use, you can reroll to a larger diameter and get your penis as far up in the system or towards the valve that you possibly can. If you don't, you have to stretch and pull the condom and half the time your penis might be part way in and half the time it may be all the way in. Any time it's not all the way in the condom, you can almost invariably end up with urine residual in there that has to be cleaned up in one way or another, in spite of the fact that you tend to push it through the valve like it's recommended. The entire system still needs improvement.

EVANS — It still needs improvement and in my case the condom was too small. In other words, I anticipated a shrinkage and the shrinkage did not occur.

CERNAN — I did too but I think what I just mentioned would tend to solve that. It's getting it over the head really.

EVANS — That's right. That's right. We said everything we want to say about waste management. We stowed everything in the waste stowage compartment except for two feces. The CO_2 absorbers - No problems this time. Nothing sticking.

CERNAN — Telecommunications - The whole thing was nominal.

EVANS — The high gain worked great. There wasn't any problem with that. DSE operation - Ground handled most of it and where the ground did not and the CMP was required, we just said configured and we did it that way. Tunnel and hatch probes - All operations were nominal with the exception of the things that are noted on the air-to-ground tapes about the docking latches.

19.0 LUNAR MODULE SYSTEMS OPERATIONS

19.1 PGNCS

CERNAN — PGNCS inertial subsystem performed exactly as advertised with the initial powerup and with the lunar surface powerup. We did not get a restart light on the initial powerup. That was the only thing we did not get on the initial powerup. The ground said it was a GO. I was a little disappointed in the AOT. It really performed like the simulator did. I could split the image on the reticle, both on the XY axis and the spiral cursor. By a slight movement of my eye, left or right or up or down, I could place the star within the reticle optical line of sight. I had to try and find a neutral position for my eye on the eyepiece so that I could be consistent in every one of my marks. That bothered me a little bit on the initial alinements until I got a constant position. That's something that, if I flew a lunar module again, I'd certainly like taken care of. The alinements came out good. But they came out good because I found an eye position where the star was focused and where the reticle was focused. I could put the star within the confines of the reticle for a good solid alinement by simply eye movement and not spacecraft or spiral cursor movement. Rendezvous radar power up and checkout was outstanding in performance during the rendezvous. No anomalies. The landing radar was not only without anomalies, before we started our 70° yaw we started to pick up some indications of radar lockon. I was about halfway through the yaw to the 34° position during the landing when the radar locked on solid. Don't remember exactly what altitude that was, but it was far in access of 35 000 feet.

SCHMITT — I think I did two or three PGNCS landing radar checks starting at about 2000 feet. They were within the motion of the tape, exactly on with each other. No anomalies there.

CERNAN — Computer subsystem - I utilized the computer exactly like the simulator in terms of verbs and nouns during descent and ascent. Every one of them came up in what I called the prescribed amount of time. We never had any overload master alarms. We never had any program alarms. We never had any anomalous program alarms. It was a duplicate and repeatable of the way I handled the computer in the simulator. Exactly.

G&N controls and displays - The DSKY speaks for itself. The displays that came up on it were exactly what were called for both in the power up in the descent and the ascent. The other two primary displays are the needles and the crosspointers. During descent, the P64 needles again were nominal in terms of what the simulator told us they would be. The P66 needles in terms of fore and aft velocity were again exactly what I'd seen in the simulator in terms of fly-to needles to null the lateral velocities.

Crosspointers - I matched the crosspointers in terms of forward and lateral velocity with what I saw out the window in P66 for final landing. That is lateral velocity on the crosspointer was effectively zero, and I agreed with that out the window. The forward velocity was probably 1 to 3 ft/sec on the crosspointer. The best estimation of my forward velocity out the window is that I had some. So, again, they were nominal.

Procedural data - The checklist, in terms of the flow through the power-up through the descent, and through the ascent, were well written. The PGNCS performed exactly as advertised in every respect.

SCHMITT — As we said about the CSM checklist, I don't think we had any changes to checklists that reflected any procedural errors prior to launch. I don't think we had any changes to the checklist that I can remember. The checklists worked perfect as far as I am concerned.

CERNAN — Let me back up and say something more about the PGNCS. The only time that the PGNCS surprised me was after TPI, when we had large residuals that I had not seen in the simulator. We had residuals in the area of 7 ft/sec in X and 4 plus or minus a few tenths in Y and Z at the end of the short nominal TPI burn. The simulator residuals were always much less than that. We had no problem. We just nulled them out with the RCS. But, nevertheless, they were there and it was about a 3-,1/2 to 4-1/2 second burn. That surprised me just a little bit. We don't have the exact numbers written down because I rolled the residuals right away and went right into P35. I do know they were

7, 4 and 4 plus or minus a few tenths. Prior to descent, they gave us a zero gyro drift compensation. They said the PGNCS was right on. However, right after orbit insertion, it looked like we might have had some g sensitive drift in Y.

SCHMITT — You did. The AGS saw it.

CERNAN — For rendezvous navigation and the short rendezvous burn, we did not see any effect of it at all.

SCHMITT — The ground tweak was 7 ft/sec and the AGS would show 9 in the same direction.

19.2 AGS

SCHMITT — Modes of operation - Nothing off nominal was used. Initialization went perfectly nominal. Calibration, the sighting with the exception of gyro cal, I think it was Z on the first activation, was slightly out of spec. Not out of limits but just slightly out of spec numbers that were given to me by Jerry Thomas, which was 0.3° spec limit. I think it was 0.5 or 0.6. Rendezvous radar navigation on the AGS was done in AUTO with the exception of the post midcourse 2, where we put in three sets of marks manually in order to maintain the AGS state vector as close to docking as possible. The AGS state vector did just that - maintained itself within 2 ft/sec and was right with the PGNCS on range at the initiation of braking. Actually it was better than that at the initiation of breaking about 500 feet it was still within two feet per second. Engine commands - All the engine discretes seemed to get into the AGS. The ground did not mention a single anomaly and I saw none on board. There were no electronic anomalies. Burn programs were perfectly nominal. In monitoring the DOI 2 and the midcourses, the AGS, as expected, did not see the short burst trim pulses that Gene made with the TTCA. The acceleration levels did not seem to be high enough to be sensed by the AGS external Delta-V. Controls and displays were excellent. After every 400 plus 3 X, PGNCS alined. There seemed to be about a quarter-degree constant bias between the AGS and the PGNCS alinement in pitch and yaw. I think it was a combination of pitch and yaw. A little bit of motion on the ball switching from PGNCS to AGS. One farther thing on calibration - there seemed to be an accumulating accelerometer bias in X that was well below any significant problem. Probably something like 0.1 ft/sec. I noticed this after the first cal and then after insertion. After insertion I did do an accelerometer cal, 400 plus 7. That seemed to improve the problem although it did not eliminate it completely. It was not a serious problem with the AGS monitoring of its state vector.

CERNAN — AGS control check - I checked it out in both pulse initially. I checked all three axes out in pulse. I got the continuous rapid fire pulses. It checked out in three axes. I checked it out in rate command both for command and attitude hold. And it was a very tight system. I checked it out in min deadband only. It was GO. There was absolutely nothing wrong with the AGS system either during powerup or during the phases of checkout.

19.3 PROPULSION SYSTEM

CERNAN — The descent burn was extremely nominal in all respects. We monitored the start and attitude hold was steady. I monitored the throttle up on the PGNCS, watched the PGNCS command it, and watched the descent follow-through. It was 100 percent on the indicator, on time at 26 seconds. We saw a throttle-down again within a few seconds of that predicted from the ground, but exactly on time with that which we saw commanded from the computer. The ROD, during the last phase of descent, during P66, responded extremely well. I knew exactly what rate of descent I had simply by the number of clips I put into the ROD. Descent and ascent was a nominal operation as prescribed and as we saw in the simulator.

SCHMITT — The one thing we previously mentioned on caution and warning was that we got a descent quantity light after touchdown by several minutes, presumably due to either fuel sensing or an actual fuel leak.

19.4 REACTION CONTROL SYSTEM

CERNAN — Attitude control modes - I flew it most of the time in pulse. After rendezvous for stationkeeping, nominal operation in all modes, nominal operation attitude hold in AUTO. Translation of control was nominal for ullage and for stationkeeping.

SCHMITT — The RCS ascent feed was good. We might add here that we did have a transducer shift in the ascent helium tanks. Tank 2, I believe, was reading hot. I believe that was what the ground called it as. They seemed to think they had a mixture ratio problem. They weren't completely sure. They had us terminate ascent feed early. It must have been 5 minutes or something like that. We terminated that early but there was no significant degradation in our RCS capability.

CERNAN — Every explosive device in the spacecraft audible.

SCHMITT — Except one. That was the second landing gear. Didn't you say you didn't hear that?

CERNAN — No, I heard it too. I could feel it when I hit the switch.

SCHMITT — I thought you said you didn't hear that.

CERNAN — The first landing gear operation, we felt, of course, the landing gear go out. The second one I could feel, in the switch, the activation.

19.5 ELECTRICAL POWER SYSTEM

SCHMITT — The batteries were excellent. There were no battery anomalies. The DC monitoring was no problem. I might mention that the ascent batteries did seem to require longer than nominal warmup time, although I do not believe it was longer than expected with reference to the ground. We unfortunately got started minutes late so we flew the first part of powered descent with battery 3 off the line in order to increase the load on the ascent batteries for preconditioning. That was not a problem at all. Battery 3 was put on somewhere in powered descent without any interference with that operation. DC monitor was fine. AC monitor was fine. Power transfer CSM/LM/CSM went nominally in every case. Abort stage configuration - Nothing to discuss that would be off nominal. Main buses performed nominally and dead facing was nominal. Explosive devices in all cases seemed to perform as expected. We heard the pyros, I think, in every case except possibly the second set of pyros on the landing gear. That might be expected, not to hear those. We heard the first, but we may have been really hearing the bolts let go and the gear start to move into place. Voltages were unchanged throughout the whole flight. Lighting - There were no lighting anomalies. Caution and warning - No anomalies. There were one or two configuration caution and warning signals which will come under ECS. What was that caution and warning we got right at the end of descent?

CERNAN — Descent?

SCHMITT — Right after touchdown.

CERNAN — No. All we got was descent quantity.

SCHMITT — That was after we vented, wasn't it?

CERNAN — No, it was before we vented. The descent quantity did not come on until after we landed and when we went through all the ascent checks. The fuel side was going down all the time. We never talked about it; we never asked.

SCHMITT — We don't know why that happened. That's right. The fuel side after touchdown continued to decrease. Sometime into the post touchdown pre-vent checklist, we got a descent quantity light. That was the only caution and warning anomaly.

19.6
ENVIRONMENTAL
CONTROL SYSTEM

SCHMITT — Oxygen cabin pressure was nominal except for a leaky main A reg, which potentially was caused by having my hoses stowed at one time with the suit in suit flow. That's up to the systems people to decide. But it did reset itself on time. It was not a serious leak. After that time, we did fly with only reg B in use. It was pretty clear that A was a usable reg; it just was leaky. Cabin atmosphere was good, good ventilation, good odor clearing. The dust clearing was remarkably good, considering the amount of dust that we had. It was within a couple hours after ingress. Although there was a lot of irritation, at least to my sinuses and nostrils, soon after taking the helmet off, about 2 hours later, that had decreased considerably.

CERNAN — The LCG cooling was perfectly nominal. The LCG cooling, I think, was a mandatory requirement pre-descent and pre-EVAs. I don't think the air cooling in the spacecraft was adequate prior to descent, which I said a long time ago, back several missions. This was really a godsend. We did not wear the liquid-cooled garments out of our own choice for ascent rendezvous, and I was very comfortable during that phase of the mission.

SCHMITT — Yes. I think had you worn them and not had cooling, you would have been uncomfortable.

CERNAN — Water supply. My first impression was that, after the first several gulps of water, there was a lot less gas in the LM water than in the command module water.

SCHMITT — True. We used all our water. We essentially ran dry at ascent. We drank a lot of water and we even used some additional water on our hands. Water glycol was nominal, and the suit circuit, with the exception of what I mentioned about REG A, was nominal.

19.7
TELECOMMUNICATIONS

SCHMITT — There was no problem monitoring the comm system. Operation of S-band high gain antenna was variable. We had some initial problems on the housekeeping day of lockup. It seemed to me to be a ground problem. I don't know their final resolution of that. It seemed that the same kind of thing happened to us on ascent, and again when we came around the horn prior to PDI. It seemed to be a ground lockup problem because it happened on the omnis as well as the high gain. We were just not getting a good strong uplink signal. I don't know what else to say. On ascent, as soon as we lit off, we lost the high gain, went over to omnis, and the omnis were giving the same indications - low signal strength, lots of noise, and a high squeal. Not a real high squeal, but an obvious squeal. It wasn't until somebody else did something that we got the comm back. I did not get the comm back; it just came back. We came through the command module for a little bit. Then they instructed me to do things I'd already done as far as going to the omnis and stuff, and then suddenly they came back up. So, I'm not sure what happened. But when we had S-band comm, it was excellent. Excellent voice.

CERNAN — The VHF comm after separation and throughout rendezvous was excellent.

SCHMITT — That's right. There was a little bit of a problem close in. I think, again, it was a question of overdriving too much.

CERNAN — The EVA antenna operations were all right, but the EVA comm was excellent throughout the first two and a half EVAs. The latter part of the third EVA, I began to get some noise in the background that the LMP did not get. It did not make the comm unreadable, but the noise was very evident. That lasted throughout the closeout of EVA-3. The LMP had no significant comm problems on the EVA, and had excellent comm. Procedures and operations of the audio center throughout the LM checkout and EVA changeover setup was nominal. It worked just as advertised.

SCHMITT — Flight recorders - I have no idea. I should mention that I probably left the LM DSEA on during the third EVA, because it was barber pole when we got back in. I suspect that we ran out of tape at that time. My regrets to Don Arabian.

20.0 LRV OPERATIONS

SCHMITT — LRV deployment was nominal. Didn't we almost slip out of the hinge pins there once?

CERNAN — I think they dropped into them. The walking hinges did not drop. They were locked in, as we reported. It seemed to have fallen into the hinges. That was the only time when there was a slight jolt. Throughout the mechanical deployment, which followed the procedures as written, she came down just as advertised and broke loose from the saddle just as advertised. The setup was nominal.

SCHMITT — We did have to push the hinge pin in.

CERNAN — I went back and reset a forward hinge pin. One out of the four hinge pins was not locked in; the yellow was not flush. Mounting and dismounting was simply a case of getting acclimated as to know how to mount and how to dismount. The biggest problem with mounting and dismounting was to be able to mount without kicking dust all over the LCRU.

SCHMITT — In my case, the problem was keeping a twist out of the lap belt, which made it difficult to unbelt.

CERNAN — Mounting on a slide slope aided dismounting.

SCHMITT — Almost all in one motion.

CERNAN — Vehicle Characteristics - Power-up - when I pushed in the Bravo and Delta circuit breakers, the gages came up just as advertised. Occasionally, I could feel a little wheel slippage. To the best of my knowledge, I had four-wheel drive and fore and aft steering the entire time, nominal. The braking action was good. As a matter of fact, on some of the extreme downslopes we were on, I had to brake continuously and stay below 18 kilometers. We barely hit 18. I had in mind the fact that the brakes could fade on you. We came down some pretty steep slopes at some reasonable speeds, and I had to brake the entire time. I worked the brakes on and off. I had no indication of brake fading at all. I never felt that I was going to lose control because of lack of braking. Acceleration - Although we could never really go in a straight line very long with the Rover because of boulders, craters, or general terrain features, I drove the Rover full out a majority of the time. Apparently, we were going upslope, especially out to station 2. I was between 10 and 12 kilometers most of the time, and that was at full throttle. I was a little bit surprised that full throttle did not give me a little bit better acceleration and a little bit better top speed.

SCHMITT — I think that 1° upslope was probably there.

CERNAN — However, the acceleration when you hit a definite grade or change in grade, you could feel that the capability to climb that grade was always there. In spite of the fact that maybe you slipped down to about 8 or 10 km/hr, you always felt that you had the torque and the power required to make that grade. I never felt that there was a grade that we tried to negotiate that I didn't have the capability to getting over with the Rover. Never. Steering and Slide Slippage - In 1/6 g with fore and aft steering and four-wheel steering for you, you've got a vehicle that is ready to react the minute you think about putting the command in. Much of the time at the speeds we were driving, as soon as that steering and side slip and sharp turn command went in, you were on three wheels. The reaction was that you did get side slip. I did feel that the majority of my more rapid or sharp turns, I'd say 50 percent of my driving, resulted in losing the back end on some of my turns. I don't know whether you felt that on the right side, Jack.

SCHMITT — Yes.

CERNAN — I was comfortable in doing it because I expected it. I felt that in keeping a reasonable speed, the rear end broke loose from me on 50 percent of the turns during my entire driving on three EVAs. It's a vehicle that you have to drive to get accustomed to. It's one you approach slowly, and then you begin to peak out and you begin to live up to its maximum performance capabilities. You can avoid obstacles very easily. The only hesitancy in doing so is that it requires the same sharp turn and generally your rear end will break out. The turning response is phenomenal. I was a little disappointed or surprised at maximum speed on what looked like a relatively level surface, which may have been a 1° or so upslope. It was not quite as fast as I thought it might be. Coming down that slope, we did a lot of zigzagging going to different stations. So I didn't get the full brunt of coming back down the same slope. Basically, I felt I could get more top speed out of the vehicle, not that I needed it, but there were times I could have used it in negotiating the surface. Torque - I don't really think I required more torque. I never lost the wheels going upslope, although I did feel the vehicle working, and you could see it in the amps that you were drawing going up some of those slopes. You could also feel it in the top speed. Again, there was ample torque to negotiate the slopes that we had confronting us. Some of those slopes, subjectively, were quite steep.

Controllability - you had to learn - just like you have to learn on most other vehicles that are essentially like that - to be gentle and smooth during the control. Sharp commands would tend to leave you without the rear end on the ground or leave you with the rear end not exactly where you wanted it. So controllability was excellent, but I felt it was very sensitive.

Crew Restrictions, Limitations, and Capabilities - Displays - I could see and read all displays all the time except when we got dust on the checklist down in front of the hand controller. Then that display became effectively unreadable until I could get off the Rover at the next stop and dust it. Hand controller operations were as advertised, very similar to the trainer. I used reverse twice, and it worked. I don't recommend it as a standard mode of operation. It's much better to have the vehicle set up for forward only control capability. My seat and foot rest were, as far as I'm concerned, perfectly adjusted and comfortable as far as position. How about yours?

SCHMITT — Fine.

CERNAN — Crew Movement Within the Suits - As far as driving the Rover was concerned, I had the same right arm restriction as far as getting my arm back and driving the Rover. But I had no wrist problems as on some of the previous flights. I wore no wristlets. I did not rub my wrist raw. I had all the wrist commands. I think that's just a function of where your arms fit in the suit. I had absolutely no wrist movement problems at all. I sat in the suit high enough to be able to see down at the displays and out in front of me. The only restriction I ever had in driving the Rover, out in front, is where coincidentally the last parking angle left the high gain antenna at a planned view. Then I had to look through the high gain antenna. Then the tendency to lose the view beyond was a little bit greater.

Seat Belt Operations - On the left side, I could not have tolerated my seat belt any smaller. It kept me in tight. I felt that I would never lose the Rover. I felt that I'd stay with the Rover even if we did a 180° roll. Yet it was loose enough to get in and out of. It might be because of just generally getting a little bit more tired, but certainly during the third EVA, I found it occasionally was a little harder to release. How about your side?

SCHMITT — Much the same. I mentioned that the seat belt got twisted occasionally. I suspect that made it harder to get out. Being tired, I'm sure, had something to do with that. Let me skip back up to crew movement within the suit. The only time there was any significant movement was when we were on side hills and moving around all the contours. I noticed I was leaning against the side of the suit, which increased the impression of being on a steep slope.

CERNAN — During the lunar Rover samples, the commander was able to take the sample from the LMP and was able to reach over and drop the sample in the LMPs sample bag without any difficulty at all. This was repeatable, based upon ground training. Exactly the same.

SCHMITT — The Rover sample worked exactly as we had planned. No changes at all.

CERNAN — The vehicle suspension characteristics were outstanding. I negotiated some intentionally, some unintentionally. I negotiated some relatively good-sized rocks, 10 to 12 inches or so, head on with the suspension system and the vehicle just walked right over these rocks without any difficulty at all. I tried to straddle the smaller craters so that we wouldn't get any side slope. In driving the vehicle, the major effort is to deter yourself from side slope activities, whether they're little craters or large craters. So you try and go down through the center of the craters if they're not too deep. If they're small craters, you try to straddle them. We went through some relatively major boulder fields, and the vehicle suspension just accepted it without any difficulty at all. I never felt that we bottomed out. We never bottomed out in terms of the wheels taking a boulder. However, we did scrape bottom once or twice in going over some boulders, centering some boulders.

SCHMITT — I never went back to look, but you mentioned you looked like you'd bent a wheel. Is that right?

CERNAN — I mentioned something about a golf-ball-size dent in the left front wheel. I inspected all the wheels after that. The left front inboard wheel was bigger than a golf ball. If you took a fist and just crunched the inner side wall, just punched it and you left an impression of your fist in it, that's about what I saw in that left front wheel. The impression was probably no more than a half an inch to three-quarters of an inch deep and a radius about the size of your fist. None of the other wheels had it because I inspected them after I saw this one. As far as driving characteristics are concerned, you wouldn't know it was there. Hand Holds on the Vehicle - The hand hold I used most to get in was the low gain antenna on the commander's side to help me to get in a proper position for strap in. Any other hand holds on the vehicle were really relatively useless, particularly in adjusting the high gain and what have you, because the vehicle when it sets by itself was a very unstable vehicle. The tendency to move or shove or lean the vehicle one way or another was very great.

SCHMITT — I used the accessory stands as my hand hold for mounting and dismounting.

CERNAN — LRV Systems Operations - The nav system was excellent. I saw the same characteristic digital movement of the gyro that we saw in LRV sim. But it certainly didn't hamper the operation of the nav system. Power Batteries - The temperature on the right number 2 battery was higher at initial powerup. We started powering up at 120°, which I think surprised everybody, including me. It stayed hot, although they both cooled down relatively. It stayed hot throughout the mission. At the end of the third EVA, it was above 138° or 140° and gave us a flag.

Steering and Traction Drive - I wiped out the hand controller as we had planned to prior to the flight about 6 or 8 times before powerup to remove any lubrication problems due to thermal characteristics. The minute I powered up (and you saw it), to the best of my knowledge, I had both front and reverse steering. Voice Communications and Antenna Management - Antenna management, because of the extensive preflight planning, was excellent. I had no trouble in handling the high gain. I could pick up the Earth and center it. It was there. I just sighted it and looked through, and it was there. I tweaked it up, and there was no problem at all. The low gain antenna, except when we did 360° pans, which I did not bother to adjust at low gain antenna following on the part of the commander to keep us within plus or minus 10° to 20°, was a simple task. It did not require any undue attention.

TV/TCU - Up until the time it failed after lift-off, the TV/TCU worked very well. Electrical and Mechanical Connections - The only connection I really had trouble with, electrical/mechanical connection, was the SEP connection to the LRV. I had to support the connector bracket with my left hand in order to get enough force on the SEP connector to mate it and lock it to the LRV.

SCHMITT — That's the standard EMU connection.

CERNAN — That's the standard EMU connection. The only other thing I'd like to mention about the LRV is it's about 99-percent required effort. Even to take a drink of water from the suit drinking bag during LRV driving could put you in some very embarrassing situations as far as following your terrain, craters, and what have you. It was almost 100-percent requirement.

SCHMITT — Geology Science Site Response - You've covered pretty well how the Rover performed on various kinds of terrain. Gene, why don't you describe the fender? That was the major dust problem.

CERNAN — With the loss of one of the fender extensions, any one of them, the dust generated by the wheels without fenders or without fenders extensions is intolerable. Not just the crew gets dusty, but everything mechanical on the Rover is subject to dust. Close to the end of the third EVA, all the mechanical devices on the gate and on the pallet in terms of bag holders and pallet locks and what have you were to the point that they would refuse to function mechanically even though the tolerances on these particular locks were very gross. They didn't work because they were inhabited and infiltrated with this dust. Some could be forced over center. Others just refused to operate even after dusting, cleaning, and a slight amount of pounding trying to break the dust loose. I think dust is probably one of our greatest inhibitors to a nominal operation on the Moon. I think we can overcome other physiological or physical or mechanical problems except dust.

SCHMITT — What we're really saying is that in any future operation, mechanical joints or levers and this sort of thing are going to have to be protected.

CERNAN — They should be sealed or protected. We had absolutely no dust problem with the wheels, and those are sealed units. Dust accumulated on the radiator.

SCHMITT — That goes for tools too. The only-tools we had locks on were the scoop and the rake, and those were getting stiff and wouldn't lock. They wouldn't relock once you adjusted them.

CERNAN — The period of time when we had lost the rear fender just put a solid coat of gray dust over everything. Once we got the fender repaired, the dust problem was at a minimum. After the long traverse rides, the radiators all required a good amount of dusting. That required X amount of time. That's going to be required again any time we have a lunar surface operation.

Payload Stowage - Jack, do you have anything? Initially, during EVA-1 prep, I think everything fit under the seats or on the pallet. The pallet fit on the Rover exactly as advertised. The SEP, the deployment of the SEP, the setup of the charges, and the charges on the pallet all fit.

SCHMITT — I'm sure we'll get into this in the system experiments, but as a general comment for any radiator surfaces that need to be protected, you need to have more than just a cursory design on the protection of those radiators. The SEP is the case in point, and that was a completely inadequate design to protect those radiators. If we ever do it again in a dust environment, you must have clear and very tight protection of your mirrors and radiators for driving.

CERNAN — Something else that dust penetrates that I don't think has been mentioned before is that it penetrates and deteriorates the capability of Velcro. I could see it on the LCRU covers and the SEP covers. The Velcro pulled off to keep the SEP covers closed, but the Velcro that kept them open didn't pull off but it was deteriorating. If you want to use tape on the lunar surface after what you're taping has been exposed to the dust, you first have to clean that surface off with a piece of tape or something and get the mirror dust off before the tape will even begin to adhere to the surface you're trying to apply it to.

SCHMITT — We ought to mention here that the gray tape in general is not very good. It will stick to itself, both inside and outside the spacecraft.

CERNAN — I had the impression that the gray tape has been sitting around for 10 years. That's the kind of adherence you had.

SCHMITT — The tape on the food bags is what we finally used whenever we needed to really tape something. It is much better tape.

CERNAN — The gray tape is very poor tape. We covered the stowage, which went exactly as planned. We had no fit problems with stowage or anything on the Rover.

CERNAN — PGA Fit and operations - The CDRs suit fit perfectly, including gloves.

SCHMITT — The LMPs was an excellent fit.

CERNAN — Doffing and donning were just as we expected. The CMP may update this, but as far as we're concerned, he had no gripes or qualms getting his suit on and off.

SCHMITT — I think he will have some comments.

EVANS — On item 21.0, EMU Systems, everything was normal with the exception of the CMPs prelaunch drink bag. Try and try as I might to get water out of it, I couldn't. After finally getting the suit off in the spacecraft, the drink bag in suit donning had somehow become stuck sideways underneath the neck ring bending the little rubber hose that we drew the water through. It did not allow any water to come through. The drink bag was filled, and it did not expand noticeably from any air that may have been in it. The problem was that it wasn't in vertically. It was kind of wedged in crossways around the neck ring. Everything else from the CMP suit worked adequately.

CERNAN — Biomed instrumentation - I think to varying degrees of individuality that we all had sensor skin problems.

SCHMITT — Yes, those are documented by the medics. Let me just say that I wore a set of connectors for the whole descent through ascent time frame, and when I took those off in the command module, the electrolyte from a couple had completely disappeared. It obviously reacted with the skin and left sort of a semiscab. It wasn't a bloody scab.

CERNAN — The commander had that too. We both had that problem.

SCHMITT — If you have the time to change them out each day, it's probably not a bad idea. We just didn't take that kind of time.

CERNAN — The LCG operation was nominal. We doffed the LCGs after the EVAs, slept in CWGs, and donned them for the EVAs. It was a very comfortable mode of operation.

SCHMITT — I really am surprised that other missions have slept in their LCGs. It just seems to me that this would have been very uncomfortable.

CERNAN — Helmet operation - The CDR's was nominal.

SCHMITT — The LMP's was fine.

CERNAN — You had your visor stuck.

SCHMITT — LEVA operation - I did have the sticky visor problem, and it was dust. We could force it closed, once we got it off. We tried once on the surface, and we couldn't get it closed.

CERNAN — That was the hard Sun visor. Lifevest - No comment; nominal. The gloves fit well and tight, and I don't have any gripes.

SHEPARD — Did you use the extra set?

CERNAN — The extra set is brand new and sitting on the surface right outside the descent stage. Neck seal - We had no problems sealing the LEVAs, helmets, or anything. UTCA operation - The CDR used his at every opportunity. I always had a bagful.

SCHMITT — I suggest that considerable thought be given to the size of condom that you pick. Mine was too small, and it inhibited the operation of the UCTA. It was a very uncomfortable situation on both EVAs, until I was able to force a urination. After the second one, I apparently popped a blood vessel. There was blood, but it disappeared after 24 hours.

CERNAN — I ended up with an external scab during the lunar surface EVAs from the sweat and the condom. It went away.

SHEPARD — Was it an abrasion problem?

CERNAN — I think it was an abrasion problem. I could not have a larger one because I don't think it would serve the purpose. It was just a lot of work and a lot of walking, and that's all there was to it.

SCHMITT — For the third EVA, I stretched the condom and it worked fine.

CERNAN — The EMU maintenance kits were fine. We used them as required, as planned.

SCHMITT — The drink bags were excellent.

CERNAN — Let me say something about the drink bags. We rotated that nozzle 90 degrees. We said it would work in training. I didn't know that drink bag was there until I wanted to get a drink of water. It never interfered with the mikes. I wore it on descent to the surface, on the surface, and drained the bag on all three EVAs. Ron's bag at launch was doubled up. He can talk about that. They just put it in wrong.

SCHMITT — Before you leave the drink bag, there's something down here for the food stick. Neither Gene nor I used all our food stick. I think it was a good idea having it there.

CERNAN — I used about half of mine most of the time.

SCHMITT — I never felt an extreme desire to eat at all. Every once in a while, I would take a little chunk off of it. Antifog was fine.

CERNAN — There was no fog problem. The PLSS PGA operations were again nominal as planned. Pressurization and ventilation were good. Liquid cooling was excellent. I never worked for any long duration in high cooling, with maybe one or two exceptions. And generally, I used high cooling only when I was hot and wanted a spurt of cold water. Probably 90 percent of the mission, I worked in medium cooling.

SCHMITT — I never went to high, not once.

CERNAN — Is that right?

SCHMITT — I was in intermediate-intermediate, which is a little better than intermediate.

CERNAN — Communications on the surface were good even before we got our antennas extended. I didn't notice any difference after the antennas were extended.

SCHMITT — I did, just a little bit. It was a little less scratchy.

CERNAN — Connectors and controls were good on the PLSS throughout the flight. They are the one thing that did not seem to get affected by the dust. They might have gotten a little stiffer, but I could not tell it. The RCU was good. The RCU fit and operated well. We did not use the OPS.

SCHMITT — Let me comment on mine. After the third EVA, we reset the regulator, and that's why we brought the CDR's PLSS back rather than the LMPs.

CERNAN — I think Jack activated his OPS with the hose free for just a moment. I think that reset the regulator. Instead of regulating at 39, it then started regulating at 43.

22.0 FLIGHT EQUIPMENT

CERNAN — The event timers and controls worked excellently. There were no anomalies, no problems.

22.1 CSM

SHEPARD — Did you have an LEB timer?

CERNAN — The LEB timer worked fine the whole flight. During launch, somehow the mission event timer on the main display panel got off. We reset it and it was fine throughout the mission. I don't know what happened during launch to cause that. We never received an explanation.

SCHMITT — That's good recollection. You're right.

CERNAN — Crew compartment configuration - As far as I'm concerned that is stowage and it was exactly as advertised. We had a few bags that blew up. Once we opened the compartments we couldn't get them back in. I had to stab holes in them. They happened to be the OPKs. I had to take my scissors and punch a hole in seven or eight of the OPKs in order to restow them. When Ron opened the compartment they just went, "plonk." Every one of them blew up. I took the scissors and went klonk, klonk, klonk, to let the air out. The only problem we had on stowage was the OPKs. The mirrors worked fine. The IV clothing and related equipment worked fine. If Jack and I had one request, it would be to carry one more CWG for cleanliness. Particularly when you come back from the LM as dirty as you are, we could have used one more CWG throughout the entire flight. We had no problems with the IV pressure garments and connecting equipment. Ron may want to mention the g-suit.

EVANS — The g-suit was a looser fit than it was when I took off which surprised me. I thought my legs would be fatter.

CERNAN — The couches - We got the center couch out and in with no problems. We got the YY struts connected and disconnected many times with no problems. The restraints worked fine. The inflight tool set was really never used except for tool B for the hatch work and tool E for all the continued panel work.

EVANS — The ones we used were good.

CERNAN — There are a lot of data collection systems. Every one of them that we used, whether it be pen and pencil in the Flight Plan or the DSE, appeared to be working nominal. Thermal control of the spacecraft - Because of the lack of the PTC control on the way home due to the UV and the IR requirements, the spacecraft was in attitudes where I think it got very cold at times.

EVANS — It got cold and damp on the inside.

CERNAN — We transferred water from the overhead hatch to the forward hatch and the forward hatch started perspiring. We warmed the spacecraft up by manual setting of the temperature control inlet valve and putting the number 3 inverter on the line. It became very comfortable and we were that way for about 36 hours. Then went back to normal for the entry.

EVANS — Camera equipment was nominal. Everything worked real fine. I haven't seen any pictures yet. SIM bay equipment - The only problem we had was extension and retraction of HF antennas. In all cases, we ended up eventually getting them fully extended and fully retracted. We never did get the retract barber pole on HF I throughout the flight. The ground was watching the motor currents and were able to tell when it was retracted. The ground never did get the barber pole indication either on the full retract of HF I.

SCHMITT — I heard Gene say it took 2 or 3 days to get squared away on how to take care of yourself and your personal items. I didn't think it took that long, I think about a day is what you require. I think you should not completely program the first 2 or 3 days. You ought to build up to a full flight schedule, for example on Skylab, over the first few days because of the variability in adaptation to the new environment. I think that organizing your own personal items does not take more than a day to really get into the swing of things. I tended not to wear the coveralls the first few days of flight. The first couple days or so I just wore the constant wear garment, but I gradually got in to where I wore the trousers and that was partly to have available the pockets for odds and ends like PRDs. I felt no thermal discomfort just wearing the CWG until transearth coast when it got much cooler because of the variable PTC attitudes.

The lightweight headsets, I did not use very much until the last couple days of the flight, but when I used them they seem to be perfectly adequate. Ron has probably talked about the problem he had with the headset which I subsequently ended up using only as a cover for my head and used the lightweight comm carrier attached to it.

Medical data seem to go fairly well; it was just a matter of keeping up to date. I did most of it on the translunar leg and Ron did most of it on the transearth leg. It varies whether we use negative reporting on food or positive reporting. It depends on what we've eaten and how much we've eaten.

Camera equipment, to our knowledge, in flight functioned very well. We understand that we may have had one jammed magazine EE, but at least it transported film for at least half a mag. At the window where most photography is being done, it's useful to have a camera configured for the anticipated type of photography that you would want.

Lens configuration, f-stop, and shutter speeds - For the most part we kept the dark slide out of the camera for rapid access to pictures. That was, in both transearth and translunar orbit operations. The kitchen timer, the interval timer I guess it's called, was a very useful item. I had the feeling that it needed a little better time calibration on it. But in using it for the SIM bay operation, sometime you would like to have a little more accurate timing. There's also, I think, usefulness in certain places in the spacecraft to have hook Velcro as well as pile, because hook is useful for hanging up washcloths and other items that in themselves represent a pile configuration. We made considerable use of the spring bungees stretched across the switch panels in order to not only control the data books that we're using but also to aid in biomed donning by putting all the gear in one spot with the bungee. Also, it was used during eating and to hold the various food packs. We tended to put the food Velcro next to the flight Velcro because it was of a superior quality. The gray tape in the CSM is really useful only when you stick it to itself. It does not stick to spacecraft or anything else very well. We tended to save the food pack tape which is much better gray tape than the stowed tape. We had an adequate amount of tissue, but I think had we had any more problem than we did with the loose bowel movements that we would have run out of tissue. I think you ought to consider that if there's any concern that we may not solve our problems of loose bowel movements that, in Skylab, there should be some way to stow a considerable amount of extra tissue. I think we had just the right amount of towels. It gave us the option of cleaning up at times and not being concerned about using dirty towels several times.

22.2 LM

CERNAN — Crew compartment configuration on the LM was as the mockup configuration and as advertised. There were no problems, throughout storage or unstorage. Restraint systems were used for descent and ascent. They worked fine. We used no tools in the LM that I can recall. Our camera equipment had only one anomaly that I know of. A 16-millimeter camera failed to start during ascent. The LMP tried to start the camera in 12 frames per second. He couldn't start it. He had to hold it and it would run in 24 frames per second. He'll have to describe the details. It did run by itself 12 frames per second later. So we might have to go back and make a check and pick up with Jack on that camera.

SCHMITT — The LM crew compartment was fine. We had no trouble except in one incidental case in finding the gear. The restraint system I used during descent and ascent worked fine. We had all the tools that we needed; of course, we didn't need any to speak of. We got by with one pair of scissors both for the cabin and surface operation for the obvious reasons that one had disappeared in the command module. Again the same comment is that in the LM the gray tape was not adequate. Camera equipment in the LM was more than adequate. We brought back the CDR surface camera and that was used during the rendezvous for air-to-air pictures and also for air-to-ground pictures and it was used for LM magazines in the command module until we had used up all the LM film that had not been used on the surface for lunar orbit operations. Only half a mag of black and white film was available for use during the postrendezvous period. I think we used just about every frame of film that was reasonably available to us in the flight. We used all the LM film but maybe half a mag of black and white and half a mag of color prior to lift-off. By the end of the rendezvous sequence, we had used up all the color; and then by the end of our TEI, we used up the rest of the black and white for target-of-opportunity pictures. I think it's a serious mistake not to do everything possible to stow more film than you need. We had just the right amount of film. We were conservative about film usage but not generally conservative. We took all the pictures we wanted. The crew just shouldn't be reluctant to take pictures because that's the prime mode of documentation.

23.0 FLIGHT DATA FILE

CERNAN — The Flight Data File was in tune with the flight; it was complete; it was followed; and it was in excellent shape.

EVANS — It was in outstanding shape. There was absolutely nothing wrong with it.

CERNAN — We had minimal updates to the crew cue cards and minimal updates to the Flight Plan. Flight planning did a super job getting it all together. There's nothing we can add to it. The Lunar Surface Checklist was the one that had most real-time updates and that was simply based on mileages and bearings because the LM position was slightly east of where we had set up the Lunar Surface Checklist. The Checklist was changed in real time because of the time allotted at each station. That's to be expected as the checklist is only a guideline anyway.

23.1 CSM

SCHMITT — Generally, I have nothing but praise for the Flight Data File, both vehicles. One comment on the Flight Plan Supplement. We had split pages for medical and food logging. That was probably a mistake. We tended to only use the book as a whole and it was a good place to keep them. If you had wanted the pages split they were too thin to maintain. Furthermore, they tended to fall out of the book. I recommend not splitting the pages or having heavier paper if you want them split. I had an extra cue card built for panel 229. I think it was an excellent card that summarized the circuit breaker functions both on 229 and on panel 8. It was not used because we had no systems anomalies of any significance that would relate to that card. But I would strongly recommend its availability if only for training. It's a good quick review of what you lose or retain for those two panels. In the Flight Plan, I added some pen and ink cues along the margins for certain observational targets that I particularly wanted to look at. These are independent of any designated experiment and I entered them as a function of time. That seemed to work very well for me. I think it is the easiest way to go, since it shouldn't concern any large number of people. Gordy Fullerton fixed up the circular orbital cue card for me with similar designation of craters as a function of time. I did not use that. Not because it wasn't a good idea but because of familiarity with the Moon, which came very quickly after a couple of orbits. You could recognize your position on the Moon fairly easily as a function of each rev, either timing the rev, approximate time since sunset, or just because you could look out the window and could tell where you were.

23.2 LM

SCHMITT — The same comments apply. I think all the Flight Data File items were excellent. We logged most of our specific items such as alinement data and comparable kinds of things in the checklist and at the point where they were collected rather than in the Data Book. Cuff checklist - We talked about that in the surface items. I thought the cuff checklist was excellent. We had the right kind of photo maps and they were useful for reference when we were around a given station. I don't think we used them as much as I had anticipated. Navigation was no problem as the points that we had selected previously were excellent points for investigation. There was no need to try to decide on an alternative point to try to study in the vicinity of a given station. The list of items to be accomplished at each station were mindjoggers to read at each station. They were not used as much as I thought they would be initially. That was mainly because we had become so familiar with the items that each station was in itself easy to recall as a result of having created the checklist. So, the checklist was turned out to be more of a learning item rather than a reference item for use on the surface. I wouldn't have done it any differently. I particularly want to compliment Chuck Lewis on the Timeline Book. The book was very very well done and we had no problems with it at all. That of course applies to every checklist we had. There were no procedural errors in any of the books. Fortunately, we did not have to use the Malfunction Book. Only once did I pull out the

Systems Data Book to check on a systems problem and I can't remember what that was now.

EVANS — I never did use the sun compass. I didn't have time to get it squared away or to figure out where it was and follow it around. I never used it as much as I thought I was going to prior to the flight. After you've been up there for a while you could look out the window and tell essentially where you were, so you really didn't need it. You kind of guess pretty much on the settings for the cameras and hopefully there wasn't any problems in that respect. We used the orbit monitor charts. They are not as good as they could be but were useful in finding out where you were and looking up a few of the craters. I did not use the contingency chart at all.

CERNAN — Let me comment about the LM landing site monitor chart and ascent monitor chart. The only time we used these charts was in observing the landing site from the command module on a day prior to landing. Because of the operation of the PGNS and the ascent guidance systems we did not have to use the ascent monitor chart at all. Lunar surface maps - We used in the cockpit after the EVAs but only pointedly toward trying to relocate our traverse and make sure we were aware of the craters we saw and where we had been. They were sort of a resume-type post-EVA rather than pre-EVA planning guide. Let Jack comment on the EVA traverse maps. I think we used them far less than Jack planned. I used my cuff checklist for all my navigation and for my traverse even though I was told that it would be relatively useless. My cuff checklist was a very vital part of my lunar surface navigation.

WARD — What did you use to make the fender with?

CERNAN — In that respect, they were very useful.

SCHMITT — I thought I would use the orbit monitor charts in the CM, so I had an extra one put on so it wouldn't interfere with the planned activities of the CMP. I did not use that very much. I eventually did some sketching on it post-TEI. I think I labeled about five specific points as areas A, B, C, D, maybe E, and these are referenced in my crew notebook for a specific observation. One item - that chart should have been identical to the CMP's chart. There were a few pen and ink changes left off such as exposure settings for certain photo targets that caused some confusion. The CSM lunar landmark maps that the LMP had added in the rear of that book, again, were not used. As I was observing a specific point or area such as Gagarin I would not take the time out to sketch on the photo. I tended to look at the first opportunity and to take notes in the notebook rather than trying to sketch on the photograph. I think having selected them and studied them preflight made it worth having them around. The necessity for flying them was probably less than the necessity for having reviewed them and studied them. I still would want to have that kind of data available in the spacecraft. I think the CMP used his visual target maps considerably. I did on a couple of occasions. For the most part, that was post-TEI and I made some notes and sketches on some of those maps. I think that function was because there was a lot of time to look at the Moon make a sketch, and then look back and fix it up post-TEI. In orbit, the time just did not exist. As Dick Gordon said a couple years ago, "Once you start flying, the clock is relentless."

EVANS — Outstanding.

CERNAN — This may be an appropriate place to comment on how we handled the lunar orbit phase with three men in the spacecraft. After the first 2 hours in the spacecraft, we figured out the most expeditious and efficient way of handling it and it worked that way throughout the rest of the flight. The CDR got in the left seat, where he belongs anyway and took the Flight Plan. Windows 5, 4, and 3 point towards the lunar surface so I let the orbital geologist and the surface geologist look out the windows and make all their finds. I kept them honest on SIM bay, made all the attitude changes that were required, and kept them up to date on all SIM bay switch changes. I ran the spacecraft, they did the orbital geology, and I kept them honest. I kept out of their way. I stayed on my side of the spacecraft, kept the systems and the world honest and they cut loose. That's the way I'd recommend doing it all the time. You didn't bump into each other. Occasionally I would sneak a peek and say they were right or wrong and let it go at that.

EVANS — Level of details provided in onboard documentation/recommendation changes - Solo phase - I have no recommended changes. It was in outstanding shape. We had gone through it before and checked it out in the simulator.

CERNAN — I want to comment under miscellaneous. I think the Flight Plan carried just enough detail to tell you what was going on and what was going to happen. If you were not familiar with the details of the operation of that particular system or what you were going to do in terms of going into PTC or any other phase in the mission it would refer you to the Systems Book or G&C Dictionary. You did not have to repeat them in the Flight Plan. I like this way of doing things. We generally had the Flight Plan plus two other documents out. One was a Systems Book and the other was probably a G&C Dictionary.

EVANS — We kept the G&C dictionary out. In the solo phase, I had the Experiments Checklist out, too.

SCHMITT — The Flight Plan was excellent. We had no problems with it at all that I'm aware of. Tommy Holloway and his people are to be complimented. The number of different requirements and experiments and general operational items that were required to be integrated was very very high. It was done in an extremely competent and usable way. I can't think of anything that I would change in the way that the Flight Plan was written.

23.5 PREFLIGHT SUPPORT

EVANS — Good.

CERNAN — Updated properly. This is both LM and CSM.

EVANS — The CSM had no problem. We would give the information to the people at the Cape and they would make the changes in the CSM and quarters copies. We had them in a timely manner. Coordination was good between the Cape and Houston. Change propulsion system - There were very few changes in the checklists themselves once you came out with the primary book. There were few changes after that. Real time procedures changes - They were quite nominal and from the CSM standpoint easily taken care of.

CERNAN — As far as I'm concerned, the LM preflight support on the Flight Data File was excellent. If there's anyplace that I'd make a comment on it was the fact that somehow the latest changes we thought were in a system somehow never got in until the morning of the sim. They were always there for the sim. Real-time procedural changes in the LM - There were really none except for the EVA. In the command module, they were so minimal that it was no problem updating the Flight Plan as we went along. I might add that the clock sync was so smooth that you wouldn't even believe it. It went "zap," we updated our clocks, and we were on our way. That put us right on the Flight Plan and that's probably one of the smoothest ideas anyone ever had.

SCHMITT — Excellent in the Flight Data File area. One specific item that I had was two or three briefing sessions on portions of the lunar orbit during which I was in the CSM spacecraft. We went over in detail the attitudes, maneuvers, and the window availabilities so that I was able to plan in a very short amount of time with minimum effort my part for my own personal observations of the lunar surface. I appreciated that extra above and beyond the call of duty on the part of the flight planners. I appreciated their taking time out to do that for me. I think it was useful to have the sessions where the flight controllers, the crew, and the flight planners met and went over those portions of the Flight Plan which were not normally simulated. I was a little bit disappointed in that some of the people who would be eventually intimately involved in the mission were not at the Flight Plan review.

24.0 VISUAL SIGHTINGS

CERNAN — Countdown - It was dark and we didn't see anything until S-IC ignition. The CDR and the CMP could see out their small windows in the BPC the glow of ignition prior to lift-off. Powered flight - During the actual powered flight of the S-IC you could not see anything at all. You couldn't see out the cockpit, as we had the lights up fairly bright. At staging, the S-IC shut down, something that you don't see in the daylight is that the fireball overtook us.

EVANS — It sure did.

CERNAN — When the S-II lit off, we literally for a nanosecond flew through the bright yellow fireball that was left over from the S-IC. Tower jett was very evident. You could see the flash and I could see the entire BPC. I could see underneath it. It was lit up underneath. The whole thing was lit up. I could see nothing on S-II until S-II shutdown. I could see the glow of S-IVB ignition. I saw the glow of S-IVB ignition, it very easily could have been the fireball of S-II which tried to overtake us but couldn't quite make it. But there was a glow right during the period of S-II shutdown to S-IVB ignition. During the S-IVB burn, you could see the glow of the aft engines throughout the burn and throughout the orbital operation. Earth orbit - I might comment that the availability of stars for a mode II or mode IV abort was pretty poor for two reasons. Number one: night adaptability because we had lights very bright. When we turned the lights down in the cockpit, I could not pick out distinct constellations such as Orion, which I was planning on using for a mode IV abort. If we would have had an SCS and G&N problem it would have been very difficult to pick out stars for that abort.

EVANS — I should mention in Earth orbit you couldn't see the stars in the telescope in the daylight but they showed up nice and bright and clear in the sextant. I think that is probably a typical thing.

CERNAN — When we burned out of Earth orbit, we started the burn in darkness and flew right on through a sunrise during the TLI burn. This was pretty spectacular. We shut down in daylight and had no other visual sightings at that point in time. Translunar/transearth - After CSM separation from the booster and docking with the LM several hours later, we could see something which may have been the S-IVB or SLA panels. As soon as we turned around for docking I could see three of the four SLA panels tumbling slowly in space. This is not unusual. That's been seen before.

EVANS — I never did see a SLA panel.

CERNAN — There seemed to be an awful lot of particles with us continually throughout the flight, both in transearth and translunar coast and in lunar orbit. These particles were obviously residue from the RCS. Others were from dumped residues. They seemed to be hanging around the LM as a result of pulling in and out of the S-IVB and they were always small particles. Some, initially, were pieces of Mylar from the S-IVB

LM separation. The others were just like small dump crystals or residue. On the LM, particularly, when you fire the RCS you could see the RCS residue.

EVANS — That residue from the RCS didn't look a lot different than a waste-water dump.

CERNAN — That's right, except that it's less dense.

EVANS — Entry - Just the fireball, and the fireball is a lot brighter than I thought it was going to be. I almost wish I would have had sunglasses. It was really bright out of the rendezvous window just shortly after the .05g when you start picking up the greatest portion of the fireball. That brightness only lasted for maybe 30 to 40 seconds. Then either you became accustomed to the brightness or the brightness decreased. From that point on, I could see the instrument panel. Long after the brightness of the fireball decreased, I could look back up through the rendezvous window and see what to me was kind of like a tunnel with a bright spot in the middle of the tunnel. Way down the tunnel, way back behind, I could see the fireball.

CERNAN — The only unusual sighting I can recall during landing or recovery is when the CMP looked out the window and saw the superstructure of an aircraft carrier and said, "Oh, we've got a tin can with us."

EVANS — Well, it was kind of foggy on the windows.

SCHMITT — Transearth we had only a small crescent of an Earth and it was not feasible to do any extensive weather observations. We had light flashes just about continuously during the whole flight when we were dark adapted. I had one which I thought was a flash on the lunar surface. That one period of time when we had the blindfolds on for the ALFMED experiment there were just no visible flashes, although that evening, that night, before I went to sleep I noticed that I was seeing the light flashes again. So, it just seemed to be that one interval either side of it where the light flash was not visible to myself or to the other two crewmen.

25.0 PREMISSION PLANNING

CERNAN — Mission plan - A lot of work went into the mission plan, with the right people. I think we came out with a mission plan and a Flight Plan which was not just a suitable one that would accomplish a purpose, but was a suitable one to be able to fly.

EVANS — The mission plan was taken care of by a lot of people. The flight planning crew insured that everything in the mission plan was taken care of. And I did not have to participate in that part of it at all.

CERNAN — Procedural changes - I think in my experience on past flights, procedural changes were held to a minimum. I think they were held to a minimum because we resisted a great many of them, particularly in terms of the lunar surface activities. Procedural changes if we would have allowed them to infiltrate the system, would have been with us right up to launch date. We put a cutoff on those several weeks before launch, accepted a few of them, and then forcefully would not accept any unless absolutely mandatory after the last 3 weeks. That was the key to keeping that Flight Plan and the lunar surface procedures intact. Mission rules and techniques - I don't think these changed from any of the previous flights.

EVANS — I don't think so either.

CERNAN — They were in good shape and followed quite well. The only place we exercised a slight different approach in mission rules was the fact that we had a DOI-1 which did not take us down to the minimum altitude that we had gone to in the past and then a DOI-2 in the LM.

EVANS — Let me go back to the Flight Plan here. I think this was the first time we've ever tried this from the command module standpoint anyhow. This was that each person responsible for a section of the Flight Plan, whether it was from LOI to DOI or DOI to circ, was brought down to the Cape and utilized on the simulator console while I or the backup crew ran through the preliminary Flight Plans before they were even in the print stage. We essentially were debugging their part of the Flight Plan. They could see the problems involved and we worked together to get a good plan. This took a day to a day and a half at a time. I think it was well worthwhile.

SCHMITT — There were periods of some difficulty, preflight particularly, in the area of medical requirements and in some last-minute possible scientific requirements particularly on the samples, but everything seemed to get resolved satisfactorily. I can't think of anything that was not handled very well by the support crew, Bob Parker in the science area, and Gordy Fullerton and Bob Overmyer in the operational areas. I guess the biggest single area that took time was the CMP's dealings with the lunar sounder. Most of our ALSEP changes were all taken care of prior to our training. We had a few minor suggestions, but they were taken care of early in the training cycle. Mission rules and techniques were fairly well defined very early by Phil Shaffer and his crowd in the techniques area. The mission rules as defined by Jerry Griffin and his people were all in the right direction in that they enhanced the probability of making a landing in a successful mission. We really never had to exercise any of the mission rules in an abnormal way. I think that the one time that a mission rule tended to be a controlling factor was in the limitation on the work at station 4, Shorty Crater. We were up against the walkback constraint and terminated that work after only 35 minutes. Another 30 minutes there would have been extremely valuable. I hope that we got enough information on the phenomena exposed at that crater that can be understood.

26.0 MISSION CONTROL

CERNAN — I think the GO/NO GOs and the performance of the CAPCOMs was outstanding. They gave us each GO, both CM and LM. There were no NO GOs, so we received all GOs. Everything was nominal. We received our updates on time. I don't think there was any concern or problem there. Consumables in both vehicles were nominal or better than nominal. Oxygen - We had plenty of oxygen in both vehicles. Electrical power - We had plenty of electrical power in both vehicles. The RCS fuel in the service module was well above the red line for the entire first part of the mission and at or above the red line the last half of the mission. We went in on double ring in the command module and we couldn't have used very much.

EVANS — I used more fuel than I would have in the simulator because there were always some rates; cross coupling in pitch and yaw in the command module RCS.

CERNAN — LM RCS - We landed with more RCS than I'd ever seen in a simulator, well over 80 percent, which made me feel good. DPS propellent - We landed with between 7 and 9 percent, which is far more than I'd ever seen in the simulator. SPS fuel - I think came out just about right on the money. We did not make any SPS midcourses on the way home and we had about 3 percent in each side. The key to the Flight Plan and the key to a smooth operation of the SIM bay in lunar orbit with all three individuals in the spacecraft was the fact that real-time changes were held to a minimum. The Flight Plan was so well thought out and was working so well that real-time changes were very simple, explicit, and not time consuming.

EVANS — It was an outstanding way to run a flight. Communications were always good.

CERNAN — The only communications problem we had in the LM was right after ascent when we lost the high gain where the ground could hear everything we said. We had a lot of noise and static in the background and we could not hear anything that the ground said until about 3 minutes into the flight.

SCHMITT — Typically outstanding support. The number of extracurricula hours the LM people and the EECOMs for the CSM in particular put in with me on Saturdays and other times just talking over systems, techniques, and mission rules were a major factor in helping me understand and keep up to speed. The help that they gave me in designing the emergency cue cards for the LM was a major contribution although we did not use them. Had we required them I think it would have gone very well. I want to point out that Dick Thorson was instrumental in organizing the LM and joint CSM/LM sessions. He was a major organizer for the creation and the updating of the emergency cue cards. It was my understanding that

some of the things I had hoped could be done during the flight were not possible because of real-time discussions in the Mission Control. Specifically, one of those things was to have a summary of the thinking of the science personnel in the back room given to me while in flight. The thinking was to be based on the data that we had transmitted to them verbally and visually through the television camera. I had hoped that I would have the benefit of their thinking, but apparently this was not possible. I would like to think that in the future we can look at ways of using the team approach to science investigations in space rather than depending solely on the observational capability and the interpretative capability of the men who are performing the job. There is no reason that I can see not to use all the brain power that is available at any given task, and part of that brain power is on the ground.

27.0 HUMAN FACTORS

27.1 PREFLIGHT

CERNAN — Preflight health stabilization and control, program - I guess it was adequate. We stayed healthy; we came back healthy. Medical care was adequate. Preflight time for rest, exercise, and sleep is something that only the crew can provide for because there's never enough time to train. When there's not enough time to train, then there's never enough time to get adequate sleep, rest, and exercise. It requires a certain amount of scheduling and crew discipline to get it all in. I believe that we got it in. Adequately prepared physiologically for the flight at lift-off. Medical briefing and exams - After extensive preflight

work for several months prior to the exams, I think that they eventually ran relatively smooth with the exception of a few misconceptions over the use of the lower body negative pressure in the CMP's G-suit garment, which eventually got resolved. Eating habits and amount of food consumption. Preflight - I think the crew has no gripes. We were satisfied.

EVANS — It was good.

CERNAN — The medical department was satisfied and we were satisfied.

SCHMITT — I personally did not find any great difficulty working out or adhering to the requirements. Medical care, although a very limited requirement, was good. I had a couple of sinus infections that reacted as they always had, and we were able over a period of 10 days or two weeks to get those cleared up. Time for exercise was probably less than it should have been, although I was able to get a good workout about every other day in addition to the workouts we got as a normal course of our EVA training. Eventually Tex started scheduling a pretty normal schedule time in the late afternoon for exercise. That helped as a reminder to see that that exercise was obtained. It is hard, at least in the lunar training program, to get in exercise periods during the day. Quite frequently, the exercise was done in the KSC gym at night. Rest and sleep is an individual thing. I made a particular effort to always get as much as I possibly could and never get behind the power curve on rest. My personal experience is that I tend to get colds and resulting sinus infections. Medical briefing was good. The exams seemed to go very well in my estimation. They were as expeditious as possible under the circumstances. I think the operational medical personnel who carried out the exams are to be complimented in their efforts to see that the exam was as painless and as efficient as possible. This should also include the postflight exams on the Ticonderoga. Eating habits and the amount of food consumption were normal except during those periods when we were on the inflight food prior to launch. Those times tended to decrease my appetite, although the food was certainly tolerable. It was not possible for me to eat the amount of food that was provided for me. This also applied to space work in the case of the inflight eating. Although I did not eat everything that was available in my food packages, I apparently needed to if I wanted to avoid losing weight. My appetite was down and I had a loss of weight. At the time of this recording, my weight has not yet gone up to preflight levels, which may have been a little high.

27.2 FLIGHT   CERNAN — Appetite and food preference.

EVANS — Appetite in-flight versus 2-week preflight - I don't think there was any change for me after the first day and a half.

CERNAN — My appetite normally inflight, based upon past experience, will decrease markedly versus nominal preflight activity. Everything else being consistently nominal, I would probably have to force myself to eat because the requirement is there to have that energy and food, but the appetite would not necessarily be there. This is just typical of me. In this particular case, the food as expected, produced a great deal of gas. For the first part of the flight, it was unpassable gas which resulted in a big football-like knot in the stomach which ranged in degree from inconvenience to annoyance to disturbing and downright painful in some cases. That also degraded the appetite because every time I ate it just stimulated this particular problem. Difference notable in food tastes inflight versus preflight - I think the greatest difference is that there is always some gas in the water. One of the biggest problems in preflight is that you always have someone like Rita Rapp to prepare them and the inconvenience part of the job is done.

EVANS — I didn't notice any real difference in food taste in flight versus preflight. To me, the wet packs taste like canned food, and the rehydratables had a better taste.

CERNAN — It is obviously a very individualistic thing.

EVANS — That's right; it's an individual thing, and I don't think it makes any difference whether from the first day on.

CERNAN — Your food preference changes as to how you feel that day. If you see a package of shrimp cocktail that doesn't look very good, you don't rehydrate it; you don't want to eat it. And, as far as I'm concerned, I could eat 10 wetpacks to every rehydratable pack.

EVANS — The first days when I really didn't feel like eating, I really didn't want the wetpacks. I would rather rehydrate something, because, to me, it had a better taste, because it didn't smell I guess. The smell of the wetpacks, when I really wasn't hungry, and still acclimating to zero-g flight on the first day there, didn't appeal to me at all.

CERNAN — I'm going to say again that the size of food portions and meal portions is a subjective individual thing. As far as I'm concerned, the food portions are entirely too great. The entire meal portions were too great and too large. I lost my 9 pounds. I predicted I'd lose somewhere between 8 and 10 pounds, and I just don't feel like I can eat as much food as there was to eat on that flight. Even though I ate more and more as the flight progressed, I very seldom consumed the entire meal that was presented at any given time.

EVANS — In my case, I think, most generally, I ate just about everything. However, if you didn't have time, which, in my case, it seemed to me like there were times when I really didn't have time - once you get all that stuff ready and it gets on through there, you really don't have time to eat everything. So you, hopefully, try to get everything made and have time to go ahead and finish everything except maybe one package.

CERNAN — Food preparation and consumption - Programs with rehydration (mixing and gases) - There is gas in all the water, and there appears to be a little more in the hot than in the cold. We added more water than was called for to make up for the amount of gas in the food. When I ate the food or drank the beverages with gas in it, I could expel it through my teeth, and not drink the gas, and still get the liquids out of the rehydratables. I don't really feel I drank or ate much gas out of the water, by comparison to what I could if I just swallowed everything that was in the package.

EVANS — In my case, I swallowed everything. I never consciously tried to separate the bubbles and the water. I guess if you got an obvious bubble on the thing, you normally don't swallow bubbles; at least, I don't think I did. In the spoon-bowl packages, if you had a bubble in there, when you opened it up the bubble was gone, so there's no problem with the spoon-bowl packages. In spoon-bowl packages, the bubbles all developed in one big bubble, and you could break that one and have no problem at all with the gas in it. The hot water had more gas bubbles than the drink gun. Maybe they were smaller bubbles; let's put it that way.

CERNAN — Food temperature - There's no question that the hot water, and it was always hot, was an excellent way of preparing the rehydratables that are desirable hot. In the LM, you did not have the privilege of hot water, and the difference in eating LM food with cold water, after eating some of that nice hot-water food in the command module, was very evident. There are some of those wetpacks (hamburgers, beef steaks, what have you) that, some way of heating them up, an oven or some other way, would increase the palatability of the food immensely.

EVANS — Let me second that statement, for sure.

CERNAN — Effect of water flavor and gas content on food - I don't think the flavor was any different than what we've seen in the past. The rehydratables are very closely attuned to the taste and flavor of regular table food, but they're never quite the same.

Use of spoon-bowl packs. We used those quite frequently, and I thought, for the most part, they worked out very well. We did something different with the soups or the more liquid spoon-bowl packs. Instead of actually opening them up and eating soup with a spoon, we just cut off the end from which you rehydrate them and suck them out. The lumpy foods, like the potatoes, we of course ate with a spoon.

EVANS — Let me make one other comment about the spoon packs, the cereal portion of it. I didn't feel like I wanted to suck the cereal out through a little hole in the thing, so I always opened those up. In general, those seemed to be the ones, in order to get them properly rehydrated, you ended up with the biggest bubble in the middle of it. So every time you tried to open one of those things, it was a messy operation. When you first opened it, it had a tendency to run right up the edge. If you tried to open it slowly, it would definitely run up the edge. If you tried to open it fast, you'd almost push it out.

CERNAN — The other thing about spoon-bowls - when you were finished and you put the germicidal tablet in them and you tried to seal them up, they'd have a tendency to produce waste along the sealing edge. Then you'd have to suck it off or wipe it off, or you'd have bubbles of soup or whatever was in the spoon-bowl floating around. Opening a can - puddings and nonliquids were fine, but when you come to the mixed fruits or the peaches, or this type of thing where you have liquid, those cans are great once you get them open. The thing about it is when you break the seal, you break it into the can, and it's not; that the can is pressurized. You break it into the can, you reduce the amount of volume, and you force liquid out. It comes off the can in bubbles and you have a mess. You end up having to be very careful. If those cans would open entirely outward instead of inward, you'd reduce that mess.

EVANS — I finally got to where I could stick the can in my mouth and open the can with my teeth.

CERNAN — And he sucked the juice out as he opened it, which is a very poor way of doing it.

EVANS — Poor, and kind of hard on your teeth.

CERNAN — Food bar usage during EVA period - The CDR and the LMP found their use very gratifying, very easy to use. The CDR, for the most part, ate from half to three-quarters of his food bar each EVA. The LMP really hardly ate any of his at all. Food waste stowage. Function of germicidal tablet pouch. We always had a couple of pouches around the spacecraft, stuck in corners, so that we could get to them every time we wanted to. We had the Skylab waterbags on board, and we had these small little valves in the waterbags. So I think from about the second day of the mission on, all three crewmen, (when we drank juices, regularly prepared juices), would rehydrate them with that same valve that we had on the water package. Rather than cut the other end of the juice bag off and suck the juice out through a flat plastic end, we'd use that Skylab waterbag valve, put it in the same hole which we rehydrated the juices out of, and drink it that way. We found it was less messy, easier to drink, and convenient to consume. And, I recommend that Skylab waterbag for all the juices. It's a neat little bag.

EVANS — It sure is. It's a neat bag; it's one less cut on the thing; and it's great.

CERNAN — Food waste stowage - We used germicidal tablets in most all the foods. We passed up a few juices sometimes, because in drinking juices the way we just described, you had to cut off another corner of the juice bag in order to get the germicidal tablet in.

EVANS — I never did put one in the juice bag.

CERNAN — I put some in some of them. Undesirable odors - I don't think there were any undesirable food odors. They were overwhelmed by the urine and feces odors and the gas odors in the spacecraft. Quantity of food eaten on lunar surface - I think probably the appetites on the lunar surface were very good. It did not appear that the food was packaged individually for CDR and LMP, so we broke out the food for that day, laid it on the table, and had a family dinner out of what was there. Quantity of food discarded on lunar surface prior to lift-off - I don't think we discarded any uneaten food on the lunar surface, although we did leave some uneaten food in the lunar module prior to jettison, but not very much. Water - Chlorine taste and odor - We never had a chlorine taste in the water, and we did chlorinate every night prior to going to bed. Iodine taste and odor - The LM water was good. We did not take a filter. Physical discomfort of gas in water - I think the LM water had quite a bit less gas after the first cupful taken out of the LM descent water. I think we had much less gas in the LM water than in the command module.

EVANS — The only comment I want to make is that the gas/water separator that you stick on the food preparation bags always leaked like a sieve. You had to keep the cap on it at all times; otherwise, you'd end up with a big blob of water on it. The cap was quite effective, though, in stopping that blob of water from forming.

CERNAN — Intensity of thirst on mission - I think the known need for staying hydrated was there. Although I drank more and more water as the mission went on, I think that there was never really a strong intensity of thirst that plagued me, except on the lunar surface after we got back in the lunar module and I found myself drinking water continually out of the water gun, prior to EVA and post-EVA. So, I think I consumed, on an average, probably twice as much water in the LM per day as I did in the command module.

EVANS — I was hustling around doing a lot more physical activity in preparation for the EVA, and at that point in time I was thirsty. I wanted to have a drink all the time. But the rest of the time I really wouldn't get that thirsty. But, you just felt like you ought to have a drink of water, and you'd drink as much as you could.

CERNAN — Work-rest sleep - Difficulty in going to sleep - I think all three crewmen experienced varying degrees of difficulty with going to sleep, and all three crewmen utilized Seconal at one time or another. The commander probably had three Seconals, and they were all taken prior to hitting the lunar surface. The CDR did not take any Seconal, to the best of his recollection, on the lunar surface or any time thereafter. I got excellent sleep when I did take a Seconal; when I did not take a Seconal, sometimes I got excellent sleep, sometimes I got marginal sleep.

EVANS — I think I probably took Seconal more than anybody up there. I'd have to look in the Flight Plan to find out for sure how many days I did not take. But I'm guessing 4 or 5 days I probably did not take Seconal, the rest of the time, I did. And, most of the time when I did take Seconal, I would sleep for 3 to 4 hours straight - just go to bed and go to sleep right off the bat - sleep for 3 or 4 hours and then kind of wake up, off and on, from that point on. Normally, when I'm around here, I require about 7 hours of sleep per night, and every once in a while 8, in order to feel real good. Up there, it seemed to me like I could get by quite adequately on 6 hours with no problem at all.

CERNAN — The sleep restraints were used every night in the command module by two crewmen when there were three men in there. The third man slept up in the couches. I thought they were adequate to help keep the temperature comfortable.

EVANS — When I was solo, I always slept in the sleep restraint in the capsule. The first night I put the lap belt on just to keep me from floating all over the thing. The rest of the time I didn't even bother with the lap belt; just jumped in the bag and floated wherever I happened to float around the spacecraft.

CERNAN — That's the same thing that I did when I was sleeping up in the couches with three men in the spacecraft. I just put the restraint on and floated around. It didn't bother me at all. In the LM, although my hammock was as tight as it would stretch, it still rested upon the suits in 1/6g, which was no problem. I thought it was extremely comfortable sleeping and, as far as I know, the LMP had no problems with the sleep restraints. A very good way to sleep in the LM, considering the tight quarters. The best thing we did was shorten that first day after launch, particularly in light of the delay, and put us on a reasonable Houston work-sleep cycle. It gave us sort of an extra day to a mission, but it put us on a very compatible work-sleep cycle, and it kept that first day to a minimum and set the rest of the days keeping pace with it. I felt very strongly that we get 8 hours of sleep period. When there were days when we were late getting to sleep about 30 or 40 minutes, I requested and got an extension of sleep period the next day. When the day had to be something very critical, like PDI day, I made sure that at all costs we did get to bed and we did get our full 8 hours of sleep. I think this probably paid off more than anything else. On the way back, after the more important aspects of the mission were complete and we got to sleep half an hour or an hour later, I made a decision that we'd just stick with the Flight Plan and getup on time. So it may have shortened our sleep period to 7 hours, 6-1/2 or 7-1/2 hours in some cases, but that worked out fine also. Prior to the major objectives of the mission, I felt very strongly about preserving that 8 hours, and we did.

EVANS — I was going to say I agree completely in that you need to get started as rested as you can.

CERNAN — Disturbances - Typical spacecraft disturbances when one man moves or one man sneezes or when one man does something else - every man does it because you just live in that kind of an environment. The CMP is the only guy that can sleep through master alarms, crew alerts, buzzers, anything. About 3 nights out, prior to coming home, he starts talking in his sleep. He was up on the couch on duty that night, and I heard "Houston, Roger" and "Houston, this is America." Then I realized you were talking in your sleep. And then it dawned on me, supposing he decides to make an SPS burn? I stayed awake all night long listening to him.

CERNAN — Exercise - I think the frequency of exercise on the way out was more consistent than on the way back. We exercised every day on the way out, for periods ranging from 10 minutes to 30 minutes on an individual basis. The quality of exercise I thought was pretty good. In some cases, the heart rate could be monitored on the ground; in other cases it was not. But in measuring your own heart rate on board, I think for the most part we all got up, consisting for periods of time, to 110 to 120 beats per minute. The exerciser - I think Ron and I used the exerciser for a couple of days on the way out. We talked about the exerciser in the spacecraft. It was used on the way out to the Moon on just a couple of separate occasions by the CMP and the CDR and it worked adequately. I didn't find any problems with it. But I think I exercised more efficiently by sitting in the commander's couch, holding on to the arm struts very tightly, and holding myself against the LEB bulkhead and running in place to produce artificial-g. I was working my arm muscles and I was running against the bulkhead, which produced force against my legs. I could really run at different speeds and for long durations, and that's the way I did all my exercise. The LMP did his exercise that same way and he did it on the right-hand couch. I don't know how he found room to move his knees, but that's where he did it.

SCHMITT — I wasn't too concerned about my legs, so I just kind of let the legs go. I really didn't exercise the legs at all except for one time. The rest of the time I was essentially trying to exercise the arms, so I used the exerciser twice, I think, by grabbing hold with one hand and pulling it through as much as I could, back and forth, that way. Then, as Gene did, I figured out I could grab hold of the struts on the spacecraft, put my feet up in the tunnel and squeeze the struts in one way and turn the hands around and push them out. I could really work up an exercise by just kind of shaking the strut as much as I could. I could actually work up a sweat just doing that type of operation.

CERNAN — Muscle soreness during or after flight - There was none on anyone's part after the flight. As far as I know, certainly on my part, there was none during the flight with one exception that I'll mention. I felt (and I don't know what the metabolic assessment is from the data that came down) that when I got to the Moon there was little or no degradation in my physical capability to do the job. I felt neither short of breath nor short of muscle response or anything when I got to the surface. The only comment I want to make, and maybe it's not muscle soreness, is that after the first EVA, both the LMP's hands and my hands were extremely sore from all that particular type of hand-dexterity labor that's required with the ALSEP and with the drill. There were no particularly abrasive areas. The fingers and the hands were just sore from continual movement, to the extent that we both had, on both hands, several blood-blister-like formations under the fingernails. We both felt a discomfort after that first EVA. The second EVA, it was less; the third EVA, it was less; and by lift-off my hands were perfect. They did not in any way hamper anything I had to do during that first EVA. I didn't even know they were sore during the first EVA until after I got out of the suit. The second EVA, they didn't hamper anything, and the third EVA, they got better still. It's like maybe having a muscle that needs work and that's what it amounted to. And the LMP's were the same way. His were bothering him the first EVA, and the second EVA they were better. It had nothing to do with the fit or size of the gloves. If I have to get a pair of gloves, I would get them fitted and sized the same way I got these fitted and sized.

EVANS — I'd like to mention one thing on muscle soreness that I heard about in preflight. I think the soreness in the back is not the lower back. It's just the muscles in the upper part of the back. I think this is from sleeping. I don't think it's from trying to hang-on to something. I think it's trying to get a relaxed position in the first part of flight, because I think it's hard to pick a relaxed position with your legs. You tend to kind of hold them in one position or another. Later, on the flight, I didn't notice that at all, and I didn't really recall trying to hang on to something. The only time I can recall trying to hang on to something is during the solo orbit periods when I was trying to focus on the camera out the window. During this type of an operation, I really used my feet to hang on to things and my back wasn't sore then.

CERNAN — Perspiration during nonexercise periods - I don't think there was any on anyone's part. During exercise periods, I found myself right at the threshold of beginning to perspire, but never really felt like I was. Inflight oral hygiene - I had no mouth discomfort. I brushed at least once a day and probably twice a day, once in the morning and once at night. I never used dental floss, and the toothbrush and toothpaste were certainly adequate.

EVANS — It felt like you needed to brush your teeth every now and then. If I have to do it again, I would get my teeth cleaned prior to going on the flight. I didn't this time for some reason.

CERNAN — Do you think that has anything to do with your smoking?

EVANS — Maybe it did. I don't know. But it felt like I really needed to have my teeth cleaned. I felt that way before flight so the flight has nothing to do with it. Brushing frequency - probably at least two or, most of the time, three times a day. Never did use dental floss. The toothbrush was great; I had no problems swallowing the toothpaste. Tasted pretty good as a matter of fact.

CERNAN — Sunglasses or other eye protective devices - For some strange reason (I would never have believed it) but I took the sunglasses out of my pocket once, put them on for about 20 minutes, and never used them the entire flight.

EVANS — The first day I didn't use them at all, and my eyes felt a little bit like maybe they were getting a little bit red or something, just a little bit tired, so I put some eyedrops in. The next day I got my sunglasses out, looked out the window a couple of times, and then needed to look at the map or something back inside. I put my head down and couldn't see the map, so I had to take the sunglasses off. So I finally said heck with it. I didn't wear the sunglasses the rest of the time. In a good portion of the visual observations, I felt that color was an important part of it. If I had sunglasses on I couldn't get a true picture of what the color is, so I didn't wear them.

CERNAN — Unusual or unexpected visual phenomena or problems experienced - I focused during rapid acceleration or deceleration with no problems.

EVANS — No problems with me.

CERNAN — Visual details - Sunlit versus down-Sun areas - Let me talk surface, and you can talk about orbit. Driving the Rover down-Sun into the west was a very degraded operation. There was no way that you could do any shadowing. We did it for a great part of the time. You just had to sort of look through the down-Sun effective zero-phase area to make sure you could see what was coming up. Driving up-Sun, again, was a degraded mode of driving. It was very bright. Everything that you were looking at was effectively washed out. But when you drove up-Sun you had a capability of either shielding your eyes with the hard-cover visor or your hand. As soon as you did that, you had absolutely distinct and perfect vision as to what was ahead of you. It was a case of being able to have the right geometry of the Sun versus your direction of driving.

CERNAN — Vision without outer visor during EVA - In effect, I never used mine. I used the protective visor and the gold visor almost the entire time except when I was in the shade and I lifted my gold visor. I hardly ever, except for occasionally driving into the Sun mode of operation, used the hard-cover visor at all. I never used the side hard-cover visors and just very seldom used the center hard-cover visor. Distance judgment versus aerial perspective during EVA - The size and distance you certainly had to multiply by a factor of 2, and maybe I would go so far as to say a factor of 5 in many cases, because there are no references on which to base size or distance. Well, I think the Moon horizon and the Earth horizon at sunrise and sunset have been discussed in detail in the past, but there is nothing unusual experienced which I didn't expect. Eye irritation during photos at window and EVA - I never wore sunglasses and in effect never had any. I take that back. One day, prior to PDI day getting in the LM, my right eye started watering for some reason but did not impair vision. I put some eyedrops in it and it seemed to soothe it. I never did anything else to my eye again until, one day on the way home, I got some chlorine in it, and I washed it out and put some more eyedrops in it. Beyond that, I had no eye irritations at all. We had chlorine all over that spacecraft. That's the way I washed my hands everyday, chlorinate the spacecraft. Helmet visor reflections - I had no particular problems with the helmet. My gold visor got very dirty and dusty and scratched up very early in the first EVA, and I cleaned it as the ground prescribed before each EVA, but it really didn't do much good. I just learned to live with it, and it really didn't degrade the operations much at all.

EVANS — Well I think most of these are pretty much not applicable to lunar-orbit-type stuff. I used your LEVA and I didn't even notice any scratches on the thing while I was out. Unless you want to talk about the eye irritations during the photos at the windows. This was essentially on the first day in lunar orbit, I think. And, for some reason, I never even noticed it from then on. I never did use the eyedrops from that point.

CERNAN — Medical kit - An adequate quantity of medications was supplied. I think there were certainly adequate quantities of medication in both vehicles. We brought the medical kit back from the LM. Why I don't know, but we did bring it back. Yet, in spite of bringing it back and having the command module kit there, we ran out of biomed sensor electrolyte sponges on the last day of the flight.

EVANS — The sponges themselves are packed in packages of six, so you throw one away every time.

CERNAN — I will say one thing. I did change my sensors one time on the lunar surface. The sponges in the LM medical kit were about half the thickness or a little bit better than half the thicknesses of the sponges in the command module medical kit. So when I put those sensors on, I put two sponges under my sensor instead of one because I didn't feel that one sponge would do the job. Two sponges were just a little too much, but I did use two sponges. Packaging of the kit was fine. Adequate instruction for use - As far as I'm concerned, there is no instruction for use on anything in the medical kit. If you want to take a Lomatil, there is nothing that says diarrhea. You don't know whether you can take one, you don't know whether you can take two. If you take two, you don't know whether you can take a Seconal with it. So, effectively, there are zero instructions. Even if the instructions were there, I'm sure you would have to talk to the ground before you take them away.

EVANS — We'll talk about the EKG on the thing. Let me make a comment. To me, we changed those things way too often. You had it on for 12 hours and you took it off and the next day you put the crazy thing back on again. If you've got it on, keep it on for 24 hours, something like that. Then let the other guys have a 2-day break on it. If, you're going to cycle it that way, don't keep changing the thing every 12 hours.

CERNAN — One thing about the sensors - Sometimes it's more inconvenient to change the sensors than it is to keep them on. Much of the time, where the guy was going to take them off rather than go through the inconvenience of taking them off and cleaning them up and getting them prepared and putting them on, whenever he had to put it on later, he'd just leave them on throughout that period. Housekeeping continues to be the major operation of space flight, particularly in spacecraft as small and as requiring as the command module and LM. Maybe in Skylab it's going to be more so, because the spacecraft is bigger. Changing sensors, for instance. As soon as you change one sensor, you've got about four or five small, loose articles in your hand. You've got to contain them. You've got to put them in a small garbage bag and then eventually put them in a big garbage bag, and every time you have a loose article with no place to put it, it's a housekeeping problem, automatically. I don't know what else you can say about that. The thing that was good as far as the command module is concerned was that we had an extra temporary stowage bag that we put up in the tunnel. That was kind of a temporary jettison bag that we filled up. As soon as it got full we would stick it into a big jettison bag and shove the jettison bag underneath the couch somewhere. It's an effective way to keep track of the junk and the trash because it's got a spring-loaded door and you shove this stuff up in the bag.

The thing about housekeeping is that it takes you anywhere from 1 to 3 days to effectively unstow the spacecraft to get at those things you need on a cyclic-type basis. Those things you need to keep living, eating, sleeping, and working with. And you have to find convenient temporary secure stowage locations for all these things. No one can really dictate whether it's going to be particularly convenient to you; but, once you do this, your housekeeping problems begin to minimize. But it's just a case of setting up those living accomodations which are compatible with three individuals who are trying to live compatibly together, both in taking care of their personal items like spoons and toothbrushes and taking care of spacecraft operational items like cameras and chlorine packages and filters and what have you. It's too inappropriate to put a lot of those things back in their original launch stowage configuration position. Shaving - I shaved once before PDI, once after PDI, and once before reentry, and I think it's one of the most clear feelings a guy can get in the spacecraft.

SCHMITT — It's great. I could only shave about a third of the face at a time, maybe a fourth, so that's the way you do it. You put a little bit on and shave that part off and start again. I've got a recommendation on the razors. And Gene didn't have that problem. guess my beard is a little thicker or something, but I couldn't use a two-bladed razor. could get one scrape out of the thing and it was full. There is just no way to clean it out and it just wouldn't cut anymore. The singleblade razor is the one that evidently has enough room in there. Even though it got plugged up with the shaving cream, it still worked okay.

CERNAN — Dust - I think probably one of the most aggravating, restricting facets of lunar surface exploration is the dust and its adherence to everything no matter what kind of material, whether it be skin, suit material, metal, no matter what it be and it's restrictive friction-like action to everything it gets on. For instance, the simple large tolerance mechanical devices on the Rover began to show the effect of dust as the EVAs went on. By the middle or end of the third EVA, simple things like bag locks and the lock which held the pallet on the Rover began not only to malfunction but to not function at all. They effectively froze. We tried to dust them and bang the dust off and clean them and there was just no way. The effect of dust on mirrors, cameras, and checklists is phenomenal. You have to live with it but you're continually fighting the dust problem both outside and inside the spacecraft. Once you get inside the spacecraft, as much as you dust yourself, you start taking off the suits and you have dust on your hands and your face and you're walking in it. You can be as careful in cleaning up as you want to, but it just sort of inhabits every nook and cranny in the spacecraft and every pore in your skin. Although I didn't have any respiratory problems, I think the LMP, which he can comment on later, had some definite local respiratory problems immediately after the EVAs due to the dust in the cabin. In sputum - I didn't spit up anything. I didn't feel any aerosol dust problem at all until after rendezvous and docking when I took off my helmet in zero-g and we had the lunar module cabin fan running the whole time. I did all the transfer with my helmet and gloves off, and I'm sorry I did because the dust really began to bother me. It bothered my eyes, it bothered my throat, and I was tasting it and eating it and really could feel it working back and forth between the tunnel and the LM. Ron, did you feel any effects of the dust when we docked and rendezvoused, particularly?

EVANS — Only when I stuck my head up in the LM. When I climbed up in the tunnel could definitely tell there was a lot of dust up in the LM and you could smell it. It's a difference, so I think you noticed it from that standpoint, but there never really was dust in the command module. The only time you ever got any dirt in the command module was when you touched something that had dirt on it. But as far as dust floating around in the command module - I don't think it ever did.

CERNAN — After rendezvous and docking - After the CDR and LMP had been living with this dust for 3 days on the lunar surface, there was a compelling urge on both of our parts to get clean. We spent about 2 or 3 hours prior to going to bed doing nothing but effectively taking soap and water and trying to wash as much of our body as we could to get free from what is really sort of a dirty feeling due to the dust. Even with soap and water it was sometimes very difficult to get clean, and the dust would get under your fingernails and other places on your body.

Radiation dosimetry - personal radiation dosimeters - Were the PRDs worn for the entire mission? Yes, with the simple exception that after rendezvous and docking, when the LMP and CDR stowed their suits, we did not transfer the PRDs. The CDR's was in the suit PGA bag for I day when it was retrieved. The LMP's was in a PGA bag for 2 days when it was retrieved. Radiation survey meter - Was it activated at any time? I thought about it, but what good would it do?

Personal Hygiene - Adequacy of wipes, size and numbers - As far as I'm concerned, the wipes might just as well be thrown off the spacecraft. They are too small to do any good. I never cut open a wipe bag. Now I think the CMP may have a change of heart.

EVANS — I used them all the time. Whenever I had one with a meal, I would cut one open and I'd just use it to wipe off my hands and mouth. When you dip out of a spoon-bowl, part of it gets on your fingers. So you'd lick your fingers and then wipe it off at the end of the meal. That's the only thing you could use them for.

CERNAN — I think the tissues, and it turned out there were plenty although the way we were using them for a while we weren't sure, and the towels are the two most important items of personal hygiene. In use of the potable water, both hot and cold, for personal hygiene - Yes, we used it and we used it effectively just like you'd wash with a washrag in your bathroom. We used it with soap and/or water and used two or three towels, one with soap, one with plain water to rinse, and one to dry. And it turned out that there were plenty of towels also. And that closes that. But I'd like to make one comment about personal hygiene and eating habits and defecation and urination habits in a spacecraft like the command module. I just personally feel very strong that we have a long, long way to go to make space a convenient, comfortable, habitable area in terms of defecation devices, in terms of urination devices, and in terms of personal hygiene to keep adequately clean and feel adequately clean. I think from what I understand of Skylab that we're taken some major steps in the right direction in terms of defecation capability, in terms of showering capability, and in terms of one other very important thing, the ability to exercise. I think if we can handle those types of living habits and learn how to handle them in Skylab, I think that one of the major modes of operations in space is going to be upgraded greatly. You do them in the command module because you have to, but because of the size or because of the facilities that you have at hand, it's a messy and sometimes a dirty and almost an unsanitary operation. But you make the best of what you can and the best certainly works. But I think Skylab is a step in the right direction. I don't know all the details of their hygiene facilities, but the thought that's going in to it I'm sure is based upon the same comments we've made here.

EVANS — I'd like to make one comment on the urine busses, as we call them. First of all, the little check valve in there is ineffective to me. You may as well have an on/off valve on this thing, because the check valve creates such a back pressure that every time I wanted to urinate I felt like I had to force it. If there is some way to get rid of that back pressure that you have to overcome in order to urinate, it would make it a lot more pleasant operation.

SCHMITT — The appetite inflight versus preflight was less again except when we were testing the preflight food, when I also had a low appetite. No notable differences in the taste of food. The things I liked in preflight I also liked in space. The things I didn't like in preflight I also didn't like in space. I didn't notice any differences. I tended to start to prefer to eat the wet packs in preference to any of the other solid foods. I would strongly recommend that the wet packs be used in preference to the rehydratable. You probably will get a different opinion from the other crewmen. The juices were good. After the one and only period of difficulty with loose bowel movements I did cut out the potassium-indicated foods. I can't say that had any effect or not, but I did not have any other loose bowel movements before the end of the flight. The first bowel movement after flight, on the Ticonderoga, was normal, the second was very loose, the third was normal, and the fourth and fifth were very loose.

The size of food portions and the meal portions - My appetite was very low the first day and gradually increased over the next 2 or 3 days. It remained essentially the same after about the third day. The most acceptable foods were the wet packs and the juices. The fruit cake was good. It was possible to eat too much or to get to the point where you didn't want any more. The chocolate was good. Of the dry crackers or cookies, the graham crackers were probably the most tasty. The peanut butter and jelly sandwiches were quite good. Food preparation and consumption. Rehydration went nominally. The nominal gas was present.

Food temperature - I tended to prefer the foods that were warm or hot, and the hot water was quite adequate for warm foods. We actually missed the warm foods in the LM where hot water was not available. I did not notice a water flavor. The water was reasonably tasty. I did not notice a high chlorine tase of any kind. All of the gas content did make it a little bit uncomfortable to eat at times. Thimble packages worked pretty well. Those that were divided I tended to cut off the other end of the package, the water insertion end, and use them as a squeeze package. Spoons worked perfectly adequate. I tended not to use the fruit in the cans because of the messiness of opening those. I think the technique that Ron worked out of opening it in or near your mouth is a good one. Puddings and this kind of thing were very good. It was only the canned fruits that I tended to avoid because they were inconvenient to use. Food bars during the EVAs I think were good to have, although I never ate more than half of one. It wasn't because it was untasty. It's just because of maybe a lack of interest in eating and using that time during the EVAs. Before and after EVAs, in the LM, I ate very well. There were some things we avoided. In my case, after having corn chowder once, which stimulated a major bowel movement, although not a loose one, I did not, thereafter, eat the corn chowder. I did not eat the cocoa because I tended to feel I got a little more gas from cocoa and an aftertaste. I did not eat the sea food items, shrimp and the lobster bisque and these sort of things, because in preflight I had noticed they tended to have a long aftertaste. Otherwise, I think all the other foods were certainly acceptable. Many times I did not eat potato-base foods because they were very filling.

Food waste stowage - I don't know how the germicidal tablet worked. The pouch was okay. It would have been nice to have had a little dispenser that was easier to use than the pouch. I don't know whether that would be possible to do or not. Seems to me it would - a little tube dispenser of some kind, where it came out more easily. We generally cut the corner off the pouch and squeezed them out. It was a little inconvenient, nothing major. We used the germicidal tablets in all the juice bags, the food bags and the wet packs. I did not use them in the tea and coffee.

Undesirable odors - Undesirable odors were at a minimum except for the occasional passing of gas. I generally had almost continuous passage of gas most of which apparently was not with significant odor. Only occasionally it seemed to be objectionable to the other crewmen. I think most of that was a water gas. Upon starting to eat, there would be an increased desire to pass gas. An increased pressure in my stomach apparently was transmitted almost immediately into the bowels. After eating I would pass gas for a couple of hours.

Quantity of foods eaten on the lunar surface - I think it was high, although probably no more than half of the food that was available. It's hard to say exactly. I think that could be worked out maybe with a detailed look at the menus. To estimate the quantity would be very difficult.

Fecal container - We used a blue bag, which is not a bad way to defecate unless the stool is loose. If it's loose it's just about impossible to use. The best thing you can do is to work out some prevention of loose stools rather than trying to handle them. Loose stools is one of the major hygiene, sanitary and operational problems that you can have on a flight. I can't emphasize that more. If it happened on a daily basis, you would eventually cut the efficiency of the crew member as much as 30 percent. I think it's important to try to understand why Apollo 17 was different than Apollo 16 in the delay of the problem of loose stools till about the 11th or 12th day. The CDR had no problem with loose stools. My personal opinion at this point, based on very little information other than observation in flight and thinking about levels of electrolyte intake, is that with the electrolyte quantity down from that imposed on Apollo 16 we did not reach an electrolyte saturation problem until the 11th or 12th day. When that saturation level was reached, I suspect that the electrolyte we were eating was dumped or concentrated in the intestines and tended to act pretty much as a laxative, an epsom salt type laxative, concentrating water in the stool. I think it's important that we reduce the electrolyte intake so that saturation is never reached. Water-Chlorine taste and odor was not apparent to me except during chlorination. Iodine taste and odor was very slight, apparent in the LM water, but not of any significance to the Physical discomfort - No physical discomfort for the LMP other than tiredness on occasion and sore muscles and the bruises under the fingernails in the case of EVA work. Gas/water separator didn't work very effectively and I'm sure that's been discussed elsewhere. Intensity of thirst during mission - Never really was thirsty, even during the EVAs, although I did stop to take a drink of water occasionally. But I never drank all the water in the insuit drink bag.

Work, rest, and sleep - The difficulty in going to sleep is variable. When Seconal was used, there was generally no difficulty in going to sleep. When it was not used, I guess there was a tendency to stay awake a little bit longer. On other occasions, the action of Seconal did not seem to affect the rate of going to sleep. There was a tendency a couple of nights to go to sleep and wake up fairly soon after going to sleep, within an hour. The second time it took a little longer, sometimes an hour to go back to sleep. But, I feel that the medical log reports for the LMP were valid and probably an average of 5-1/2 to 6 hours of sleep per night was good. I don't think, except for maybe one night, that I went much below that. The sleep was never continuous for more than 3 hours without waking up. I feel that 6 hours is adequate sleep for the kind of work we were doing. The programming of 8 hours is necessary in order to get 6 hours because of the periods of wakefulness and for the difficulty in getting the cabin organized and everybody to bed at the programmed time. So maintaining an 8-hour sleep period is mandatory in order to obtain the 6 hours required to perform the mission without getting tired or getting behind the power curve with respect to sleep.

Restraints - I had the feeling that I wanted to have my head and limbs restrained in order to get a good sleep, although I did sleep at times without that restraint. My personal opinion is to make them smaller with a somewhat more feeling of restraint. When I slept in the couch, I tended to put a shoulder strap over my head and cinch it down very lightly so that I had that feeling of head restraint. Probably one of the biggest things that made sleep difficult was the loss of sensory perception of limb position in zero gravity. When they were not being moved, you lost that perception. It came back immediately upon moving them. In general, the other crewmen did not disturb my sleep. I'm not sure why I would awake when I did. It did not normally seem to be the activities or the restlessness of the other crewmen. In one or two cases, I think it was the other crewmen, but most of the time I don't think it was.

Exercise - I ran maybe a mile and a half on the afternoon of the day of launch, keeping up to the daily running program that had continued for several months prior to launch. In flight, every day except PDI day and rendezvous day the LMP did some kind of exercise. Particularly running in place against the LEB, using the arms and shoulders on the Y-Y strut of the seat in order to provide an artificial gravity of sorts. And that seemed to be the best way that I could find to get significant heart rates. I think the medical people should have the information on those heart rates. The heart rate that I was capable of generating before my arms got tired tended to decrease, I think, with mission duration. On the day before entry, it got back up to 120. I'm not sure how much motivation had to do with that - motivation versus deconditioning. After some isometrics under the right-hand couch for 5 to 10 minutes, then I would run in place for 5 to 7 minutes, something like that. I did not use the exerciser. I found these other methods seemed to be better for my own personal needs. Muscle soreness during or after flight. The only muscle soreness that I can say I recognize was the very extreme soreness post EVAs, but that had disappeared by the next morning. And that was in the hands, soreness in the hands. After the bicycle exercises on the Ticonderoga, the next morning after the first exercise my calves were sore, and they remained sore after the second exercise on the bicycle. Within 24 hours, there was no noticeable soreness in the calves. At the conclusion of each of my running-in-place exercises, I was perspiring, not to a drippy extent but to a damp extent. Never got any real visible drops of perspiration, but I did feel damp, particularly around the head. After a few minutes of just floating quietly, that perspiration generally evaporated.

Oral hygiene - I brushed about every other day and had no discomfort in the mouth. I did not use the dental floss and the toothbrush was perfectly adequate. The toothpaste seemed to me to be a little less abrasive than you might desire, but it did freshen your mouth and seemed to clean the teeth adequately.

Sunglasses - I used the sunglasses most of the time to look at the Moon in particular. I wore them in the cabin during PTC when the Sun was coming in the windows, up through PDI day. After that I didn't use them in the LM except occasionally to look out the window at the lunar surface. After rendezvous I didn't find the desire as great to use the glasses. Initially, it seemed as if some of the moderate to light headaches that I had might have been the result of the sharp contrast of lighting that we were exposed to as much as it was to any kind of vestibular disorientation. So I'm not quite sure whether which was which, but the headaches did disappear by the third day. By post rendezvous, I did not feel the need even to look at the surface through the sunglasses. It was as if my eyes had started to self-compensate for the increased brightness that we were exposed to. Partly, I used my glasses because they do have a small correction for my astigmatism, and that did increase the resolution with which I could view the surface. Looking at the Earth and translunar coast with the sunglasses, I often did that for the correction. I used the binocular and the sunglasses and it did seem to help the resolution of viewing cloud patterns and geographic locations. When I used the sunglasses they seemed to be very adequate in terms of the level in which they reduced the brightness. As soon as I looked in the cabin to look at instruments and this sort of thing, the glasses did inhibit the observation of those instruments and the lettering on the panels, and I would push them up on my forehead for cabin work.

Unusual and unexpected visual phenomena problems experienced - Let me reference you to the description I tried to make of the sunrise color-banding in the Earth-orbit portion of the flight. We talked a lot on the tapes about the orange, yellow, and red hues to the gray in lunar orbit around the edge of Serenitatis Basin. That is also on the tapes and most of that orbital descriptive work was in the post rendezvous timeline. My solar corona sketch is in my crew notebook and I'll have to get that for reproduction. And I think the only other thing I would add is that with the sunset corona, I was able to see very strong linear streamers very close to the Sun. With the sunrise, I don't recall ever seeing strong streamers or bright streamers down close, within a solar diameter or two. But the most diffuse and broad streamers were quite obvious and are covered in the sketch and I think in some verbal descriptions on the tapes. I noticed no eye focus problem during rapid acceleration and deceleration. The best viewing Sun angle for viewing lunar topography was the low Suns, and the best Sun angle for seeing albedo and color differences was directly down Sun or zero-phase. Often, during the EVAs, I would have the gold visor down three-quarters to protect most of my face from the Sun. But for close-in detail I would look through the lower one-quarter, where I'd just have the clear helmet available in order to see more detail without looking directly into the Sun. When we were driving up-Sun with the Sun on the visor (having had some problems with the hard-shell visor movement), I mainly used my arms to shade the helmet or the LEVA so that I could see up-sun. And that worked fairly well.

Distance judgment versus aerial perspective - The distances and sizes I used were compensated by some early estimates of crater size based on the size of the LM and ALSEP distances and items like that, although I never did feel comfortable with the numbers I used. I was doing it on a subjective basis as a result of those early observations rather than on what the crater really looked like. The craters always seemed to look smaller than I felt I knew they were, although probably never by more than a factor of 2 or 3. Distances would have to be the same, or the same way through judgment of how far away from something you are. It generally results in an underestimate. You always think you're closer than you really are. I think the tapes cover some comments on the Earth illumination at the horizons. Briefly, right at the terminator horizon of the Earth, you get sharp shadow definition of cloud features. At the sunlit horizon from lunar distance that's a very clear definition between the black of space and the upper portion of the Earth. In Earth orbit and near Earth, you can see the gradation of that horizon caused by the atmosphere. At night around the Earth, there's a very clear horizon glow all around the Earth. Air glow, I guess you would call it. And the horns of the crescent Earth are much sharper and elongate compared to those of the crescent Moon, as if light was being defracted into the atmosphere and in extending the length of the horns of the crescent. With the setting and rising Sun around the Moon, you would get a - in the case of the setting Sun - a few reflections off of the high peaks some significant amount of time after the Sun had set. And the same would apply conversely to the rising Sun. The first indication of sunrise, in addition to the solar corona brightening, was a few bright areas on the peaks near the terminator that were high enough to catch the first morning rays and reflect them back around toward the spacecraft.

Eye irritation during photos - I did not notice any. Helmet visor reflections I guess have been very well covered. With the dust and scratches on the helmet, of course, you needed to shade the helmet more and more in order to see with the Sun directly on the helmet. Medical kits certainly seemed adequate. We did run out of electrolyte and some more should be packaged, I would think, for the comparable amount of time we had because we actually did not change sensors out according to the Flight Plan. We generally wore sensors longer than the Flight Plan required, which meant had we done it according to the Flight Plan we definitely would have run out of electrolyte early. I think it is a mistake not to have a fairly clear summary of instructions for use of each of the drugs, if for no other reason than the no-comm case when a drug might be required. For

most of those drugs, they would essentially be of little use to us in a no-comm situation because we would not know exactly what they were for and which drugs could be taken in combination without an adverse effect. Housekeeping was relatively easy, in general, except for the waste management portion. Within a day, the routine of where to put things to keep track of them and how to eat and all the normal and more mundane aspects of living were fairly clearly defined in my mind and did not present any serious problem.

S7255937

Shaving - I did not shave until the day before entry and after the press conference. I felt no significant discomfort from the beard during any of the time in orbit. There was a little bit of stickiness involved with wearing the chin strap but that was insignificant. I think that having a beard or not having a beard has to be purely a personal item. It cost me about an hour to shave it off, but I think that's comparable to the amount of time it would have taken to stay clean-shaven. It was difficult to shave off. I went through about three of the double-edge blades. And although none of them were seriously degraded, it just seemed that with a new blade the whole shaving process was easier. One thing to do prior to shaving is make sure you set yourself up with a good light. I might have been able to cut 10 or 15 minutes out of the shaving if I'd had better lighting. I also recommend that, prior to using the brushless shave cream, you get a lot of hot water on a rag and soak your beard with it. I also washed the beard with soap and hot water before applying the brushless shave cream. In spite of the difficulty in shaving, there was no pulling or discomfort associated with it. It was mainly a problem of the clogging of the razor and I think the dulling of the razor but there was no pulling of the beard at all.

72H1561

S7338346

Dust - We'll just talk about in-cabin dust. After the first EVA, there was considerable dust in the cabin. It would be stirred up by movements of the suit and the gear that we had. Almost immediately upon removing my helmet, I started to pick up the symptoms that you might associate with hayfever symptoms. I never had runny eyes or runny nose. It was merely a stuffiness in the nose and maybe in the frontal sinuses that affected my speech and my respiration considerably. After about 2 hours within the cabin, those symptoms gradually disappeared. By morning of the next day, they were gone completely. After the second and third EVAs, although I'm sure the dust was comparable, the symptoms were not nearly as strong as after the first EVA. That was as if I either developed a mucous protection of the affected areas or had some way or another very quickly developed an immunity to the effects of the dust. Let me mention the PRDs. The first couple of days, my PRD resided in my temporary stowage bag because I did not wear the coveralls. After the second day, I wore the coverall pants and the PRD was in the pocket of the pants. After rendezvous, the PRD inadvertently was left stowed in my suit and so it resided in the PGA bag for two days before I had it available to put in the coverall pocket. That PRD was in my PGA pocket during the CSM EVA.

Personal hygiene - I think the LM feces bags are superior to the CSM's in that they have a goodly quantity of tissues cut to size and are quite good. I see no reason why those couldn't be the same kind of blue bags in the CSM. Although we ended up having plenty of tissue, there was some concern initially whether we would. And I think had we had any greater problem in loose bowel movements we probably would have had an inadequate supply of tissues. Tissues are extremely useful in all kinds of personal and cabin hygiene and there should never, if at all possible, be any concern over not having enough tissues. Particularly, if you are using the BUSSes, you tend to use a tissue every time you use the BUSS - at least one. You tend to use one during the meals, and of course a lot of them in the use of the blue bag. Potable water was used for personal hygiene. I washed several times with soap, and post rendezvous I actually washed my hair quite adequately by putting a lot of water on a towel and wetting the hair quite well. Then, just in a normal terrestrial way, I rubbed soap into it and then washed the soap out again with a couple of wet towels. The soap on board seemed to be quite good. It did a good job of cleaning but also was not overly sudsy and seemed to wipe off or wash off very well. It did not leave any noticeable residue that was uncomfortable.

~EASY TO USE!~

CD-ROM

The attached CD-ROM requires no installation.

It is designed to leave no footprint on your computer's hard-drive and requires no special drivers to run.

All of the files are programmed as a web page and only require a web browser to view*. All files are HTML, JPG, MPG and Quicktime MOV.

MAC Users open the file "apogee.htm" in your Web browser to access the disc. It is located in the folder called "Files".

The disc includes:

The entire Television broadcast from the Apollo 17 landing site. ~Over eleven hours of video!~

Twenty-Two Quicktime* Panoramas taken at Taurus-Littrow which allow the user to scroll around and view the landing site in 360°. Just click on the images and drag to view.

Watch for Apollo 17 The NASA Mission Reports Volume Two with an exclusive interview with Commander Gene Cernan!

---

* Quicktime is a free program which can be acquired from http://www.apple.com